Nanomaterials for Wound Healing

This book provides a comprehensive examination of wound healing, starting with the foundational knowledge, including the factors affecting it, and the principles guiding wound healing. It delves into the molecular underpinnings of wound healing, particularly in the context of diabetes, and discusses the challenges of chronic wound healing, including the limitations of existing treatments. It explores the current status and future prospects of nanomaterials, including innovative approaches like smart 3D-printed hydrogels, drug-loaded nanomaterials, and natural bioactive compounds encapsulated within nanomaterials. The book also introduces bioinspired nanomaterials and their application in wound dressings and tissue regeneration, emphasizing the diagnostic-to-therapeutic strategy continuum. Further chapters cover the specific roles of nanobiomaterials in diabetic wound healing, skin tissue engineering, and the antibacterial properties of nanomaterials that promote wound healing. It looks into engineered nanomaterials, green synthesized nanomaterials for wound healing and cell growth stimulation, and the advanced application of nanocellulose. The role of the immune system in wound healing and strategies for modulating immune responses to enhance healing processes are explored. Last, the book addresses the mechanisms of nanomaterials in wound healing, their recent advancements, limitations, future perspectives, and the importance of conducting toxicity studies on nanomaterials used for wound healing. This book is an invaluable source for researchers, clinical practitioners, and students of biomedical engineering, human physiology, and nanotechnology.

Nanomaterials for Wound Healing

Edited by

Mohammed Gulzar Ahmed, Sumel Ashique,
Arshad Farid, and Gokhan Zengin

CRC Press
Taylor & Francis Group
Boca Raton London New York

CRC Press is an imprint of the
Taylor & Francis Group, an informa business

Designed cover image: Shutterstock

First edition published 2025

by CRC Press
2385 NW Executive Center Drive, Suite 320, Boca Raton FL 33431

and by CRC Press
4 Park Square, Milton Park, Abingdon, Oxon, OX14 4RN

CRC Press is an imprint of Taylor & Francis Group, LLC

ISBN: 9781032825144 (hbk)
ISBN: 9781032825151 (pbk)
ISBN: 9781003504948 (ebk)

DOI: 10.1201/9781003504948

Typeset in Times
by Deanta Global Publishing Services, Chennai, India

Contents

Editors

Mohammed Gulzar Ahmed serves as the Principal and Dean at Yenepoya Pharmacy College & Research Centre, part of Yenepoya (Deemed to be University), located in Mangalore, Karnataka, India. With a career that spans over 22 years in teaching and an additional 3 years dedicated to research, he has made significant contributions to the field. His scholarly output includes more than 150 publications in internationally and nationally accredited journals. His expertise encompasses novel drug delivery systems, nanotechnology, targeted drug delivery, and stability studies. Dr Ahmed has secured seven patents and has authored three book chapters and four books. He has also obtained a DST-SERB CRG grant, along with several seed grants from Yenepoya and Rajiv Gandhi University of Health Sciences (RGUHS). He has been recognized by the university with awards for Best Researcher, Best Paper, and Good Teacher.

Sumel Ashique currently holds the position of Assistant Professor in the Department of Pharmaceutical Sciences at Bengal College of Pharmaceutical Sciences & Research in Durgapur, West Bengal, India. He brings three years of teaching experience to his role. His academic contributions include over 60 publications in both international and national journals of repute. His areas of expertise are in drug delivery, nanotechnology, and targeted treatment strategies. He has been awarded four patents from IP Australia and has published six book chapters in international books. Additionally, he has published fifteen book chapters and edited five books with international publishers.

Arshad Farid has been serving as an Assistant Professor at the Gomal Center of Biochemistry and Biotechnology, Gomal University, D.I. Khan, Pakistan, since April 2014. With approximately 16 years of combined teaching and research experience, he has published more than 90 research articles in journals of international and national repute. His research is markedly interdisciplinary, covering areas such as medicinal plants, natural product chemistry, herbal drug development, ethnopharmacology, nanotechnology, drug delivery, and microbiology. He has contributed more than 20 book chapters and holds 15 international patents from countries including Germany, the UK, and South Africa. Additionally, he has presented over 15 abstracts at national and international conferences. Dr Farid actively contributes to the scientific community through peer review of papers for more than 20 respected journals. He was a fellow of the British Council–Charles Wallace Fellowship program at the University of Aberdeen, Scotland, UK, in 2023, and previously received a fellowship in 2007 for training in environmental, agricultural, and analytical chemistry laboratory techniques at the Department of Chemistry, University of Glasgow, Scotland, UK. He has also been a research fellow at INTI International University, Malaysia, since June 1, 2023.

Gokhan Zengin is an alumnus of Selcuk University's Faculty of Science in Turkey, where he also earned his PhD from the Natural and Applied Sciences Institute's Department of Biology. Presently, he holds a professorship at Selcuk University within the same faculty and department. His research focuses on biological activity, chromatographic techniques, and enzyme inhibition. Dr Zengin has contributed 800 publications to his field of research, amassed over 18,000 citations, and delivered numerous presentations at scientific congresses.

1 Introduction to Wound Healing

Factors Affecting, Phases, and Principles of Wound Healing

Sindhu Priya E S[1], *Tahreen Taj*[1], *Fatima Fakrudheen*[1], *Sahana T G*[2], *and Mohammed Gulzar Ahmed*[1*]

[1] Yenepoya (Deemed to be University), Karnataka, India
[2] University of Toronto, Toronto, Canada
[*] Corresponding Author

1.1 INTRODUCTION

Wound healing is a vital physiological response to tissue injury, reflecting the ability of the body to repair and regenerate damaged areas. While most tissues can repair and regrow after injury, the skin, being the outermost barrier, is particularly vulnerable to constant injuries. Therefore, in this chapter, we will focus specifically on skin injuries. The skin is the largest organ, acting as a critical barrier against the external environment and playing a significant role in the exchange of fluids, gases, and heat. Skin injury can be open or closed (for example, hematomas, contusions, blisters, surgical incisions, lacerations, abrasions, punctures, and avulsions). When injury occurs, the structural and functional integrity of the skin is compromised, and the skin tries to restore the structure and function of damaged tissue by initiating a complex and intricate process referred to as wound healing.

The wound healing process is continuous and can be categorized into four overlapping phases: hemostasis, inflammation, proliferation, and remodeling (Broughton et al., 2006b; Centre, 2009; Davidson & Dipietro, n.d.; P. Martin, 1997; Young et al., 2016). Each phase involves the coordinated action of resident and migrating cells, cytokines, growth factors, and the extracellular matrix to achieve wound closure (Barrientos et al., 2008; Regan & Barbul, 2012; Singer & Clark, 1999). The extent of wound healing depends on the extent of tissue damage. Most acute wounds typically progress sequentially through these phases, restoring functional tissue within four to six weeks. However, some chronic wounds fail to heal efficiently (Järbrink et al., 2017). These wounds are stalled in one of the healing phases due to unresolved factors that prevent the timely progression of events. Various factors, including age, nutrition, tissue hypoxia, growth factor imbalance, and other underlying conditions like diabetes and an impaired immune system, affect the rate and duration of wound healing (S. Guo & DiPietro, 2010a; Thomas Hess, 2011). Chronic wounds with a compromised healing process pose a significant challenge as they can increase the risk of infection and negatively impact the quality of life. Uncontrolled events during the proliferation or remodeling phases could also lead to scar tissue or necrosis, resulting in nonfunctional tissue.

DOI: 10.1201/9781003504948-1

Wound healing is essential for restoring damaged tissue's structural integrity, allowing it to resume its normal physiological functions. Efficient and timely wound healing is crucial for maintaining overall health and well-being, as prolonged healing times often indicate other underlying conditions. Understanding the mechanisms and factors involved in wound healing is fundamental for developing effective treatments and improving patient outcomes.

1.2 FACTORS AFFECTING WOUND HEALING

1.2.1 Local Factors

1.2.1.1 Oxygenation

The presence of oxygen plays a role in cellular functions such as energy production, collagen synthesis, and angiogenesis. Insufficient oxygen supply, known as hypoxia, can hinder these processes, resulting in delayed wound healing. Hyperbaric oxygen therapy has been utilized to boost wound oxygenation and enhance healing outcomes (Gottrup, 2004; Sen, 2009).

1.2.1.2 Infection

Infections pose a barrier to wound healing by prolonging the phases and potentially leading to chronic wounds. Effective management of wounds necessitates infection control measures, including the application of antiseptics, antibiotics, and debridement (Bowler et al., 2001; Edwards & Harding, 2004). The presence of biofilms can complicate infections further, making them more resistant to treatment and prolonging inflammation (James et al., 2008).

1.2.1.3 Moisture

Maintaining an optimal moisture balance is essential for wound healing. Excessive dryness may lead to scab formation and hinder epithelial migration, while excessive moisture can cause maceration and elevate the risk of infection. Advanced wound dressings have been developed to regulate moisture levels effectively (Atiyeh et al., 2003; Junker et al., 2013). Hydrocolloid, hydrogel, and foam dressings are examples of modern wound care products that help in maintaining the proper moisture balance (Okan et al., 2007).

1.2.1.4 Foreign Bodies

Foreign bodies in the wound bed can trigger a prolonged inflammatory response, leading to delayed healing and chronic wound formation. Thorough wound cleaning and debridement are necessary to remove foreign materials and facilitate healing (Jones et al., 2015; Kamata et al., 2019). The necrotic tissue also acts as a foreign body and needs to be removed to prevent infection and promote healing (Wysocki, 2002).

1.2.1.5 Wound Temperature

Wound temperature can influence the rate of biochemical reactions and cellular activities. Lower temperatures may hinder healing processes while maintaining a slightly elevated temperature could potentially enhance cell proliferation and migration (Leaper, 2006).

1.2.2 Systemic Factors

1.2.2.1 Age

As people age, their cell functions tend to decline, resulting in a reduced immune response and slower wound healing. Older individuals often experience delayed wound closure and an increased risk of chronic wounds. Age-related changes in skin structure, such as reduced collagen and elastin production, contribute to a delayed healing process (Gosain & DiPietro, 2004).

1.2.2.2 Nutrition

Adequate nutrition is crucial for wound healing as it provides essential nutrients such as proteins, vitamins, and minerals. Malnutrition can hinder collagen production, immune function, and overall cellular activities, leading to delayed healing (Cereda et al., 2015). Specific nutrients like vitamin C, zinc, and arginine play vital roles in collagen formation, immune function, and cell proliferation (Stechmiller, 2010).

1.2.2.3 Chronic Diseases

Chronic disease conditions such as cardiovascular diseases, diabetes, and peripheral vascular diseases can have a significant impact on wound healing. For instance, diabetes is associated with impaired leukocyte function, reduced collagen synthesis, and poor vascularization (Falanga, 2005; Maruyama et al., 2007). Cardiovascular diseases result in reduced blood flow to the wound site, delaying the delivery of essential nutrients and oxygen (Mustoe et al., 2006).

1.2.2.4 Medications

Certain medications, such as nonsteroidal anti-inflammatory drugs (NSAIDs), corticosteroids, and chemotherapy agents, can impede the wound-healing process by impairing immune responses and cellular activities. Careful management of these medications is necessary for optimal healing (S. Guo & DiPietro, 2010; Robson, 1997). Corticosteroids, for example, can inhibit the inflammatory response and collagen synthesis, delaying healing (Busti et al., 2005).

1.2.2.5 Smoking

Smoking has a detrimental effect on wound healing due to its impact on oxygen delivery, resulting in inflammation and decreased cellular function. Nicotine and other toxic substances in cigarettes reduce collagen deposition, hinder fibroblast function, and decrease capillary growth, leading to delayed healing and increased complication rates (Manassa et al., 2003; Mosser & Edwards, 2008). Carbon monoxide from cigarette smoke binds more effectively to hemoglobin than oxygen, reducing the amount of oxygen supplied to tissues (Silverstein, 1992).

1.2.2.6 Stress

Psychological stress can have an adverse effect on wound healing by triggering the release of stress hormones like cortisol, which is responsible for suppressing the immune response and altering inflammatory processes. Chronic stress is often associated with delayed wound healing and increases the susceptibility to infection (Broadbent et al., 2003; Walburn et al., 2009). Stress management techniques and psychological support can have a positive impact on improving wound healing (Ebrecht et al., 2004).

1.2.2.7 Obesity

Obesity can impede wound healing by reducing skin perfusion, delaying immune function, and increasing the risk of infection. Excess adipose tissue can lead to a state of chronic inflammation, which may thereby negatively impact the healing process. Surgical wounds in obese patients are more prone to complications such as dehiscence and infection (Waisbren et al., 2010).

1.2.3 Genetic Factors

Genetic predisposition also plays a role in influencing wound healing. Variations in genes related to collagen production, growth factor synthesis, and inflammatory responses can affect individual healing rates and susceptibility to chronic wounds (Nwomeh et al., 1998; Sood et al., 2014). Ongoing studies are looking into factors aiming to create customized wound care plans tailored to genetic profiles (Broughton et al., 2006a).

1.3 PHASES OF WOUND HEALING

In the human body the biological process that occurs in response to injury is interesting and complicated, and in simple terms, it is called wound healing. Whether it is small damage to the skin or a major injury, the human body not only repairs it but it starts rebuilding the damaged tissues through different phases that play fundamental roles in the healing process. Furthermore, the immune system, as well as clotting factors, play extensive roles in various healing mechanisms (Singh et al., 2017). Wound healing comprises numerous phases that are activated by different pathways, such as intracellular and intercellular biochemical pathways. Additionally, several cells are also involved in the wound healing process (Tottoli et al., 2020).

Wound healing comprises different phases:

1. Hemostasis phase
2. Inflammatory phase
3. Proliferative phase
4. Remodeling phase

1.3.1 Hemostasis Phase

Hemostasis is an important physiological process that initiates promptly following an injury to halt bleeding and initiate the process of healing. The word *hemostasis* precisely describes the mechanism by which the body discontinues the flow of blood at an injured place by conjoining the terms *hemo*, meaning "blood," and *stasis*, meaning "to halt." It involves numerous simultaneous actions, each of which is essential to avoid extreme loss of blood and possible injury (Ellis et al., 2018).

Four distinct phases make up the progression of hemostasis: (1) narrowing of the blood vessel, (2) development of a transient "platelet plug," (3) coagulation cascade activation, and (4) development of a "fibrin plug" or the final clot (LaPelusa et al., 2023).

1.3.1.1 Narrowing or Constriction of the Blood Vessel

The first reaction to a skin injury is the hasty constriction of the blood vessel at the damage site. Vasoconstrictors, such as endothelin released from the injured endothelium, cause the vascular smooth muscle to contract, narrowing the artery. Other chemicals such as epinephrine, prostaglandins, norepinephrine, and catecholamines also play a part in this temporary reduction in bleeding. However, this constriction will only stop blood loss for a short period of time, as the muscles immediately relax again due to rising hypoxia and acidosis, which may allow bleeding to resume (LaPelusa et al., 2023).

1.3.1.2 Platelet Activation and Plug Formation or Formation of a Temporary "Platelet Plug"

Platelets are crucial for clot formation, discovered in the 19th century. The anucleate cells are derived from megakaryocytes and those are very important in the immediate consequences of trauma. Platelets are called a natural source of progression factors and also may discharge a lot of other materials, including fibronectin, vitronectin, and sphingosine 1-phosphate, which are essential in retaining healthy tissues and confirming restoration as well as repair (Locatelli et al. 2021). The platelets will come into contact with a damaged blood vessel, then attach to that particular area and get activated. This causes additional platelets to be formed, creating a "platelet plug." This platelet plug acts as a temporary seal to stop the flow of blood from the injury. Additionally, platelets release constituents like platelet-derived growth factor (PDGF), which stimulate the tightening of nearby smooth muscles and stabilize the initial clot. There is a higher chance of bleeding and a slower rate of wound healing caused by the effects of microgravity on platelet aggregation and function (Ji et al.,2022).

1.3.1.3 Activation of the Coagulation Cascade

The coagulation cascade is a complex series of events where multiple blood proteins are involved, resulting in the fibrin formation, which are substantial proteins that strengthen the platelet plug. The coagulation cascade leads to the accumulation of platelets at the site of the wound and the development of a fibrin clot that stops the flow of blood in order to minimize the amount of blood lost.

The various coagulation factors, which include several zymogens, get activated sequentially, such as prekallikrein and prothrombin. This cascade confirms a strong barrier against further blood loss by strengthening the first plug and accelerating the clotting process (Ji et al.,2022).

1.3.1.4 Development of the Fibrin Plug or the Final Clot

The ultimate stage of hemostasis is critical in converting a temporary platelet plug into a stable fibrin clot. After an injury, this conversion generates a strong network that immediately seals the injured blood vessel and stops the flow of blood. This process is not just vital for preventing excessive blood loss; it also kicks off the healing of the wound. A fundamental component of this process is fibrinogen, a fibrous protein with α-helical coiled-coil shape. It is united by proteins such as myosin, epidermin, and keratin. This structural identification of fibrin comes from its specific patterns observed by using wide-angle X-ray diffraction studies. The architecture of this fibrin network plays an important role in wound healing efficacy. Some of the important characteristics of fibrin include the number of connections between the fibers, the thickness of the fibers, and the overall porosity and permeability of the clot (Rodrigues et al., 2019).

This structure of fibrin not only stops bleeding but also supports the repair process by interacting with various cells, for example, fibroblasts, endothelial cells, smooth muscle, keratinocytes, and leukocytes. These connections are essential as they facilitate the migration and function of cells involved in healing. In addition, the type of fibrinogen present will affect the healing process. There are two natural variants of fibrinogen: high-molecular-weight and low-molecular-weight fibrinogen. These vary in their effect on the speed of clotting, cell growth, and how they polymerize to form fibrin. Each difference has an exceptional function in adapting the development of novel blood vessels, a process essential to the effective repair and healing of damaged tissue. However, the transformation of a platelet plug into a fibrin clot not only serves to stop bleeding but also establishes a framework critical for wound healing. The form of fibrinogen present and the structural characteristics of the fibrin clot have a significant impact on how well the healing process works (Weisel et al., 2017).

1.3.2 Inflammatory Phase

After the body achieves a pause the bleeding, it calls in its defense team known as inflammation. In this phase, in order to stop the bleeding, the blood vessels will narrow first. Later, they widen to allow the entry of reinforcements such as white blood cells and antibodies. It's like a military operation where the troops rush to the site of injury to start the healing process. There may be discomfort, swelling, and some bleeding, which happens because it is really helping in the removal of any potentially dangerous materials (Laurens et al., 2006).

Inflammation isn't just about controlling bleeding; it also acts as a shield against infections. The swelling that takes place progresses the environment in which the cells can restore and perform their function. The redness and warmth show that the body is genuinely working hard to repair the damaged cells. Here, two particular cells play a major role, namely, neutrophils and macrophages. Under the guidance of signaling molecules in this inflammatory phase, the neutrophils and macrophages rush in and clean up the mess by getting rid of any foreign invaders or damaged tissue (Hayat et al., 2023).

Initially, neutrophils rush in to fight off any bacteria, while macrophages come later to clean up the mess. It's a well-coordinated effort to ensure that the wound heals quickly and efficiently. Once they've completed their task, the repair process is started by other cells such as fibroblasts and epithelial cells. These inflammatory cells settle down on a fibrin scaffold, which is a protein formed when blood clots. Later on, they release chemical signals that promote the development of new blood vessels and clotting, along with the growth of fresh tissue (Davis et al., 2016).

The reaction of inflammatory cells is accurately standardized, and the healing process is hindered by both too much and too little inflammation. Here, in particular, macrophages are essential players in tissue repair, displaying versatility and adaptability in response to different stimuli. Their nonexistence or malfunction can cause tissue repair to be hindered. They release a variety of cytokines and growth factors (Eming et al., 2007).

The body's primary method of defense against invasive microorganisms in wounds is innate inflammation. It is triggered by signals released from injured cells as well as bacterial components, commonly referred to as damage-associated molecular patterns (DAMPs) and pathogen-associated molecular patterns (PAMPs) (L. Chen et al., 2017). Through attributing to particular receptors, these DAMPs and PAMPs stimulate resident immune cells, including mast cells, macrophages, Langerhans cells, and T cells, which in turn trigger inflammatory pathways. White blood cells are attracted to the injury site due to the release of pro-inflammatory cytokines and chemokines (L. Chen et al., 2017; Krzyszczyk et al., 2018).

In addition, these pro-inflammatory molecules promote blood vessel widening, aiding in the adhesion and movement of immune cells. Selectin is a type of adhesion molecule that plays a crucial role in this process, as blocking it significantly impairs immune cell infiltration and wound healing (Baum & Arpey, 2005).

Macrophages play an essential role in tissue repair. As they consume debris and pathogens, they first promote inflammation before switching to an anti-inflammatory function to aid in the healing process. Traditionally, wound macrophages were categorized into M1 and M2 subsets; however, this categorization is currently regarded as being too basic (Rodrigues et al.,2019). Human and murine macrophages exhibit a spectrum of responses to stimuli, governed by tissue status and environmental cues. They can be referred to as pro-inflammatory (M1-like) and anti-inflammatory (M2-like) groups for ease (Xue J et al., 2014). Their nonappearance during different stages of healing leads to numerous complications, highlighting their importance in scavenging debris, bacteria, and inflammatory cells while promoting repair practices (M. Li et al., 2021).

By phagocytosing pathogens and releasing antimicrobial compounds, neutrophils are drawn to chemical signals, which are essential in eliminating necrotic tissue and infections. Histamine released by mast cells helps in neutrophil recruitment during the initial inflammatory phase, while T cells are essential for controlling the immunological response as well as promoting tissue repair (de Oliveira et al., 2016). In the early reaction to tissue damage, the resident T cells are essential, and circulating T cells will be beneficial in the resolution of inflammation. Generally, creating efficient wound healing techniques requires responsiveness to the intricate relationships between various immune cells (Keyes et al., 2016).

Furthermore, pro-inflammatory macrophages are triggered by stimuli like LPS and IFN-γ, promoting inflammation by releasing growth factors, inflammatory cytokines, and reactive oxygen species. As they phagocytize apoptotic neutrophils, they become the main inflammatory mediators (Marrocco et al., 2022). Changes in cytokine levels and phagocytosis are two examples of the processes that drive the shift in inflammation toward anti-inflammatory activity. Alternatively, activated macrophages express cytokines that aid in resolving inflammation and promoting wound repair, along with enzymes like arginase. They also release growth factors to support tissue rebuilding and stabilize blood vessels (Koh et al., 2011).

1.3.3 Proliferative Phase

The proliferative phase becomes a critical step in the healing of wounds once the inflammation subsides. It is characterized by a synchronized attempt to rejuvenate the vascular network, which covers the wound surface by re-epithelialization, and to produce granulation tissue. Growing more vigorous collagen, more blood vessels, and surface epithelium healing are all examples of the remodeling that occurs in the scar as the proliferative phase continues. Myofibroblasts play a fundamental role in wound contraction by gripping the wound edges and pulling them together, similar to the mechanism observed in smooth muscle cells (Landen et al., 2016).

This phase can be outlined into three primary stages (Wallace et al., 2023):

a. Filling the wound with fresh connective tissue and blood vessels
b. Contracting the wound edges toward the center
c. Covering the wound with proliferating epithelial cells to form a complete closure

This proliferative phase witnesses the reconstruction of the wound along with fresh tissue composed of collagen and extracellular matrix. Granulation tissue is generated when the temporary wound matrix following hemostasis is replaced. It is made up of a variety of cells, such as blood vessels wrapped in collagen bundles, granulocytes, fibroblasts, and macrophages. This process partially reestablishes both the structure and function of the damaged skin, while concurrently, the wound contracts as new tissue is synthesized (Wilkinson et al., 2020). Moreover, by providing adequate oxygen and nutrients, a new vascular network must be established to ensure the strength of the granulation tissue (Darby et al., 2014).

At the cellular level, the proliferative phase coordinates the stimulation of endothelial cells, fibroblasts, macrophages, and keratinocytes, which results in matrix formation, angiogenesis, and wound closure. Activation of keratinocytes can be seen early post-injury, which results in a partial epithelial–mesenchymal transition and the acquisition of an invasive phenotype that facilitates migration, especially during re-epithelialization. This progression encompasses the rearrangement of cell adhesion and the release of matrix metalloproteinase (MMP) to facilitate migration while depositing new extracellular matrix proteins to rebuild the basement membrane (Rodrigues et al., 2019).

In addition to cellular processes, with neuropeptides such as substance P influencing proliferation and angiogenesis, innervation also plays an essential role in wound healing. For the most part, in diabetic wounds, the restoration of nerve fiber regeneration is crucial, as deficits in nerve regeneration hinder healing. During wound repair, glial cells are activated and express components that are crucial for chemotaxis, which aids in efficient healing as well (Um et al.,2017).

Simultaneously, epidermal cells that help in wound resurfacing are produced by the proliferation of hair follicle stem cells. In response to diverse signaling molecules, fibroblasts replace the temporary matrix with extracellular matrix-rich granulation tissue, which promotes angiogenesis and aids in the movement and differentiation of wound cells (Morgun et al., 2020).

The main cause of angiogenesis is hypoxia, which causes microvascular endothelial cells to proliferate as well as move toward the wound area and build novel capillaries to meet the metabolic needs of the healing tissue. The overall goal of the proliferative phase of wound healing is to restore tissue integrity along with function through a complex interaction of cellular and molecular mechanisms. Macrophages also play an important role in this process, secreting proteases and chemotactic factors to aid endothelial cell behaviors and guide vessels to the next phase called remodeling (Everts et al., 2015).

1.3.4 Remodeling or Maturation Phase

The remodeling phase is the ultimate phase of wound healing. Here, the newly formed tissue undergoes a process of strengthening and gaining flexibility. Cells that played a crucial role in healing the injury eventually undergo programmed cell death, or apoptosis (Nourian Dehkordi et al., 2019).

This phase involves the remodeling of collagen from type III to type I, which leads to the cessation of the wound. The proliferative phase's first disordered collagen deposition causes the wound to thicken. However, during the maturation phase, collagen becomes aligned along tension lines, and excess water is reabsorbed, which allows the collagen fibers to lie closer together and cross-link (Riwaldt et al., 2021).

After the injury, remodeling typically begins around 21 days and may continue for a year or more. This process includes the whole response to injury, starting with the formation of a fibrin clot and culminating in the formation of a mature scar with a high concentration of type I collagen (J. Li et al., 2007).

By substituting fibronectin, hyaluronan, and proteoglycans for the original fibrin clot and ultimately generating mature collagen fibrils, fibroblasts are essential to the remodeling of the extracellular matrix. The composition of collagen in scar tissue evolves predominantly between type I and type III collagen in granulation tissue as healing progresses, thereby increasing the tensile strength of the scar (Potekaev et al., 2021).

Scar tissue, however, will not be able to perfectly mimic the structure and integrity of skin that is not injured. Whereas collagen fibrils in normal skin have a basket interlace orientation, they adopt enormous parallel bundles in scar tissue. Consequently, only around 80% of the pre-wounding strength is restored by scar tissue following an injury (Witte et al., 1997).

For extracellular matrix remodeling, the balance between collagen synthesis and degradation must be maintained, and this is accomplished by temporally regulating important metalloproteinases. The breakage of natural helical collagens during repair is assisted by these collagenases, which are expressed by a variety of cell types such as anti-inflammatory macrophages, keratinocytes, and fibroblasts (Wang et al., 2018).

Additionally, myofibroblasts are important for wound contraction, which is aided by mechanical strain and increased transforming growth factor beta (TGF-β) expression. These cells, characterized by an abundance of alpha-smooth muscle actin (α-SMA), can result in localized adhesions and high contractile forces. A process known as contracture, where the pseudopodial extensions help in myofibroblast contraction by enabling them to stick together and draw the matrix together (Darby et al., 2014). Eventually, the response of wound healing reduces as endothelial cells, macrophages, and fibroblasts go through apoptosis or exit the area of injury, leaving behind a mature scar (Zhou et al., 2023).

1.4 PRINCIPLES OF WOUND HEALING

Wounding and wound healing are universal processes in all body tissues and organs. These repair processes, while familiar to all tissues, are not uniform. The healing process, though continuous, is divided into distinct phases to facilitate comprehension of the physiological processes occurring in the wound and surrounding tissue. Understanding these phases is crucial as it forms the basis for effective healthcare practices (Broughton et al., 2006b). Healing is a complex course involving coordinated interactions between immunological and biological systems. It consists of a cascade of carefully and precisely regulated steps and events that correlate with the appearance of various cell types in the wound bed during various phases of the healing process. Timing and interactions between the components in the wound healing process vary for acute and chronic wounds. However, the main phases remain the same. The various processes of acute tissue repair that are triggered by tissue injury may be unified into an order of four time-dependent phases: (a) coagulation and hemostasis, beginning immediately after injury; (b) inflammation, which begins shortly after that; (c) proliferation, which starts within days of the injury and encompasses the major healing processes; and (d) wound remodeling, in which scar tissue formation takes place, and which may last up to a year or more (Witte & Barbul, 1997).

On the other hand, chronic wounds, while initiating the healing process, undergo prolonged inflammatory, proliferative, or remodeling phases, leading to tissue fibrosis and nonhealing ulcers. The wound-healing process involves various specialized cells, such as platelets, macrophages, fibroblasts, and epithelial endothelial cells. These cells cooperate with the extracellular matrix. Healing is impacted by the activity of proteins and glycoproteins, including growth factors, cytokines, chemokines, inhibitors, and their receptors, in addition to the different cellular connections. Each stage of wound healing has certain milestones that must occur for routine healing to progress. To truly grasp the differences inherent in chronic wounds that prevent healing, it is crucial to delve into the process of healing in normal wounds.

Injury triggers an organized and complex cascade of cellular and biochemical events that result in a healed wound. For didactic purposes, the wound healing response can be divided into three distinct and overlapping phases: (a) hemostasis and inflammation, (b) proliferation, and (c) maturation or remodeling. Failure or prolongation in one phase may delay healing or lead to nonclosure of the wound. Wound-healing failures remain a significant clinical problem with an enormous impact on healthcare costs. The fundamental physiology of healing is essential for a clearer understanding of the pathophysiologic processes that impair healing.

1.4.1 Hemostasis and Inflammation

The inflammatory phase is a crucial phase of healing, characterized by increased vascular permeability, chemotaxis of cells, local release of cytokines and growth factors, and activation of migrating cells. The hemostatic mechanisms alone would be ineffective over the longer term. Hemostatic events trigger the coagulation cascade via both intrinsic and extrinsic mechanisms, leading to platelet aggregation and clot formation to limit blood loss. As blood spills into the injury site, the platelets and blood components come in contact with exposed collagen and other extracellular matrix components. This contact triggers the release of clotting factors from the platelets. It develops a blood clot composed of fibronectin, fibrin, vitronectin, and thrombospondin (Gantwerker & Hom, 2011). The blood clot and platelets trapped within it are essential for hemostasis. They provide a provisional matrix for cell migration in the subsequent hemostatic and inflammatory phases. The cytoplasm of platelets contains α-granules filled with cytokines and growth factors, like TGF-β, PDGF, epidermal growth factor, and insulin-like growth factors (Locatelli et al., 2021). These molecules act as promoters in the wound-healing cascade by activating and attracting neutrophils, macrophages, endothelial cells, and fibroblasts (Weisel & Litvinov, 2017). Platelets also contain vasoactive amines, such as serotonin, that are stored in dense bodies and cause vasodilation and increased vascular permeability. This leads to fluid extravasation in the tissue, which results in edema, which, in turn, potentiates itself during the following inflammatory phase.

Platelets get activated when exposed to extravascular collagen (like type I collagen), which they detect via specific integrin receptors. These cell surface receptors mediate a cell's interactions with the extracellular matrix (ECM). Platelets release soluble mediators (cyclic AMP and growth factors) and adhesive glycoproteins after contact with collagen, which signals them to become sticky and aggregate. The essential glycoproteins released from the platelet alpha granules include fibronectin, thrombospondin, fibrinogen, and von Willebrand factor. As platelet aggregation proceeds further, clotting factors will be released. Subsequently, a fibrin clot is deposited at the injury site. The fibrin clot serves as a temporary matrix (Locatelli et al., 2021). The aggregated platelets will be trapped in the fibrin web to provide the clot's bulk. Their membranes offer a surface on which inactive clotting enzyme proteases are bound, activated, and accelerate the clotting cascade.

Platelets, the tiny powerhouses of the blood, play a crucial role in the wound-healing cascade. They release a variety of growth factors from their alpha granules, including PDGF, transforming growth factor alpha (TGF-α), TGF-β, insulin-like growth factor-1 (IGF-1), vascular endothelial growth factor (VEGF), and essential fibroblast growth factor (bFGF). These growth factors act as catalysts in the wound-healing cascade, activating and attracting neutrophils, macrophages, endothelial cells, and fibroblasts. To start the inflammatory response, neutrophils and monocytes are drawn from the vasculature by PDGF and TGF-β. F-Met-Leu-Phe, a bacterial waste product and a breakdown fragment produced from complement C5a, also offers further chemotactic cues for luring neutrophils to the injury site. Concurrently, VEGF, TGF-α, and bFGF stimulate endothelial cells to start angiogenesis. After that, PDGF activates and attracts fibroblasts, encouraging them to go to the wound site and start producing collagen and glycosaminoglycans—extracellular matrix proteins that promote cellular migration and interactions with the framework supporting the matrix. Hemostasis, platelet deposition at the site of injury, and interactions between soluble mediators and growth factors and the extracellular matrix thereby mark the beginning of the healing process and set the foundation for later healing activities (P. Martin & Nunan, 2015; Witte & Barbul, 1997).

1.4.1.1 Inflammation

Hemostasis precedes inflammation. The accompanying obligatory rupture of vessels exposes the subendothelial collagen to platelets, aggregating platelets and activating the intrinsic part of the coagulation cascade.

The humoral and cellular inflammatory phase follows next to create an immunological defense against microbes that invade. It is divided into two phases: early and late inflammatory (Sorg et al., 2017).

The following stage of wound healing, inflammation, starts within the first 24 hours of the damage and can last for up to two weeks in nonhealing wounds and much longer in chronic wounds. The granules secreted by mast cells include enzymes, histamine, and other active amines; these are the cause of the telltale symptoms of inflammation, such as rubor (redness), calor (heat), tumor (swelling), and dolour (pain) around the wound site. Neutrophils, monocytes, and macrophages are the critical cells during inflammation. They clear the wound of debris and infection and release soluble mediators, including growth factors, such as, PDGF, TGF-β, TGF-α, IGF-1, and FGF, and proinflammatory cytokines, such as IL-1, IL-6, IL-8, and TNF-α, that aid in the recruitment and activation of fibroblasts and epithelial cells to prepare the wound for the next stage of healing. Cytokines are crucial for controlling inflammation during the healing process of wounds.

In addition to the contraction of chemoattractant cytokine(s), growth factors, and cytokines, a third significant class of tiny regulatory proteins has been discovered. These proteins are collectively referred to as chemokines. The initial lack of recognition of the structural and functional similarities among chemokines resulted in a unique nomenclature made up of numerous acronyms based on biological functions (e.g., monocyte chemoattractant protein-1 [MCP-1], macrophage inflammatory protein-1 [MIP-1]), source of isolation (platelet factor-4 [PF-4]), physiological characteristics (interferon-inducible protein of 10 kDa [IP-10]), or controlled upon regular T-cell expression and secretion upon activation (RANTES). It was discovered that the roughly 40 chemokines could be divided into four main groups according to the arrangement of cysteine residues found close to the N-terminus when their biochemical characteristics were determined.

A recent trend has been reestablishing a more organized nomenclature system based on these four significant classes. Chemokines are primarily involved in two processes in health and development: (1) controlling leukocyte trafficking and (2) controlling the recruitment and activation of neutrophils, lymphocytes, macrophages, eosinophils, and basophils during inflammation (B. Guo et al., 2021).

1.4.1.2 Neutrophils

Neutrophils are the initial inflammatory cells that react to the coagulation cascade and soluble mediators generated by platelets. By eliminating foreign objects, devitalized tissue, and phagocytosing and destroying germs, they act as the body's initial defense against infection. Meaningful interactions between adhesion molecules (selectins, cell adhesion molecules [CAMs], and cadherins) and receptors (integrins) associated with the plasma membranes of circulating leukocytes and vascular endothelial cells occur during the process of extravasation of inflammatory cells into a wound. Leukocytes first attach themselves flimsily to the endothelial cell walls by their selectin molecules, which results in them decelerating and rolling on the surface of endothelial cells. While moving, chemoattractants (growth factors, cytokines, or bacterial products) can activate leukocytes. Following activation, leukocytes attach themselves securely to endothelial cells by binding their integrin receptors to ligands produced on activated endothelial cells, such as VCAM and ICAM.

Leukocytes move into the injured tissue by squeezing between venule endothelial cells and using their integrin receptors to recognize and bind to extracellular matrix components. Chemotactic signals outside the venule trigger this process. The inflammatory cells release the enzymes collagenase and elastase to aid in their migration past the endothelial cell basement membrane and into the extracellular matrix at the wound site. Additionally, inflammatory mediators like TNF-α and IL-1 are produced and released by neutrophils, which attract and activate epithelial and fibroblast cells.

After migration into the wound site, neutrophils produce oxygen-free radicals that kill phagocytized bacteria. They also release large amounts of neutrophil collagenase and neutrophil elastase, two proteases that break down damaged extracellular matrix components. Bacteria that remain in a wound over time can make it more chronic by attracting neutrophils and causing them to generate reactive oxygen species, cytokines, and proteases. Due to apoptosis, neutrophils in the wound often become depleted after two to three days, and tissue monocytes take their place (Lux, 2002).

1.4.1.3 Macrophages

Activated macrophages play pivotal roles in regulating healing, and the healing process does not generally proceed without macrophages. About 24 hours after the damage, circulating monocytes are drawn to the wound site and eventually mature into macrophages. They spread through the same routes as neutrophils. In response to chemokines, cytokines, growth factors, and soluble fragments of extracellular matrix components created by the proteolytic breakdown of collagen and fibronectin, they are induced to develop into active tissue macrophages. Tissue macrophages play two roles in healing, much like neutrophils do. They scavenge and destroy germs while eliminating devitalized tissue with the help of elastase and releasing MMPs as they patrol the wound region. Macrophages secrete inhibitors of the proteases, which allows them to control the proteolytic degradation of wound tissue. This feature sets them apart from neutrophils. They play an essential role in mediating the shift from the inflammatory to the proliferative healing phase, in addition to their phagocytic function.

They release various growth factors and cytokines, including PDGF, TGF-β, TGF-α, FGF, IGF-1, TNFα, IL-1, and IL-6 (Frenette & Wagner, 1996). While some soluble mediators encourage angiogenesis, others draw in and prime fibroblasts for new tissue matrix synthesis, deposition, and organization. The wound is decreasing macrophage count, and the lack of neutrophils suggests that the inflammatory phase is ending and the proliferative phase is about to commence (Ellis et al., 2018).

1.4.2 Proliferative Phase

This phase consists of four processes that co-occur and depend on each other: angiogenesis, granulation tissue formation, re-epithelialization, and wound contraction. VEGF, PDGF, FGF, and TGF-β1 (Figure 1.1) modulate all these phenomena, and diverse signalling pathways are involved. Angiogenesis, the formation of vascularity, provides oxygen and growth factors to induce granulation tissue formation. BFGF, VEGF, and TGF-β signalling pathways stimulate angiogenesis. VEGF is mainly responsible for endothelial proliferation and migration. Blood vessel maturation is promoted via MAPK and PI3K-Akt-eNOS, and the later signalling pathway produces ROS. At the same time, the low generation of ROS stimulates the proliferation and migration of fibroblasts, enhancing collagen production to prepare granulation tissue formation and wound closure. Granulation tissue formation and type III collagen are promoted principally by bFGF and TGF-β and provide the structure for fibroblast and keratinocyte migration and vascular formation.

Re-epithelialization, known by the proliferation and migration of keratinocytes, promotes the closure of wounds mainly stimulated by signalling pathways such as MAPK, FAK-paxillin, PI3K-Akt-mTOR pathways of VEGF, EGF, bFGF, TGF-β, and ROS. Dysfunction of angiogenesis is present in diabetic foot ulcers and burns, highlighting this event's relevance in non-healing conditions. (R. F. Martin, 2020)The acute wound transitions to tissue repair once an ongoing injury has stopped,

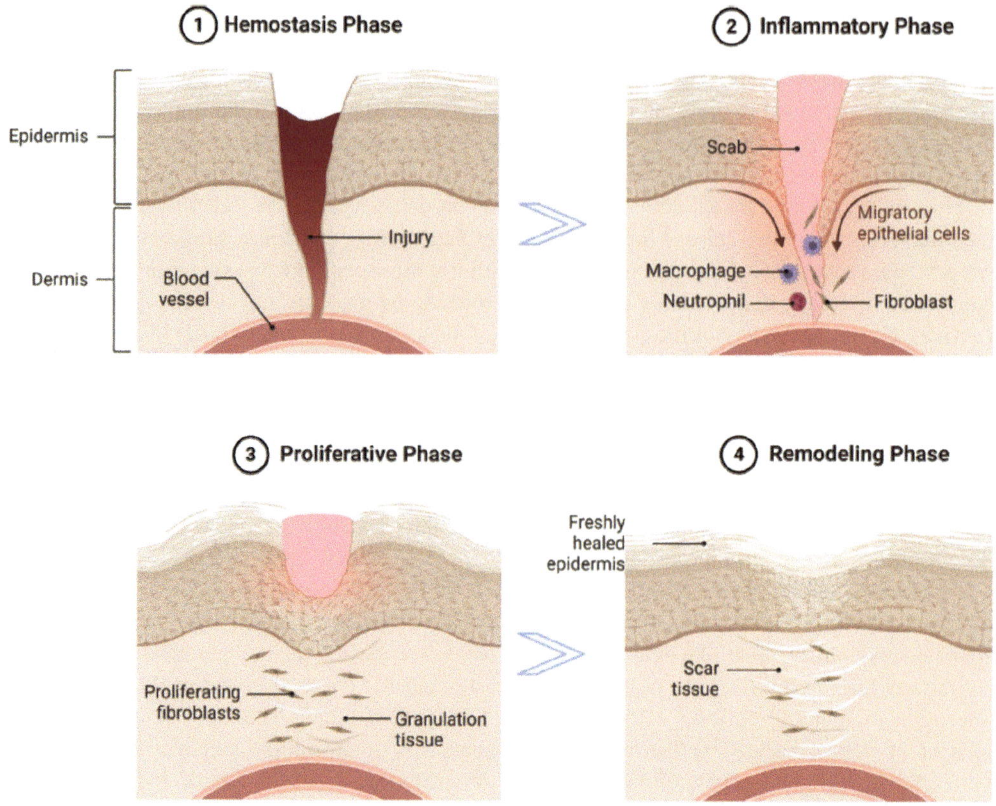

FIGURE 1.1 Illustration of the wound healing process showing the cells involved in each stage.

hemostasis has been reached, and an immune response has taken hold. After injury, the proliferative phase begins on the third day and lasts almost two weeks. It is characterized by fibroblast migration and deposition of a newly synthesized extracellular matrix, replacing the provisional network of fibrin and fibronectin. At the macroscopic level, this phase of wound healing can be seen as an abundant formation of granulation tissue. The diverse processes in the proliferative phase are briefly discussed below (Diegelmann et al., 1981).

1.4.2.1 Fibroblast Migration

For the first three days after damage, the surrounding tissue's fibroblast and myofibroblast counts are stimulated to rise. They then migrate into the wound, attracted by factors such as TGF-β and PDGF released by inflammatory cells and platelets. Fibroblasts initially show up in the wound on the third day following injury, and phenotypic modification is necessary for their accumulation. Once within the wound, they multiply rapidly and generate the matrix proteins type 1 and type 3 procollagen, fibronectin, hyaluronan, and proteoglycans. All of their products are deposited in the local milieu (Foster et al., 2021).

Fibroblasts begin moving by binding to matrix components such as fibronectin, vitronectin, and fibrin via their integrin receptors. Specific amino acid sequences (arginine–glycine–aspartic acid [RGD]) or binding sites in these matrix elements are where integrin receptors bind. The fibroblast finds another binding site by extending a cytoplasmic projection. However, one end of the cell stays attached to the matrix component. Upon locating the subsequent location, the initial site is liberated (perhaps due to local protease activity), and the cell utilizes its network of actin fibers within the cytoskeleton to advance. Proteolytic enzymes are secreted locally by fibroblasts to aid in their migration through the matrix. Fibroblasts secrete three different types of MMPs: stromelysin

(MMP-3), which has various protein substrates in the ECM; collagenase (MMP-1); and gelatinases (MMP-2 and MMP-9), which break down gelatin substrates (Rodrigues et al., 2019).

1.4.2.2 Collagen and Extracellular Matrix Production

Collagen, proteoglycans, and other granulation tissue components are synthesized and placed mainly by fibroblasts. PDGF and TGF-β are two of the most significant growth factors that regulate fibroblast activity. PDGF, predominantly from platelets and macrophages, stimulates several fibroblast functions, including proliferation, chemotaxis, and collagenase expression.

Collagens are an essential component in all phases of wound healing. Fibroblasts synthesize them, giving all tissues strength and integrity. They are essential, particularly during the proliferative and remodeling stages of repair. Collagens are a foundation for the intracellular matrix formation within the wound. Unwounded dermis contains 80% type I and 20% type III collagen, whereas acute wound granulation tissue expresses 30–40% type III collagen (Tottoli et al., 2020).

TGF-β is the primary control signal that governs extracellular matrix deposition and is released by both macrophages and platelets. TGF-β raises the total amount of matrix protein produced by inducing the transcription of genes related to fibronectin, collagen, and proteoglycans. Simultaneously, TGF-β inhibits the release of proteases that break down the matrix and promotes the production of tissue inhibitors called metalloproteinases (TIMP) to impede the matrix's breakdown even further. Recent research suggests that many of the effects of TGF-β on the formation of the extracellular matrix are mediated by a novel growth factor called connective tissue growth factor (CTGF) (Rivera & Spencer, 2007).

Once the fibroblasts have migrated into the matrix, they change their morphology, settle down, and increase and synthesize granulation tissue components, including collagen, elastin, and proteoglycans. Collagen is produced when fibroblasts adhere to the cables of the temporary fibrin matrix (Broughton et al., 2006b).

1.4.3 Remodeling Phase

The last stage of the healing process, remodeling, is when the granulation tissue turns into scar tissue, and its tensile strength rises. During the maturation process of granulation tissue, there is a decrease in cell density and metabolic activity in the granulation tissue, as well as a reduction in the number of capillaries due to aggregation into larger vessels and a decrease in the amount of glycosaminoglycans and the water associated with them (GAGs and proteoglycans). Additionally, collagen type, quantity, and organization alterations occur, improving tensile strength.

The scar is formed in the remodeling or maturation phase, the fibroblast matures to myofibroblasts, and the collagen structure is remodeled. The TGF-β1 and bFGF stay at last to enhance ECM maturing, known as replacement and degradation of type III collagen by type I collagen by the action of collagenases, metalloproteinases, and fibroblasts. In this process, ROS has an active role in enhancing bFGF expression, modulating collagen production, and remodeling the ECM. The principal activated signaling pathways in this phase are the MAPK, Smad, and β-catenin pathways (Table 1.1). The complications associated with this phase are the overexpression of MMP and collagenases that constantly destroy ECM structure in chronic wounds and the underexpression of the latter enzymes and elevated synthesis of type III collagen in excessive scarring wounds such as hypertrophic wounds, burns, and infected wounds (Attinger et al., 2006).

Signaling pathways are the mediators of cellular responses, in which redox signaling is also a critical point in all the wound healing phases. Therefore, ROS at low or controlled concentrations function as a pathogen controller and help to activate proliferation, migration, inflammation, and angiogenesis cell responses. Nonetheless, excess or uncontrolled ROS induce a chronic inflammatory reaction during the inflammation phase, occurring in an impaired wound. In this regard, antioxidants play a vital role in the efficiency and speed of the wound-healing process (Rivera & Spencer, 2007).

TABLE 1.1
Process and Time Frame of Different Stages Involved in the Healing of Wounds

Phase Number	Phase Name	Description	Duration	Key Events
1	Hemostasis	First reaction to injury: formation of a transient clot through platelet aggregation and constriction of blood vessels.	Minutes to hours	Vasoconstriction, platelet aggregation, clot formation
2	Inflammatory	To eliminate debris, pathogens, and injured cells, inflammation begins. Immune cells including neutrophils and macrophages are involved in this phase.	Days	Neutrophil infiltration, macrophage recruitment, removal of foreign material, release of growth factors, and cytokines
3	Proliferative	Angiogenesis (the production of new blood vessels), fibroplasia (the formation of fibrous tissue), and epithelialization (the regeneration of the surface layer) are examples of processes that result in the formation of new tissue.	Days to weeks	Angiogenesis, fibroblast proliferation, collagen synthesis, epithelial cell migration and proliferation
4	Remodeling/ maturation	Longest stage, during which tissue strength grows, wound contraction occurs, and collagen fibers restructure. Depending on the extent of the wound, this period may continue for a few weeks or several years.	Weeks to years	Collagen remodeling, wound contraction, increase in tensile strength, scar maturation, eventual scar remodeling, and reduction in size

1.5 CONCLUSION

Wound healing is a multifaceted but essential part of pathology that requires a number of well-organized procedures intended to preserve the integrity of damaged tissues. Chronic wounds are initiated by a variety of factors; these wounds commonly remain in the inflammatory phase and do not heal on their own without medical intervention. The first step in the healing process is immediate coagulation to stop the loss of blood. Later steps include getting rid of pathogens and debris during inflammation, proliferation to repair tissue structures, and remodeling to strengthen the tissue. Transitioning through the healing phases and diminishing problems can be assisted by effective clinical treatment, which comprises appropriate cleansing, debridement, dressing, ensuring enough perfusion and oxygenation, preventing infection, and providing appropriate nutritional maintenance. It is imperative to distinguish the complex processes and comprehend the elements influencing wound healing, and follow the principles of efficient wound management in order to promote optimum healing outcomes and minimize the impact on patients and healthcare systems.

REFERENCES

Atiyeh, B. S., El-Musa, K. A., & Dham, R. (2003). Scar quality and physiologic barrier function restoration after moist and moist-exposed dressings of partial-thickness wounds. *Dermatologic Surgery*, *29*(1), 14–20. https://doi.org/10.1046/j.1524-4725.2003.29002.x

Attinger, C. E., Janis, J. E., Steinberg, J., Schwartz, J., Al-Attar, A., & Couch, K. (2006). Clinical approach to wounds: Débridement and wound bed preparation including the use of dressings and wound-healing adjuvants. *Plastic and Reconstructive Surgery*, *117*(Supplement), 72S–109S. https://doi.org/10.1097/01.prs.0000225470.42514.8f

Barrientos, S., Stojadinovic, O., Golinko, M. S., Brem, H., & Tomic-Canic, M. (2008). Growth factors and cytokines in wound healing. *Wound Repair and Regeneration*, *16*(5), 585–601. https://doi.org/10.1111/j.1524-475X.2008.00410.x

Baum, C. L., & Arpey, C. J. (2005). Normal cutaneous wound healing: Clinical correlation with cellular and molecular events. *Dermatologic Surgery*, *31*, 674–686. https://doi.org/doi:10.1111/j.1524-4725.2005.31612

Bowler, P. G., Duerden, B. I., & Armstrong, D. G. (2001). Wound microbiology and associated approaches to wound management. *Clinical Microbiology Reviews*, *14*(2), 244–269. https://doi.org/10.1128/CMR.14.2.244-269.2001

Broadbent, E., Petrie, K. J., Alley, P. G., & Booth, R. J. (2003). Psychological stress impairs early wound repair following surgery. *Psychosomatic Medicine*, *65*(5), 865–869. https://doi.org/10.1097/01.PSY.0000088589.92699.30

Broughton, G., Janis, J. E., & Attinger, C. E. (2006a). The basic science of wound healing. *Plastic and Reconstructive Surgery*, *117*(Supplement), 12S–34S. https://doi.org/10.1097/01.prs.0000225430.42531.c2

Broughton, G., Janis, J. E., & Attinger, C. E. (2006b). Wound healing: An overview. *Plastic and Reconstructive Surgery*, *117*(7 Supplement), 1–32. https://doi.org/10.1097/01.prs.0000222562.60260.f9

Broughton, G., Janis, J. E., & Attinger, C. E. (2006c). Wound healing: An overview. *Plastic and Reconstructive Surgery*, *117*(Supplement), 1e-S–32e-S. https://doi.org/10.1097/01.prs.0000222562.60260.f9

Busti, A. J., Hooper, J. S., Amaya, C. J., & Kazi, S. (2005). Effects of perioperative antiinflammatory and immunomodulating therapy on surgical wound healing. *Pharmacotherapy*, *25*(11), 1566–1591. https://doi.org/10.1592/phco.2005.25.11.1566

Centre, L. H. S. (2009). *Introduction to wound healing*. http://www.lhsc.on.ca/Health_Professionals/Wound_Care/intro/structur.htm%5Cr

Cereda, E., Klersy, C., Serioli, M., Crespi, A., & D'Andrea, F. (2015). A nutritional formula enriched with arginine, zinc, and antioxidants for the healing of pressure ulcers. *Annals of Internal Medicine*, *162*(3), 167–174. https://doi.org/10.7326/M14-0696

Chen, L., Deng, H., Cui, H., Fang, J., Zuo, Z., Deng, J., Li, Y., Wang, X., & Zhao, L. (2017, December 14). Inflammatory responses and inflammation-associated diseases in organs. *Oncotarget*, *9*(6), 7204–7218. https://doi.org/doi:10.18632/oncotarget.23208.

Chen, L., & DiPietro, L. A. (2017, October 1). Toll-like receptor function in acute wounds. *Advances in Wound Care*, *6*(10), 344–355. https://doi.org/https://doi.org/10.1089/wound.2017.0734

Darby, I. A., Laverdet, B., Bonté, F., & Desmoulière, A. (2014, November 6). Fibroblasts and myofibroblasts in wound healing. *Clinical, Cosmetic and Investigational Dermatology*, *7*, 301–311. https://doi.org/doi:10.2147/CCID.S50046

Davidson, J. M., & Dipietro, L. (n.d.). *The wound-healing process* (pp. 59–82).

Davis, J., & McLister, A. (2016, June 29). Introduction to wound management. *Smart Bandage Technologies: Design and Application*, *8*, 1.

de Oliveira, S., Rosowski, E. E., & Huttenlocher, A. (2016, May 27). Neutrophil migration in infection and wound repair: Going forward in reverse. *Nature Reviews Immunology*, *16*(6), 378–391. https://doi.org/doi:10.1038/nri.2016.49.

Diegelmann, R. F., Cohen, I. K., & Kaplan, A. M. (1981). The role of macrophages in wound repair. *Plastic and Reconstructive Surgery*, *68*(1), 107–113. https://doi.org/10.1097/00006534-198107000-00025

Ebrecht, M., Hextall, J., Kirtley, L.-G., Taylor, A., Dyson, M., & Weinman, J. (2004). Perceived stress and cortisol levels predict speed of wound healing in healthy male adults. *Psychoneuroendocrinology*, *29*(6), 798–809. https://doi.org/10.1016/S0306-4530(03)00144-6

Edwards, R., & Harding, K. G. (2004). Bacteria and wound healing. *Current Opinion in Infectious Diseases*, *17*(2), 91–96. https://doi.org/10.1097/00001432-200404000-00004

Ellis, S., Lin, E. J., & Tartar, D. (2018). Immunology of wound healing. *Current Dermatology Reports*, *7*(4), 350–358. https://doi.org/10.1007/s13671-018-0234-9

Eming, S. A., Krieg, T., Davidson, J. M. (2007, March 1). Inflammation in wound repair: Molecular and cellular mechanisms. *Journal of Investigative Dermatology*, *127*(3), 514–225. https://doi.org/https://doi.org/10.1038/sj.jid.5700701

Everts, P. A., Lana, J. F., Onishi, K., Buford, D., Peng, J., Mahmood, A., Fonseca, L. F., van Zundert, A., & Podesta, L. (2015). *Angiogenesis and tissue repair depend on platelet dosing and bioformulation strategies following orthobiological platelet-rich plasma*. https://doi.org/https://doi.org/10.3390/biomedicines11071922

Falanga, V. (2005). Wound healing and its impairment in the diabetic foot. *The Lancet, 366*(9498), 1736–1743. https://doi.org/10.1016/S0140-6736(05)67700-8

Foster, D. S., Januszyk, M., Yost, K. E., Chinta, M. S., Gulati, G. S., Nguyen, A. T., Burcham, A. R., Salhotra, A., Ransom, R. C., Henn, D., Chen, K., Mascharak, S., Tolentino, K., Titan, A. L., Jones, R. E., da Silva, O., Leavitt, W. T., Marshall, C. D., des Jardins-Park, H. E., ... Longaker, M. T. (2021). Integrated spatial multiomics reveals fibroblast fate during tissue repair. *Proceedings of the National Academy of Sciences, 118*(41). https://doi.org/10.1073/pnas.2110025118

Frenette, P. S., & Wagner, D. D. (1996). Adhesion molecules — blood vessels and blood cells. *New England Journal of Medicine, 335*(1), 43–45. https://doi.org/10.1056/NEJM199607043350108

Gantwerker, E. A., & Hom, D. B. (2011). Skin: Histology and physiology of wound healing. *Facial Plastic Surgery Clinics of North America, 19*(3), 441–453. https://doi.org/10.1016/j.fsc.2011.06.009

Gosain, A., & DiPietro, L. A. (2004). Aging and wound healing. *World Journal of Surgery, 28*(3), 321–326. https://doi.org/10.1007/s00268-003-7397-6

Gottrup, F. (2004). Oxygen in wound healing and infection. *World Journal of Surgery, 28*(3), 312–315. https://doi.org/10.1007/s00268-003-7398-5

Guo, B., Dong, R., Liang, Y., & Li, M. (2021). Haemostatic materials for wound healing applications. *Nature Reviews Chemistry, 5*(11), 773–791. https://doi.org/10.1038/s41570-021-00323-z

Guo, S., & DiPietro, L. A. (2010). Factors affecting wound healing. *Journal of Dental Research.* https://doi.org/10.1177/0022034509359125

Hayat, M., Nawaz, A., Chinnam, S., Muzammal, M., Latif, M. S., Yasin, M., Ashique, S., Zengin, G., & Farid, A. (2023, July 1). Formulation development and optimization of herbo synthetic gel: In vitro biological evaluation and in vivo wound healing studies. *Process Biochemistry, 130,* 116–126.

James, G. A., Swogger, E., Wolcott, R., Pulcini, E. deLancey, Secor, P., Sestrich, J., Costerton, J. W., & Stewart, P. S. (2008). Biofilms in chronic wounds. *Wound Repair and Regeneration, 16*(1), 37–44. https://doi.org/10.1111/j.1524-475X.2007.00321.x

Järbrink, K., Ni, G., Sönnergren, H., Schmidtchen, A., Pang, C., Bajpai, R., & Car, J. (2017). The humanistic and economic burden of chronic wounds: A protocol for a systematic review. *Systematic Reviews.* https://doi.org/10.1186/s13643-016-0400-8

Ji, Y., Song, W., Xu, L., Yu, D.-G., & Annie Bligh, S. W. (2022). A review on electrospun poly(amino acid) nanofibers and their applications of hemostasis and wound healing. *Biomolecules, 12,* 794. https://doi.org/https://doi.org/10.3390/biom12060794

Jones, E. M., Cochrane, C. A., & Percival, S. L. (2015). The effect of pH on the extracellular matrix and biofilms. *Advances in Wound Care, 4*(7), 431–439. https://doi.org/10.1089/wound.2014.0538

Junker, J. P. E., Kamel, R. A., Caterson, E. J., & Eriksson, E. (2013). Clinical impact upon wound healing and inflammation in moist, wet, and dry environments. *Advances in Wound Care, 2*(7), 348–356. https://doi.org/10.1089/wound.2012.0412

Kamata, M., Sakamoto, Y., & Kishi, K. (2019). Foreign-body reaction to bioabsorbable plate and screw in craniofacial surgery. *Journal of Craniofacial Surgery, 30*(1), e34–e36. https://doi.org/10.1097/SCS.0000000000004945

Keyes, B. E. et al. (2016). Impaired epidermal to dendritic T cell signaling slows wound repair in aged skin. *Cell, 167,* 1323–1338. https://doi.org/doi:10.1016/j.cell.2016.10.052

Koh, T. J., & DiPietro, L. A. (2011, July 11). Inflammation and wound healing: The role of the macrophage. *Expert Reviews in Molecular Medicine, 13,* e23. https://doi.org/doi:10.1017/S1462399411001943

Krzyszczyk, P., Schloss, R., Palmer, A., & Berthiaume, F. (2018, May 1). The role of macrophages in acute and chronic wound healing and interventions to promote pro-wound healing phenotypes. *Frontiers in Physiology, 9,* 419. https://doi.org/doi:10.3389/fphys.2018.00419

Landén, N. X., Li, D., & Ståhle, M. (2016, October). Transition from inflammation to proliferation: A critical step during wound healing. *Cellular and Molecular Life Sciences, 73*(20), 3861–3885. https://doi.org/doi:10.1007/s00018-016-2268-0

LaPelusa, A., & Dave, H. D. (2023, May 1). *Physiology, hemostasis.* Treasure Island, FL: StatPearls Publishing; 2024 Jan–. PMID: 31424847.

Laurens, N., Koolwijk, P. D., & De Maat, M. P. (2006, May 1). Fibrin structure and wound healing. *Journal of Thrombosis and Haemostasis, 4*(5), 932–939. https://doi.org/https://doi.org/10.1111/j.1538-7836.2006.01861.x

Leaper, D. (2006). Effects of local and systemic warming on postoperative infections. *Surgical Infections, 7*(Supplement 2), s-101–s-103. https://doi.org/10.1089/sur.2006.7.s2-101

Li, J., Chen, J., & Kirsner, R. (2007). Pathophysiology of acute wound healing. *Clinics in Dermatology, 25,* 9–18. https://doi.org/doi:10.1016/j.clindermatol.2006.09.007

Li, M., Hou, Q., Zhong, L., Zhao, Y., & Fu, X. (2021, June 16). Macrophage related chronic inflammation in non-healing wounds. *Frontiers in Immunology, 12,* 681710. https://doi.org/doi:10.3389/fimmu.2021.681710

Locatelli, L., Colciago, A., Castiglioni, S., & Maier, J. A. (2021). Platelets in wound healing: What happens in space? *Frontiers in Bioengineering and Biotechnology, 9.* https://doi.org/10.3389/fbioe.2021.716184

Lux, C. N. (2022). Wound healing in animals: A review of physiology and clinical evaluation. *Veterinary Dermatology, 33*(1), 91. https://doi.org/10.1111/vde.13032

Manassa, E. H., Hertl, C. H., & Olbrisch, R.-R. (2003). Wound healing problems in smokers and nonsmokers after 132 abdominoplasties. *Plastic and Reconstructive Surgery, 111*(6), 2082–2087. https://doi.org/10.1097/01.PRS.0000057144.62727.C8

Marrocco, A., & Ortiz, L. A. (2022, October 21). Role of metabolic reprogramming in pro-inflammatory cytokine secretion from LPS or silica-activated macrophages. *Frontiers in Immunology, 13,* 936167. https://doi.org/doi:10.3389/fimmu.2022.936167

Martin, P. (1997). Wound healing--aiming for perfect skin regeneration. *Science, 276*(5309), 75–81. https://doi.org/10.1126/science.276.5309.75

Martin, P., & Nunan, R. (2015). Cellular and molecular mechanisms of repair in acute and chronic wound healing. *British Journal of Dermatology, 173*(2), 370–378. https://doi.org/10.1111/bjd.13954

Martin, R. F. (2020). Wound healing. *Surgical Clinics of North America, 100*(4), ix–xi. https://doi.org/10.1016/j.suc.2020.05.012

Maruyama, K., Asai, J., Ii, M., Thorne, T., Losordo, D. W., & D'Amore, P. A. (2007). Decreased macrophage number and activation lead to reduced lymphatic vessel formation and contribute to impaired diabetic wound healing. *The American Journal of Pathology, 170*(4), 1178–1191. https://doi.org/10.2353/ajpath.2007.060018

Morgun, E. I., & Vorotelyak, E. A. (2020, November 9). Epidermal stem cells in hair follicle cycling and skin regeneration: A view from the perspective of inflammation. *Frontiers in Cell and Developmental Biology, 8,* 581697. https://doi.org/doi:10.3389/fcell.2020.581697.

Mosser, D. M., & Edwards, J. P. (2008). Exploring the full spectrum of macrophage activation. *Nature Reviews Immunology, 8*(12), 958–969. https://doi.org/10.1038/nri2448

Mustoe, T. A., O'Shaughnessy, K., & Kloeters, O. (2006). Chronic wound pathogenesis and current treatment strategies: A unifying hypothesis. *Plastic and Reconstructive Surgery, 117*(Supplement), 35S–41S. https://doi.org/10.1097/01.prs.0000225431.63010.1b

Nourian Dehkordi, A., Mirahmadi Babaheydari, F., Chehelgerdi, M. et al. (2019). Skin tissue engineering: Wound healing based on stem-cell-based therapeutic strategies. *Stem Cell Research & Therapy, 10,* 111. https://doi.org/https://doi.org/10.1186/s13287-019-1212-2

Nwomeh, B. C., Yager, D. R., & Cohen, I. K. (1998). Physiology of the chronic wound. *Clinics in Plastic Surgery, 25*(3), 341–356. https://doi.org/10.1016/S0094-1298(20)32468-8

Okan, D., Woo, K., Ayello, E. A., & Sibbald, G. (2007). The role of moisture balance in wound healing. *Advances in Skin & Wound Care, 20*(1), 39–53. https://doi.org/10.1097/00129334-200701000-00013

Potekaev, N. N., Borzykh, O. B., Medvedev, G. V., Pushkin, D. V., Petrova, M. M., Petrov, A. V., Dmitrenko, D. V., Karpova, E. I., Demina, O. M., & Shnayder, N. A. (2021, December 18). The role of extracellular matrix in skin wound healing. *Journal of Clinical Medicine, 10*(24), 5947. https://doi.org/doi:10.3390/jcm10245947

Regan, M. C., & Barbul, A. (2012). The cellular biology of wound healing. In *Fibrin sealing in surgical and nonsurgical fields.* https://doi.org/10.1007/978-3-642-95720-8_1

Rivera, A. E., & Spencer, J. M. (2007). Clinical aspects of full-thickness wound healing. *Clinics in Dermatology, 25*(1), 39–48. https://doi.org/10.1016/j.clindermatol.2006.10.001

Riwaldt, S., Corydon, T. J., Pantalone, D., Sahana, J., Wise, P., Wehland, M., Krüger, M., Melnik, D., Kopp, S., Infanger, M., & Grimm, D. (2021, June 17). Role of apoptosis in wound healing and apoptosis alterations in microgravity. *Frontiers in Bioengineering and Biotechnology, 9,* 679650. https://doi.org/doi:10.3389/fbioe.2021.679650

Robson, M. C. (1997). Wound infection. *Surgical Clinics of North America, 77*(3), 637–650. https://doi.org/10.1016/S0039-6109(05)70572-7

Rodrigues, M., Kosaric, N., Bonham, C. A., & Gurtner, G. C. (2019). Wound healing: A cellular perspective. *Physiological Reviews, 99*(1), 665–706. https://doi.org/10.1152/physrev.00067.2017

Sen, C. K. (2009). Wound healing essentials: Let there be oxygen. *Wound Repair and Regeneration, 17*(1), 1–18. https://doi.org/10.1111/j.1524-475X.2008.00436.x

Silverstein, P. (1992). Smoking and wound healing. *The American Journal of Medicine, 93*(1), S22–S24. https://doi.org/10.1016/0002-9343(92)90623-J

Singer, A. J., & Clark, R. A. F. (1999). Mechanisms of disease - Cutaneous wound healing. *New England Journal of Medicine, 341*(10), 738–746.

Singh, S., Young, A., & McNaught, C. E. (2017, September 1). The physiology of wound healing. *Surgery (Oxford), 35*(9), 473–477. https://doi.org/10.1016/j.mpsur.2017.06.004

Sood, A., Granick, M. S., & Tomaselli, N. L. (2014). Wound dressings and comparative effectiveness data. *Advances in Wound Care, 3*(8), 511–529. https://doi.org/10.1089/wound.2012.0401

Sorg, H., Tilkorn, D. J., Hager, S., Hauser, J., & Mirastschijski, U. (2017). Skin wound healing: An update on the current knowledge and concepts. *European Surgical Research, 58*(1–2), 81–94. https://doi.org/10.1159/000454919

Stechmiller, J. K. (2010). Understanding the role of nutrition and wound healing. *Nutrition in Clinical Practice, 25*(1), 61–68. https://doi.org/10.1177/0884533609358997

Thomas Hess, C. (2011). Checklist for factors affecting wound healing. *Advances in Skin & Wound Care, 24*(4), 192. https://doi.org/10.1097/01.ASW.0000396300.04173.ec

Tottoli, E. M., Dorati, R., Genta, I., Chiesa, E., Pisani, S., & Conti, B. (2020). Skin wound healing process and new emerging technologies for skin wound care and regeneration. *Pharmaceutics, 12*(8), 735. https://doi.org/10.3390/pharmaceutics12080735

Um, J., Yu, J., & Park, K. S. (2017). Substance P accelerates wound healing in type 2 diabetic mice through endothelial progenitor cell mobilization and Yes-associated protein activation. *Molecular Medicine Reports, 15*, 3035–3040. https://doi.org/doi:10.3892/mmr.2017.6344

Waisbren, E., Rosen, H., Bader, A. M., Lipsitz, S. R., Rogers, S. O., & Eriksson, E. (2010). Percent body fat and prediction of surgical site infection. *Journal of the American College of Surgeons, 210*(4), 381–389. https://doi.org/10.1016/j.jamcollsurg.2010.01.004

Walburn, J., Vedhara, K., Hankins, M., Rixon, L., & Weinman, J. (2009). Psychological stress and wound healing in humans: A systematic review and meta-analysis. *Journal of Psychosomatic Research, 67*(3), 253–271. https://doi.org/10.1016/j.jpsychores.2009.04.002

Wallace, H. A., Basehore, B. M., & Zito, P. M. (2023, June 12). *Wound healing phases.* Treasure Island, FL: StatPearls Publishing; 2024 Jan–. PMID: 29262065

Wang, X., & Khalil, R. A. (2018). Matrix metalloproteinases, vascular remodeling, and vascular disease. *Advances in Pharmacology, 81,* 241–330. https://doi.org/doi:10.1016/bs.apha.2017.08.002

Weisel, J. W., & Litvinov, R. I. (2017). *Fibrin formation, structure and properties* (pp. 405–456). https://doi.org/10.1007/978-3-319-49674-0_13

Wilkinson, H. N., & Hardman, M. J. (2020, September 30). Wound healing: Cellular mechanisms and pathological outcomes. *Open Biology, 10*(9), 200223. https://doi.org/10.1098/rsob.200223

Witte, M. B., & Barbul, A. (1997). General principles of wound healing. *Surgical Clinics of North America, 77*(3), 509–528. https://doi.org/10.1016/S0039-6109(05)70566-1

Wysocki, A. B. (2002). Evaluating and managing open skin wounds: Colonization versus infection. *AACN Clinical Issues: Advanced Practice in Acute and Critical Care, 13*(3), 382–397. https://doi.org/10.1097/00044067-200208000-00005

Xue, J., Schmidt, S. V., Sander, J., Draffehn, A., Krebs, W., Quester, I., De Nardo, D., Gohel, T. D., Emde, M., Schmidleithner, L., & Ganesan, H. (2014, February 20). Transcriptome-based network analysis reveals a spectrum model of human macrophage activation. *Immunity, 40*(2), 274–288. https://doi.org/doi.org/10.1016/j.immuni.2014.01.006

Young, A., Mcnaught, C., & Young, A. (2016, October). *The Physiology of Wound Healing, 9*(2011), 11–12. https://doi.org/10.1016/j.mpsur.2011.06.011

Zhou, S., Xie, M., Su, J., Cai, B., Li, J., & Zhang, K. (2023, July). New insights into balancing wound healing and scarless skin repair. *Journal of Tissue Engineering, 14.* https://doi.org/https://doi.org/10.1177/20417314231185848

2 Molecular Basis of Wound Healing in Diabetes

Nitish Kumar, Bharti Verma, Rahul Kumar, Avani Sharma, Deepika Singh, and Nidhi Tyagi*

SRM Institute of Science and Technology (Deemed to be University), Uttar Pradesh, India
*Corresponding Author

2.1 INTRODUCTION

In the 21st century, diabetes mellitus (DM), which was formerly viewed as a relatively unimportant health issue worldwide, has become a serious threat to human health (Hayat et al., 2023). DM is a chronic illness marked by high blood glucose levels brought on by either insufficient or inefficient insulin production or usage by the body. Another way to characterize DM is as a chronic illness that affects the metabolism of fats, carbohydrates, and proteins, impacting a large number of populations in the world (Patel et al., 2012). The two most common forms of diabetes are type 1 diabetes (T1DM) and type 2 diabetes (T2DM), both of which clearly show genetic variability. Type 1 diabetes, which is insulin-dependent and juvenile onset, is characterized by the rate of autoimmune pancreatic β-cell destruction that leads to complete insulin insufficiency. Just 5–10% of people with diabetes have this type of the disease (Bereda, 2021). Adult-onset type 2 diabetes, which is insulin-independent, is characterized by peripheral insulin resistance and insufficient insulin production by pancreatic beta cells. The majority of diabetics (95%) have this type of the disease (Bereda, 2021). Gestational diabetes mellitus (GDM) is another type of diabetes that is mostly linked to pregnancy. GDM includes a wide spectrum of hyperglycemia, from glucose levels indicative of overt diabetes found in early pregnancy (less than 20 weeks of gestation) to mild impaired glucose tolerance (IGT) or impaired fasting glucose (IFG) found in late pregnancy (McIntyre et al., 2019). Numerous severe signs and symptoms of DM include polyuria, polydipsia, polyphagia, and hyperosmolar coma, as well as several long-term consequences, such as diabetic foot, nephropathy, diabetic neuropathy, diabetic retinopathy, and weight loss, which is frequently accompanied by skin rashes (El Sayed et al., 2023).

Globally, 537 million individuals suffer from diabetes. About 38.4 million Americans in the US have diabetes; 8.7 million do not have a diagnosis. Furthermore, 97.6 million persons over the age of 18 and 27.2 million adults over the age of 65 have prediabetes (National Diabetes Statistics Report, 2021). The frequency of diabetes has increased during the past 20 years worldwide, and it is predicted that there will be 578 million cases by 2030 and 780 million cases by 2045 (Aschner et al., 2021). Diabetes is associated with around 5 million deaths annually and accounts for 50–70% of limb amputations (Patel et al., 2019).

Patients with diabetes have difficulties healing wounds because their glucose metabolism is compromised, which can result in hyperglycemia. Twenty to forty percent of diabetics have chronic sores that never heal. Longer wound healing periods are correlated with higher levels of hyperglycemia and HbA1c; estimates of the lifetime risk of developing diabetic foot ulcers (DFUs) range from 5% to 25%

DOI: 10.1201/9781003504948-2

(Hegazi et al., 2024). Ten to twenty percent of diabetic patients have DFUs, which carry a high risk of amputation and substantial healthcare expenses (Alexiadou et al., 2012). Within five years after an amputation, mortality rates can range from 13% to 18% (Mieczkowski et al., 2022). The startlingly high mortality rates following amputations underscore the critical need for improved diagnostic and therapeutic approaches. It is imperative to address this issue because the cost of DFUs is predicted to increase from USD 7.03 billion in 2019 to USD 11.05 billion by 2027 (Jhamb et al., 2016).

Understanding the molecular processes behind wound healing is crucial for managing and preventing diabetes. Molecular information enables the development of customized therapies that target specific dysregulated pathways and accelerate wound healing. It may be possible to improve wound healing in diabetics by concentrating on specific chemicals or mechanisms associated with inflammation, angiogenesis (the formation of new blood vessels), and tissue repair. Beata Mrozikiewicz-Rakowska et al. investigate the safety and effectiveness of allogenic adipose-derived stem cells (ADSCs) in the treatment of diabetic foot ulcers in humans. An important step toward the development of cell-based therapeutics is the correlation between the molecular background and the improvement in wound healing that ADSCs contribute to (Mrozikiewicz-Rakowska et al., 2023). When treating big, complex DFUs, Shawn Cazzell et al. assessed the safety and effectiveness of DermACEL, an acellular dermal matrix allograft. D-ADM showed promise in quickly shrinking the size of large, complex DFUs. Remarkably, D-ADM was applied just once to all wounds that fully healed (Cazzell et al., 2019).

The purpose of this chapter is to analyze the existing literature on the molecular and genetic factors involved in wound healing in diabetes. In this chapter, we provide an overview of the mechanisms underlying impaired wound healing in diabetic individuals. This chapter delves into the pathogenesis of diabetic wounds, highlighting the significance of key molecules and diverse cellular signaling pathways in diabetic wound healing, as well as the ways in which diabetes modifies these pathways. Last, we underline some potential therapeutic approaches targeting molecular pathways, such as various gene therapies that improve the wound repair process.

Interventions that have improved clinical outcomes for people with DFUs include some strategies that fall under standard of care therapies and advanced therapies, with the standard of care treatment involving wound debridement, offloading, and glycemic and infection control. Advanced therapies include hyperbaric oxygen therapy (HBOT), wound dressings, negative pressure wound therapy (NPWT), and growth factor therapies such as platelet-rich plasma, stem cells, skin replacement therapies, radio wave or electric therapy, and cell- and tissue-based products (Burgess et al., 2021). Many novel treatments are currently available, and well-conducted randomized controlled trials (RCTs) have provided evidence of their efficacy. Sahar Moeini et al.'s (2022) clinical investigation examines the effectiveness of the inforatio approach, which aids in the healing of deep fat ulcers.

2.2 MOLECULAR PATHOGENESIS OF DIABETIC WOUND HEALING

Wound healing involves a complex sequence of cellular and molecular processes, including inflammation, cell proliferation, angiogenesis, collagen deposition, and re-epithelialization (Salazar et al., 2016). Usually, hemostasis—which prevents blood loss and the entry of microorganisms into the injured area—is the first step in the healing process. This phase is quickly followed and overlapped by an inflammatory phase, during which neutrophils, which are pro-inflammatory cells, first upregulate. Macrophages then follow, clearing pathogens, debris, growth factors, and other cells and cytokines. The inflammatory phase is followed by the proliferative phase, during which the injured area is filled with new tissue, angiogenesis (the formation of new blood vessels), and matrix construction. The extracellular matrix's tensile strength is subsequently increased during the final remodeling phase, which also lowers the blood supply to the injured area (Komarcević, 2000; Sharp et al., 2011). Diabetes hinders hemostasis, inflammation, proliferation, and the remodeling phase of wound healing, which delays the healing process and has a long-term detrimental impact on morbidity, mortality, and quality of life (Singh et al., 2011).

There are several factors that might impede and impact the process of wound healing, including both systemic and local ones. The slow healing process in diabetic patients is also caused by a number of physiological factors, including increased serum matrix metalloproteinase-9, impaired collagen accumulation and variation in the ratio of collagen types, dysregulation in neuropeptide expression in the skin along with a suppressed inflammatory response, deficiency of thrombin-activatable fibrinolysis inhibitor, advanced glycation end product (AGE) modification of platelet-derived growth factor (PDGF), decreased number of epidermal nerves, impaired epidermal barrier function, and imbalance between the accumulation of extracellular matrix (ECM) components and their remodeling by matrix metalloproteinase (Patel et al., 2019). Some of the local factors that affect the wound healing process are mentioned in Figure 2.1.

2.3 ROLE OF KEY MOLECULES IN WOUND HEALING

Wound healing is a complex biological process that is initiated whenever tissue integrity is breached (Grazul-Bilska et al., 2003). The process of wound healing is dynamic, involving the activation of

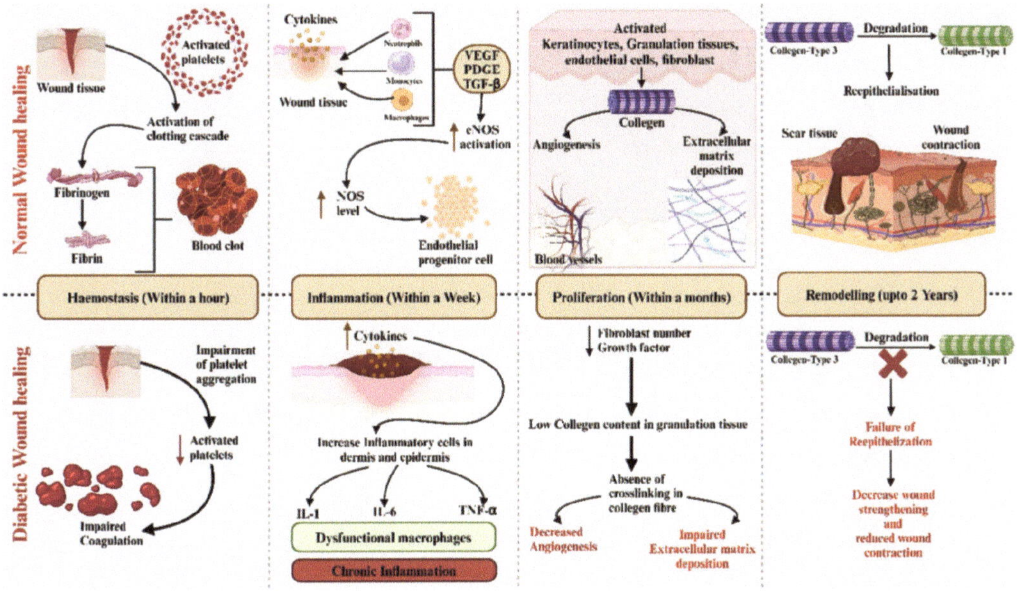

FIGURE 2.1 Pathophysiological phases involved in normal and diabetic wound healing. Hemostasis is the first phase of wound healing, where a platelet plug prevents blood loss and a preliminary release fibrin matrix is formed. In the inflammatory phase that follows, neutrophils, macrophages, and mast cells enter the wound to eradicate bacteria, remove debris, and secrete growth factors like VEGF, which phosphorylates and activates eNOS in the bone marrow, raising NO levels and causing the mobilization of bone marrow EPCs to the circulation. During the proliferation phase, keratinocytes, fibroblasts, endothelial cells, and leukocytes migrate and multiply more in the wound, leading to improved angiogenesis and re-epithelialization processes, and an increase in the synthesis of extracellular matrix components. In the final remodeling phase, myofibroblasts, macrophages, and endothelial cells undergo apoptosis. During this, type III collagen is replaced by type II collagen, which is necessary for the tensile strength and wound contraction. Diabetic wound repair involves altered processes involved in wound closure for people with diabetes mellitus, beginning with a reduction in fibrinolysis and an imbalance of cytokines. Prolonged inflammatory phases are caused by growth factors that enable a constant input of inflammatory cells. Additionally, hyperglycemia causes a reduction in angiogenesis. Reduced collagen production leads to failed re-epithelialization, which in turn reduces wound contraction and tensile strength. VEGF, vascular endothelial growth factor; eNOS, endothelial nitric oxide; EPCs, endothelial progenitor cells; PDGF, platelet-derived growth factor; TGF-β, transforming growth factor beta; NOS, nitric oxide synthase; IL-1, interleukin-1; IL-6, interleukin-6; TGF-α, transforming growth factor alpha.

different cell types, growth hormones, extracellular matrix components, and cytokines to restore lost or injured skin layers (Barrientos et al., 2008). The role of growth factor in wound healing is summarized in Table 2.1.

2.3.1 TGF-β Signaling

Transforming growth factor beta (TGF-β) is a family of pluripotent cytokines that includes three isoforms: TGF-β1, TGF-β2, and TGF-β3. TGF-β1 has a major role in the healing of cutaneous wounds (Steed, 1997). The signaling of TGF-β controls several aspects of wound healing, such as tissue remodeling, ECM deposition, and inflammation. TGF-β stimulates the processes necessary for wound closure, including collagen synthesis, myofibroblast differentiation, and fibroblast activation (Vaidyanathan, 2021). On the other hand, diabetes-related dysregulated TGF-β signaling may lead to fibrosis, excessive scarring, or poor wound healing.

TABLE 2.1
Role of Growth Factors in Diabetic Wound Healing with Their Effects

Growth Factor	Synthesis	Effects	Reference
Epidermal growth factor (EGF)	Platelets, macrophages	Promotes the proliferation and migration of keratinocytes along with the stimulation of the formation and synthesis of ECM.	Steed, 1997
Transforming growth factor	Lymphocytes, macrophages, hepatocytes, platelets	TGF-α: Chemotactic and mitogenic for fibroblasts and keratinocytes. TGF-β1 and TGF-β2: Promote angiogenesis, upregulate collagen production and inhibit degradation, and promote chemoattraction of inflammatory cells. TGF-β3 (antagonist to TGF-β1 and β2): Exhibited scarless wound healing in adults when TGF-β1 and TGF-β2 are inhibited and has been discovered to be highly expressed in fetal scarless wound healing.	Ganesh & Don, 2020
Vascular endothelial growth factor (VEGF)	Endothelial cells	Promotes formation of new blood vessels. Increases migration and functions of endothelial cells.	Diller & Tabor, 2022
Fibroblast growth factor (FGF)	Macrophages, mast cells, T-lymphocytes	Promotes angiogenesis, granulation, and epithelialization via endothelial cell, fibroblast, and keratinocyte migration, respectively.	Steed, 1997
Platelet-derived growth factor (PDGF)	Platelets, macrophages, and endothelial cells	Stimulates function of neutrophils and macrophages while simultaneously attracting macrophages and fibroblasts to the injury site. Promotes collagen and proteoglycan synthesis.	Ganesh & Don, 2020
Interleukins	Macrophages, keratinocytes, endothelial cells, lymphocytes, fibroblasts, osteoblasts, basophils, mast cells	IL-1: Pro-inflammatory, chemotactic for neutrophils, fibroblasts, and keratinocytes. Activates neutrophils. IL-4: Activates fibroblast differentiation. Induces collagen and proteoglycan synthesis. IL-8: Chemotactic for neutrophils and fibroblasts.	Bennett & Schultz, 1993
Keratinocyte growth factor	Fibroblasts	Promotes the migration, differentiation, and proliferation of keratinocytes.	Steed, 1997

It's important to note that wound healing was delayed when both TGF-β1 and lymphocytes were lacking, rather than when either one was present alone (Bennett & Schultz, 1993). It was proposed that this might be because TGF-β2 and TGF-β3 express themselves later when TGF-β1 is not present. TGF-β2 has been implicated in every phase of wound healing, including the induction of angiogenesis and the attraction of fibroblasts and inflammatory cells to the injured area, thus fostering the synthesis of collagen (Valluru et al., 2011). It is thought that the two TGF-β receptors (TβRI and TβRII) are the pathways via which all three isoforms bind and signal. Consistently phosphorylated, TβRII phosphorylates TβRI upon ligand interaction. Through the receptor Smads (Smad-2 and Smad-3) and co-Smad-4, phosphorylation of the receptor complex triggers the SMAD intracellular signaling cascade (Diller et al., 2022). The nuclear membrane is crossed by the receptors Smad-4 and SMADs, which control many genes.

2.3.2 Cytokines

With specialized functions at various phases, cytokines are essential to the healing of wounds. To attract cells required for repair, such as interleukin-1 (IL-1), interleukin-6 (IL-6), and tumor necrosis factor alpha (TNF-α), they first induce inflammation (Wang et al., 2022). Neutrophils and monocytes produce IL-6, which has been demonstrated to be crucial in starting the healing response. Cytokines such as PDGF, fibroblast growth factor (FGF), and TGF-β stimulate the development of new tissue and the expansion of cells as healing proceeds (Schultz et al., 2009).

At higher levels, especially for prolonged periods of time, TNF-α has a detrimental effect on healing. TNF-α suppresses the synthesis of ECM proteins and TIMPs while increasing the synthesis of matrix metalloproteinases (MMPs; MMP-1, MMP-2, MMP-3, MMP-9, MMP-13, and MT1-MMP). Specifically, ECM remodeling and breakdown by MMPs facilitate the formation of new tissue (Ganesh & Don, 2020). In order to ensure appropriate tissue shape and function during wound healing, this regulated breakdown is necessary. Cytokines aid in the last step by strengthening and reshaping the tissue, which promotes effective healing (Schultz et al., 2009).

Hyperglycemia is a major factor in the delayed healing of diabetic wounds. "Metabolic memory" refers to the persistence of diabetic problems in persons with diabetes, even after blood sugar control. This could be connected to changes in the cells' epigenetic makeup brought on by repeated exposure to high glucose levels (Aronson, 2008). The etiology of type 2 diabetes is attributed to a combination of hereditary and environmental variables. Epigenetics plays a significant role in many physiologic and pathologic disorders by acting as a link between genetics and environment. Recent research indicates that the healing process and cell morphologies in diabetics may be impacted by the epigenetic regulation of structural and immune cells within wounds (Negre-Salvayre et al., 2009). Precision medicine approaches to diabetic wound healing will benefit from the study of epigenetics. The study of heritable and nonheritable genetic code is known as epigenetics. A concept related to genetics is epigenetics; "the study of molecules and mechanisms that can perpetuate alternative gene activity states in the context of the same DNA sequence" is used here in accordance with Giacomo's definition (Giugliano et al., 2008). This idea centers on the role that histone modification, DNA methylation, noncoding RNA (ncRNA) regulation, and RNA methylation play in the healing of diabetes wounds (Ju et al.,2024).

2.3.3 Insulin Signaling Pathway

Insulin is essential for controlling the metabolism of glucose and accelerating the healing of wounds. Insulin signaling may be compromised in diabetes, which would reduce target cells' sensitivity to the hormone (Abdelkader et al., 2016). Due to the disruption of many cellular processes involved in tissue repair, this impediment can impede wound healing.

In the proliferation or fibroplasia phase of wound healing, the insulin receptor-mediated PI3K–Akt–Rac1 signal pathway stimulates the migration of keratinocytes and vascular endothelial cells

(Hrynyk et al., 2014). It also inhibits MIP-2 expression, advances neutrophil attenuation, and reduces neutrophil infiltration in wounds. Finally, it regulates the inflammatory response during wound healing (Oryan et al., 2017).

2.3.4 MAPK Signaling Pathway

In response to varied extracellular stimuli, the mitogen-activated protein kinase (MAPK) pathway controls cell division, proliferation, and death. MAPKs, such as p38, JNK, and ERK, facilitate cell migration, proliferation, and ECM remodeling, all of which are important for wound healing (Wang et al., 2022). Diabetes-related dysregulation of MAPK signaling can impede these processes and postpone wound healing. One of the key causes of the delay in wound healing in diabetics is impaired keratinocyte migration brought on by hyperglycemic conditions. The P38/MAPK pathway has been proposed to modulate keratinocyte migration and autophagy during wound healing. Li et al. discovered that under high-glucose conditions, the P38/MAPK pathway was downregulated and autophagy inactivation occurred concurrently. This prevented keratinocyte migration (Schultz et al., 2009). It has been demonstrated that negative-pressure wound therapy can reduce inflammation and accelerate wound healing by downregulating the MAPK/JNK signaling pathway, depicting the essential role of the MAPK pathway in diabetic wound healing.

2.3.5 PI3K/AKT Pathway

Also known as the protein kinase B or PKB pathway, cell migration, proliferation, and survival all depend on the phosphatidylinositol 3-kinase (PI3K)/AKT pathway. This pathway is triggered by insulin, which causes AKT to become phosphorylated and activated. AKT then controls a number of downstream effectors that are important for cell division and survival. When a ligand binds to receptor tyrosine kinases (RTKs), AKT activation triggers the PI3K/AKT pathway (Jere et al., 2019). The gene PTEN (phosphatase and tensin homolog), which is a tumor suppressor and represents both lipid and protein phosphatase, negatively controls PI3K/AKT signaling by dephosphorylating phosphatidylinositol-3,4,5-trisphosphate (PIP3) and converting it to phosphatidylinositol 4,5-bisphosphate (PIP2). It stimulates the phosphorylation of AKT, and some factors like mTOR and GSK-3 accord in the metabolism of glucose and wound healing in DM (Wang et al., 2022). Diabetes-related impaired PI3K/AKT signaling can inhibit wound healing by lowering cell migration and proliferation.

2.3.6 Wnt/β-Catenin Pathway

The Wnt/β-catenin pathway is one of the primary mechanisms that govern wound healing and enhances epithelial remodeling and wound angiogenesis. Wnt family members are implicated in various biological processes, such as cell division, apoptosis, proliferation, and stem cell pluripotency maintenance, according to a plethora of experimental research. There are two types of Wnt signaling: canonical and non-canonical Wnt signaling (Jere et al., 2022). Upon binding to the receptor/coreceptor, the Wnt ligand initiates the Wnt/β-catenin signaling cascade. Certain Wnt proteins that are produced artificially move to the Golgi apparatus, where they attach to Wls, a transmembrane protein that Wnt secretes, before being transported to the cell membrane by the Golgi system. Different signaling pathways are activated when the secreted Wnt ligand interacts with members of the Frizzled protein family (Wang et al., 2022).

Stable conductance of the Wnt/β-catenin signaling pathway depends on the β-catenin stability. In the absence of the Wnt signal, a complex of Axin, casein kinase 1a (CK1a), and glycogen synthase kinase 3b (GSK3b) phosphorylates the free cytosolic β-catenin. Wnt ligand binding to the receptor/coreceptor initiates the Wnt/β-catenin signaling cascade. The Wnt signal promotes the dishevelled (DVL) protein's activation, which deactivates the Axin protein complex. This leads to the accumulation of cytoplasmic β-catenin, which promotes its translocation to the nucleus and the formation of an active transcriptional complex with T-cell-specific factor (TCF) and lymphoid enhancer-binding factor 1 (LEF1) protein transcription (Zhang et al., 2018) (Figure 2.2).

The Wnt signaling pathway is enhanced by the secreted proteins that make up the RSPO protein family, which includes RSPO1 to 4. The stabilization of Wnt receptors and their coreceptors is carried out by RSPOs. It has been claimed that transplanting Wnt signaling-activated cells increases diabetic wound healing since the Wnt/β-catenin pathway is known to enhance diabetic wound restoration (Jere et al., 2022).

2.3.7 Transcription Factors (TFs)

TFs regulate inflammation, cellular proliferation, re-epithelialization, angiogenesis, and the creation of granulation tissue, all of which are involved in the change from the inflammatory to the proliferation phase. Because they regulate the expression of genes involved in vital cellular processes, transcription factors are essential in the healing of diabetic wounds. These proteins encourage resolution and tissue repair by regulating the expression of inflammatory mediators, which in turn controls inflammation. Additionally, they promote the development of new blood vessels for better tissue perfusion by stimulating the angiogenic factors that are produced.

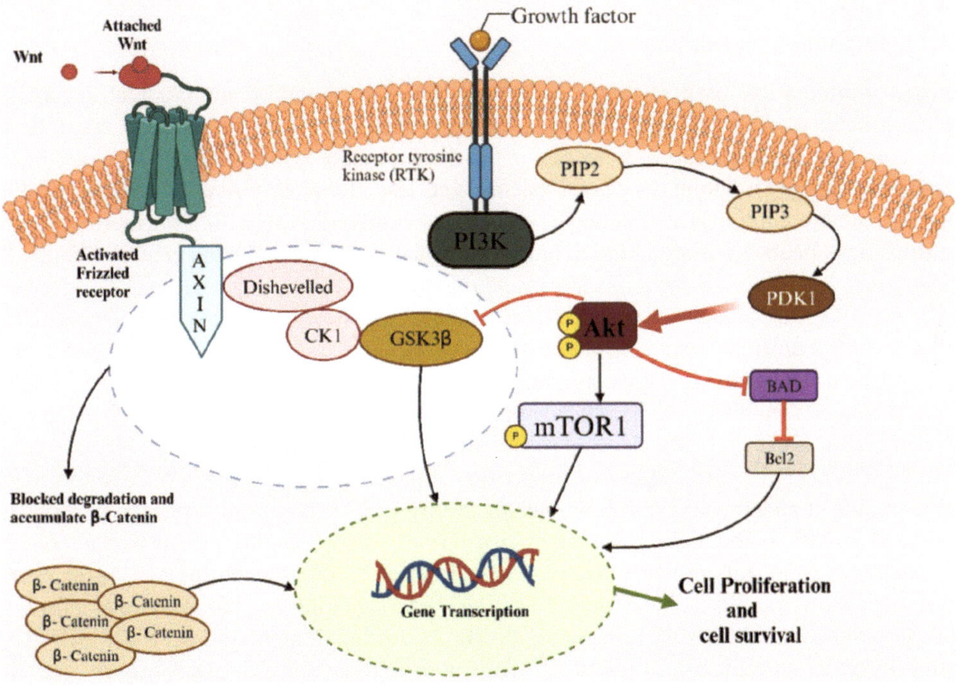

FIGURE 2.2 Mechanism of Wnt and PI3K pathways in diabetic wound healing. The Wnt signaling pathway involves the recruitment of dishevelled (DVL) protein to the cell membrane by Wnt binding to the Frizzled and LRP5/6 receptor complex. This prevents the destruction complex from phosphorylating β-catenin. As a result, β-catenin builds up in the cytoplasm and moves into the nucleus, where it triggers transcription factors that promote the expression of target genes, including Axin2, c-myc, and cyclin D1. Concurrently, transcription and protein translation are interfered with in diabetic wounds by PI3K/AKT signaling. Phosphatidylinositol-4,5-bisphosphate (PI-4,5-P2) is phosphorylated by PI3K, which then recruits AKT and PDK to produce phosphatidylinositol-3,4,5-triphosphate (PI-3,4,5-P3). Activated AKT divides into pathways that control transcription and translation, such as mTOR, Bcl2, and GSK3β. The regulation of cellular activities includes growth factor release, cell proliferation, motility, invasion, survival, and collagen deposition

Additionally, transcription factors influence the expression of ECM constituents and enzymes involved in ECM turnover, which regulate extracellular matrix remodeling and ensure appropriate tissue support and shape. Furthermore, by controlling the expression of genes required for tissue regeneration, they encourage cell migration and proliferation. In addition, transcription factors control the expression of genes related to wound closure and epithelial cell differentiation, which ensures rapid healing.

2.4 CONTRIBUTION OF REACTIVE OXYGEN SPECIES AND HYPERGLYCEMIA TO DIABETIC WOUND HEALING

Reactive oxygen species (ROS) are essential modulators of multiple stages in the healing process of wounds. In fact, the defense against external harm necessitates low ROS levels (Dunnill et al., 2017). Still, one of the main causes of diabetic wounds that do not heal is redox imbalance, which is brought on by increased oxidative stress on tissues and a decline in antioxidant activity (Cano et al., 2018). Clinical research revealed that the highly oxidizing environment linked to tissue hypoxia and hyperglycemia infiltrates diabetic wounds that do not heal, delaying wound healing. The activity of antioxidant enzymes is significantly reduced in people with type 2 diabetes who have the disease for a long time (Dworzanski et al., 2020). Through topical infection, ischemic lesion, neuropathy, and skin injury, oxidative stress may affect the healing of diabetic wounds.

2.4.1 Diabetic Skin Injury

The steps of migration–proliferation, remodeling, coagulation, and inflammation all overlap in the healing process of diabetic skin damage (Armstrong et al., 2018). For quick recovery, a balanced redox state is probably essential. A decline in the diabetic ulcer healing reserve is suggested by the appearance of more inflammatory cells, edema, and less granulation tissue formation in diabetic skin compared to normal skin (Boulton et al., 2020; Armstrong et al., 2018). Furthermore, it seems that diabetic patients are more likely to have tissue near the periwound and central wounds that is vulnerable to high levels of oxidative stress and damage, which impairs healing (Long et al., 2016; Lan et al., 2013). According to Kim et al. (2019), there is a possibility that bacteria and high levels of oxidative stress contribute to the formation of chronic wounds. Studies investigating oxidative stress and hypoxia have revealed that these conditions may cause the downregulation of genes involved in miRNA synthesis in fibroblasts grown in culture (Baldini et al., 2020). Recent studies have shown that diabetes is associated with elevated serum levels of AGEs and their epidermal receptors (Kim et al., 2018; Niu et al., 2012). ROS overproduction is induced by the binding of AGEs to receptors on the surface of endothelial cells (Goldin et al., 2006). ECM proliferation appears to be negatively impacted by AGEs in the diabetic wound microenvironment, which also affects wound contraction and exacerbates the inflammatory response (David et al., 2017). According to a recent study, in keratinocytes involved in diabetic wound healing, Nrf2 signaling activation dramatically boosted TGF-β1 and decreased MMP9 (Long et al., 2016). According to an additional study, SIRT1 (silent mating information regulation 2 homolog) activation may enhance angiogenesis in wounds and speed up wound healing in diabetic mice by protecting vascular endothelial cells from the damaging effects of oxidative stress, and it may be due to the AKT–Nrf2 mechanism (Li et al., 2019).

2.4.2 Diabetic Neuropathy

Approximately half of the population with diabetes will have peripheral neuropathy of clinical significance (Armstrong, 2000; Armstrong et al., 1998, 2017). Skin protection is lost or becomes less effective as a result of sensory nerve dysfunction. Plantar capillary blockage, local tissue ischemia, and tissue death are caused by motor neuropathy, which raises plantar pressure and directly damages the tissue (Obrosova, 2023). Diabetes patients with autonomic neuropathy of the sweat glands experience irregular temperature regulation, decreased sweating, and dry, chapped skin, all of which compromise the integrity of the skin and lower its defense against infection. Additionally, it causes changes in the blood flow to the skin as well as microcirculatory problems such as tension and loss of innervation in the peripheral sympathetic nervous system, which can result in abnormal arteriovenous shunting and voltage malfunction in the placebo response. Consequently, there is a chance of developing nutritional capillary ischemia and abnormal blood flow circulation.

The role of hyperglycemia in the production of oxidative stress has been well studied.

Increased oxidative stress appears to negatively impact the peripheral nerve's blood supply, structure, and metabolism (Rosenberger et al., 2020). This results in widespread degradation of the peripheral nervous system's components, including Schwann cells, myelinated axons, and sensory neurons found in the dorsal root ganglia. Axons are rich in mitochondria, which provide a direct nerve energy source. Insufficient supply of ATP, on the other hand, may cause them to lose their capacity to move axons, further encouraging axonal damage and the development of diabetic neuropathy. Axon oxidative stress damage in hyperglycemic conditions is caused by the loss of the ability to detoxify the excess oxidative stress with an insufficient ATP supply, which accelerates axonal damage or apoptosis (Fernyhough et al., 2014). Numerous biochemical pathways may be affected by oxidative stress, according to studies.

2.4.2.1 Activated Polyol Pathway

The polyol pathway primarily consists of sorbitol dehydrogenase converting sorbitol to fructose and aldose reductase (AR) converting glucose to sorbitol. Elevated intracellular glucose levels in diabetic patients lead to an increase in AR's affinity for glucose, which further results in enhanced sorbitol production. Frequent deposition of sorbitol leads to a decrease in Na^+/K^+-ATPase activity, which lowers nerve cells' physiological reserve and, in turn, lowers nerve conduction velocity. It is also known that the depletion of NADPH, which comes from the pentose phosphate pathway that produces GSH from glutathione, causes the hyperglycemia-induced polyol pathway to create increased oxidative stress. Concurrently, as a byproduct of metabolism, excess fructose increases glycation and NADPH consumption, exacerbating intracellular oxidative stress (Pang et al., 2020). According to diabetic mouse models, sorbitol greatly accumulates in the sciatic nerve of diabetic $AR^{+/+}$ mice. Additionally, there were traces of DNA damage and oxidative stress in the sciatic nerve (Ho et al., 2006). Till now, aldose reductase inhibitors have been employed to block the polyol pathway, which inhibits oxidative stress (Srivastava et al., 2011). This could help treat diabetic ulcers that coexist with neuropathy.

2.4.2.2 Activated Hexosamine Pathway

Fructose-6-phosphate, a metabolic intermediate of glycolysis, is transferred from the glycolytic pathway to the hexosamine pathway in an environment where hyperglycemia is prevalent. It is subsequently transformed into uridine diphosphate N-acetylglucosamine (UDP–GlcNAc). Following this, UDP–GlcNAc binds to the transcription factors' serine and threonine residues (Hosseini et al., 2013). According to studies, the hexosamine pathway is activated, which impairs neuron function. Furthermore, the hexosamine pathway has the ability to stimulate the production of cytokines like TGF-α and TGF-β1 (Sayeski et al., 1996; Kolmlitty et al., 1998).

Hyperglycemia stimulates the hexosamine pathway, which leads to an increase in transcription factor Sp1, which in turn stimulates the overexpression of PAI-1 and TGF-β1 in artery endothelial cells (Du et al., 2000; Chen et al., 1998). Moreover, the hexosamine pathway is induced by hyperglycemia, and thrombospondin 2 (TSP2) is expressed more when nuclear factor kappa B (NF–κB) signaling is present. One important matricellular protein in damage reactions is TSP2. This seems to slow down the healing of diabetic wounds (Kunkemoeller et al., 2019).

2.4.2.3 Activated AGE Pathway

Reversible early glycosylation products can be formed in DM as a result of nonenzymatic glycosylation of glucose interacting chemically with lipids, nucleic acids, and amino acids. In the end, extracellular and intracellular structure and function are changed when AGEs bind with their receptors (Chen et al., 2012).

The generation of the ECM and increased tissue stiffness can be caused by the creation of AGEs on ECM molecules, either directly or indirectly, by activating TGF-β receptors to promote cell proliferation. When AGEs attach to their receptor (RAGE) on the surface of endothelial cells, a signaling cascade begins, which results in the overproduction of ROS, p21, RAS, and MAPKs as well as the activation of NADPH oxidase. One important target of AGE–RAGE signaling is NF–κB.

The AGE–RAGE axis-induced excess oxidative stress triggers NF–κB and influences the transcriptional activation of many cytokines and adhesion molecules, such as tissue factor, endothelin-1, ICAM-1, and E-selectin, which are intimately associated with inflammation and diabetic angiopathy (Goldin et al., 2006). AGEs also increase ROS generation and decrease endothelial NO's bioavailability (Frank et al., 2005; Bucala et al., 1991). Diabetes-related microvascular and macrovascular problems have been linked to the AGE–RAGE axis (Brownlee et al., 1984). Thus, by affecting vasodilation, thickening the capillary basement membrane, and causing endothelial hyperplasia, the AGE pathway process may cause neuropathy. By enhancing apoptosis, lowering fibroblast proliferation, and decreasing the activity of growth factors such as fibroblast growth factor, AGEs adversely affect the healing of diabetic ulcers (Duraisamy et al., 2001). Moreover, through stimulating M1 polarized macrophages, AGE-induced autophagy is linked to delayed cutaneous ulcer healing (Kang et al., 2011; Guo et al., 2016).

2.4.2.4 Activated PKC Pathway

A class of serine/threonine protein kinases known as PKC has critical physiological roles that impact numerous cellular signaling transduction pathways (Freeley et al., 2011). The overproduction of oxygen free radicals in a hyperglycemic environment decreases the activity of glyceraldehyde phosphate dehydrogenase (GAPD). As a result, dihydroxyacetone phosphate (DHAP) is converted more frequently to diacylglycerol (DAG), activating the PKC pathway (Brownlee et al., 2001). Activated PKC can mediate the generation of oxygen free radicals, which can block NO action and cause endothelial function to be damaged. According to recent research, benfotiamine, a lipid-soluble vitamin B1 analog, may promote diabetic wound healing by inhibiting the activation of the AGE and PKC pathways, as well as the hexosamine pathway (Stirban et al., 2006). According to preclinical research, the molecular and cellular makeup of wound beds was considerably enhanced by activating Nrf2-mediated antioxidant mechanisms in the local regeneration environment of diabetic wounds (Rabbani et al., 2018). Acute glucose fluctuations cause the PKC pathway to become active, which in turn causes the peripheral nerve's endothelial dysfunction and experience oxidative stress. Additionally, they trigger the secretion of adhesion molecules and inflammatory factors (Monnier et al., 2006; Ceriello et al., 2008).

2.4.3 Diabetic Peripheral Artery Disease

Diabetic wounds provide a clear description of the role of diabetic angiopathy. Irreversible complicated nonenzymatic glycation, increased oxidative stress and inflammation, endothelial dysfunction, and a hypercoagulable state all contribute to diabetic microangiopathy and macroangiopathy (Goldberg, 2009).

Oxidative stress results in the production of inflammatory mediators, pericyte degeneration, thickening of the basement membrane, endothelial hyperplasia, NO reduction, impaired vasodilation, and an increase in procoagulant biomarkers, such as IL-6, TNF-α, D-dimer, and PAI-1. Oxidative stress is caused by activated biochemical pathways, such as the AGE–RAGE pathway, the polyol pathway, PKC activation, and the hexosamine pathway. However, each of these contributes to the development of diabetic microangiopathy (Balasubramanian et al., 2020; Ngo et al., 2005; Domingueti et al., 2016). Changes in the structure of capillaries or arteriolar vessels restrict the transport of nutrients and white blood cells to particular tissues, make infections more likely to occur and worsen, and hasten the development of diabetic ulcers.

Furthermore, new research suggests that hypoglycemia may be a major factor in the vascular problems of diabetes (Hanefeld et al., 2013). Even under normal glucose environments, hyperglycemia or brief fluctuations in blood sugar levels might cause long-lasting changes in the expression of proatherogenic and OS genes (EI-Osta et al., 2008). Diabetes causes tissue ischemia to worsen and the lumen of major vessels to shrink or obstruct, delaying the repair of diabetic wounds.

2.4.4 Topical Infections

Skin infections, including bacterial and fungal infections, are more common in patients with uncontrolled diabetes mellitus. In turn, infections slow the healing process of the wound. The skin microbiome and the microbiome found in diabetic wounds are intimately connected, exhibiting both individual variations and complicated time-dependent characteristics. Diabetes-related alterations are visible in diabetic wounds. The most common species involved in this is the staphylococcal species (Daeschlein et al., 2019; Spichler et al., 2015; Moffarah et al., 2016).

2.5 ROLE OF ANTIOXIDANTS IN DIABETIC WOUND HEALING

As a mitochondria-targeted antioxidant, SKQ1 raised macrophage infiltration, reduced neutrophil counts, and boosted myofibroblast counts. It also reduced lipid peroxidation levels without changing blood levels of IL-6 and TNF. Treatment with SKQ1 also enhanced cell migration, which improved the genetically diabetic mice's ability to heal their skin wounds (Demyanenko et al., 2017).

Through the scavenging of ROS, regeneration of endogenous and exogenous antioxidants, restoration of oxidized proteins, suppression of NF–κB, and modulation of gene transcription, α-lipoic acid (ALA) has been shown to slow the development of diabetic neuropathy. It is now frequently utilized in the clinic to treat diabetic microangiopathy (Park et al., 2004; Xiang et al., 2008; Ametov et al., 2003).

Coenzyme Q10 (CoQ10), an endogenously produced lipid and vitamin-like molecule that is mostly found in mitochondria, reduces apoptosis and oxidative stress (Galeshkalami et al., 2019). CoQ10 is an electron transporter in the respiratory chain of the mitochondria, which raises ATP production and intensifies the antioxidant activity of the mitochondria (Bhadri et al., 2013).

Aldose reductase inhibitors are used to block the polyol pathway from oxidative stress (Srivastava et al., 2011), which may help diabetic ulcers with neuropathy.

Naturally occurring as an antioxidant found in fruits and vegetables, ferulic acid (FA) prevents lipid peroxidation and enhances the production of antioxidant enzymes such as glutathione, superoxide dismutase, catalase, nitric oxide, and serum zinc and copper. This suggests that FA may aid in the healing process of diabetic ulcers (Ghaisas et al., 2014). Additionally, syringic acid therapy enhances wound healing by promoting migration and proliferation (Ren et al., 2019).

Dietary antioxidant chlorogenic acid has been linked to accelerated wound healing, increased hydroxyproline content, decreased malondialdehyde/nitric oxide levels, and increased reduced glutathione levels in wound beds, according to animal studies (Bagdas et al., 2015). Nevertheless, more studies are required to validate these findings.

By lowering the level of ROS and increasing the number of M2 phenotype macrophages surrounding the wound, the drug-loaded ROS-scavenging hydrogel promotes wound closure.

Furthermore, these hydrogels that develop in wounds also promote the release of colony-stimulating factors from granulocyte macrophages to deal with external bacteria and enhance wound closure (Zhao et al., 2020).

N-acetyl cysteine (NAC) is an antioxidant that has been shown to enhance angiogenesis and endothelial cell activity. Mohamed et al. found that taking NAC on a regular basis enhanced muscle fiber damage, neovascularization, perfusion, and post-amputation stump healing (Babizhayev et al., 2015; Zayed et al., 2017).

2.6 ANGIOGENESIS AND VASCULAR FUNCTION

Angiogenesis is the process by which existing vasculature turns into new capillaries (Kolte et al., 2016). A balance of pro-angiogenic and anti-angiogenic factors controls both pathologic and physiologic angiogenesis. The ability to provide nutrients to tissues in a controlled manner, as well as their refinement and integrity, are characteristics of the vessels created during nonpathologic angiogenesis (Fukumura et al., 2008). Malignancy is most frequently related to pathological angiogenesis, wherein excessive angiogenesis can terminate as a tumor grows or continue uncontrollably (Kerbel, 2008). In addition, psoriasis, endometriosis, rheumatoid arthritis, and Crohn's disease are also associated with dysfunctional angiogenesis (Nissen et al., 1998). The underlying cause of diabetic retinopathy, atherosclerosis, and numerous other comorbidities affecting diabetic patients has been linked to dysfunctional angiogenesis (Fowler, 2011).

In order to maintain a functional vascular network that is neither proliferating nor contracting, basal levels of pro-angiogenic factors like VEGF and FGF, as well as anti-angiogenic molecules like Ang-1 and pigment epithelium-derived factor (PEDF), are expressed (Demidova-Rice et al., 2012). In the damaged and surrounding vasculature, hypoxia has a significant role in activating endothelial cells (Tandara et al., 2004).

Neutrophils are the initial responders in the acute phase of inflammation, and the innate immune system brings leukocytes to the damaged site in this hypoxic environment (Larjava et al., 2012). Persistent angiogenesis is a defining characteristic of the proliferative phase of wound healing. Numerous pro-angiogenic components are created in wounds subsequent to the oxygen gradient established by the injury. These substances, with VEGF being the most prominent among them, encourage capillaries to evolve into developing juvenile loops and branches. One of the most significant angiogenic agents in wounds has been demonstrated to be VEGF, whose synthesis occurs downstream of hypoxia. Hypoxia-inducible factor-1 (HIF-1), a transcriptional activator that upregulates target genes including VEGF-A to induce angiogenesis, is activated by hypoxia after injury (Liu et al., 2008). The primary isoform of VEGF-A in the wound attaches to endothelial cells' receptors to control the formation of new blood vessels (Eming et al., 2007). Starting from the wound periphery and moving into the wound bed, VEGF and other pro-angiogenic factors direct vascular growth toward low oxygen regions (Tandara et al., 2004).

Capillary pruning in wounds is regulated by the release of a number of anti-angiogenic, vascular maturation factors, the best characterized of which are PEDF and Sprouty-2 (SPRY2) (Pries et al., 2014).

PEDF is a powerful anti-angiogenic factor in the vasculature and a member of the serine protease inhibitor (SERPIN) family (Filleur et al., 2009). It has been demonstrated that PEDF causes EC apoptosis and lowers the permeability of neovasculature that is prone to leaking (Broadhead et al., 2010). Metformin, a well-known medication for treating diabetes, was demonstrated in a study by Chen et al. (2016) to reduce prostate cancer cell proliferation, migration, and tumor formation via a mechanism involving the overexpression of PEDF. As a result of its demonstrated ability to inhibit angiogenesis and vascularity in tumors, PEDF is presently being studied as a potential anti-tumor and anti-angiogenic treatment (Becerra et al., 2013). Another component that has been shown to help with capillary remodeling in wounds is (SPRY2), which is an intracellular protein. It reduces the impact of VEGF on EC proliferation in wounds by blocking MAPK signaling (Wietecha et al., 2013).

The angiopoietins 1 and 2, which cooperate with the Tie2 receptor, are another set of factors known to affect the normal course of wound capillary growth and remodeling in addition to PEDF and SPRY2 (Fagiani et al., 2013). Strong maturation factor Ang1 keeps ECs and pericytes in capillaries stable. An antagonistic impact of angiopoietin-2 (Ang2) destabilizes vessels (Staton et al., 2010). When it comes to binding to the Tie2 tyrosine kinase receptor, both angiopoietins compete (Maisonpierre et al., 1997). Therefore, large levels of Ang2 are observed during the proliferative and pro-angiogenic phases of wound healing during the normal angiogenic process, but elevated levels of Ang1 are observed during the maturation phase (Brudno et al., 2013).

The process of neovascularization in response to ischemic conditions is significantly aided by bone marrow-derived endothelial progenitor cells (EPCs), as may be the case in diabetic wounds that are worsened by reduced circulation to the peripheral tissues. Gallagher and colleagues (2007) demonstrate in this issue of the JCI that EPCs in the bone marrow respond to ischemia by following chemokine gradients, which causes these cells to migrate to regions of hypoxia, where they subsequently take part in the construction of new blood vessels.

The process of eNOS activation in the bone marrow, which is thought to be compromised in diabetics, mobilizes bone marrow-derived EPCs and keeps them from significantly accumulating at the wound site. It has been demonstrated that in certain tissues, hyperoxia stimulates eNOS activation (Potter et al., 1999). EPC is mobilized into the bloodstream by means of a rise in bone marrow NOS activation brought on by hyperoxia. The generation of NO is stimulated by increased eNOS, and this helps the bone marrow to generate EPCs. Even though hyperoxia can raise blood levels of circulating EPCs, the cells are not efficiently mobilized to the wound site, which adds another obstacle to the healing process.

2.7 THERAPEUTIC APPROACHES TARGETING MOLECULAR PATHWAYS

It's interesting to note that medications that are frequently used to treat diabetes, such as insulin, metformin, some sulfonylureas, thiazolidinediones, and DPP-4 inhibitors, have demonstrated a wide range of diverse effects in addition to anti-inflammatory capabilities that may be helpful in the management of chronic wounds. Moreover, these drugs have been shown to have positive effects on angiogenesis, granulation tissue creation, MMP decrease, keratinocyte and fibroblast proliferation, and polarization regarding a healing-promoting macrophage phenotype (Salazar et al., 2016). Table 2.2 and Table 2.3 summarized the pre-clinical and clinical interventions respectively in wound healing.

2.7.1 Drug Repurposing in the Treatment of Wound Healing

2.7.1.1 Statins

Statins are among the first-line types of medications used to decrease cholesterol. When statins limit the formation of cholesterol, they also inhibit the synthesis of farnesylpyrophosphate (FPP), which inhibits wound healing, and cortisol. Through keratinocyte migration impairment, they both decrease epithelization by acting on glucocorticoid receptors. Additionally, statins have been shown to have pleiotropic effects that are independent of cholesterol levels. They also possess antioxidant, antimicrobial, and anti-inflammatory qualities, as well as the capacity to improve endothelial function and the results of healing. By encouraging neovascularization, mobilizing endothelial precursor cells from bone marrow, and increasing their proliferation, migration, and survival, simvastatin has been demonstrated to enhance tissue repair (Dimmeler et al., 2001). The improved healing capacity in diabetic animal models may be explained by elevated VEGF and NO content in the wound area (Bitto et al., 2008).

2.7.1.2 Phenytoin

In vitro, phenytoin selectively promotes the proliferation of fibroblasts while inhibiting the proliferation of other cell types (Moy et al., 1985). However, it has also been proposed that fibroblast proliferation can be indirectly controlled by the production of cytokines from keratinocytes (Vijayasingham et al., 1991). The harmed area appears stronger because phenytoin increases the creation of granulation tissue, decreases collagenase activity, and enhances collagen manufacture

TABLE 2.2
Preclinical Studies of Pharmacological Intervention via Molecular Approach in Diabetic Wound Healing

Pathway	Drug and Dose	Preclinical Outcomes	Reference
VEGF, TGF-β, IL-8, MMP-9	Propranolol (8 mg/kg)	Reduced inflammatory cell and MMP-9 levels and increased cell proliferation, mast cell number, collagen deposition, blood vessel density, and nitric oxide levels	Romana-Souza et al., 2009
HIF-1α, SDF-1α, VEGF	Deferoxamine (100 mg/kg)	Increased neovascularization and healing by promoting endothelial tube formation and cell proliferation through upregulation of HIF-1α	Hou et al., 2013
TNF-α, IL-1β, collagen expression	Neurotensin (50 μg/ml)	Reduced inflammatory status, improved healing, re-epithelialization by decreasing inflammatory infiltrate and increasing fibroblast migration and collagen deposition	Moura et al., 2014
TGF-β	Hyaluronic acid (30 mg/kg)	Improved healing by increasing skin remodeling proteins, TGF-β, and transglutaminase-II	Galeano et al., 2011
TGF-β	Propolis (10 mg/kg)	Enhance healing by increasing collagen production via TGF-β1 and smad2, 3 signaling	Hozzein et al., 2015
eNOS	Azelnidipine (3 mg/kg)	Accelerated healing by stimulating NO production and enhancing histologic processes	Bagheri et al., 2011
TNF-α, IL-1β, and IL-6	Naringin (80 mg/kg)	Enhanced healing by inducing angiogenesis and downregulating the expression of TNF-α, IL-1β, and IL-6, and upregulating the expression of IGF-1, VEGF, and TGF-β	Kandhare et al., 2014
MMPs and TIMP-1	Pentoxifylline (25 mg/kg)	Accelerates healing by decreasing MMPs expression and increasing TIMP-1 expression	Babaei et al., 2013
Nrf2, Ang-1, and Tie-2 signaling	Bee venom (200 ug/kg)	Enhanced wound healing by increasing collagen and BD-2 expression and reinstating the levels of Ang-1, Nrf2, and Tie-2 downstream signaling	Hozzein et al., 2018
SIRT1	Resveratrol (50 mg/kg)	Antioxidant activity	Huang et al., 2019

and deposition (Patil et al., 2015). Additionally, it exhibits antibacterial activity by lowering the wound's bacterial load. Phenytoin has demonstrated the capacity to raise FGF and VEGF at the site of the lesion, which in turn stimulates the development of new blood vessels (Pereira et al., 2010). Consequently, biopsies of wound tissue treated with phenytoin show indications of reduced inflammatory cell infiltration, enhanced collagenization, and neovascularization.

2.7.1.3 Metformin

Metformin improves glucose uptake and utilization by peripheral tissues, reduces intestinal glucose absorption, inhibits hepatic glucose synthesis, and improves intestinal glucose uptake to produce its antihyperglycemic effect. It has been demonstrated that metformin also affects other inflammatory pathways, including those mediated by mitogen-activated protein kinase (MAPK)/c-Jun NH2-terminal kinase (JNK) and NF–κB (Ursini et al., 2018). All of these findings suggest that the potential of metformin for the treatment of wound healing may be supported by its immunomodulatory and anti-inflammatory effects. In fact, AMPK activation via mTOR and NOD-like receptor protein 3 (NLRP3) inflammasome inhibition may modify macrophage polarization toward the M2 phenotype, working together to resolve wound healing. After receiving metformin systemically for 14 days, db/db diabetic rats showed improved angiogenesis and faster wound healing. Additionally,

TABLE 2.3
Clinical Data of Pharmacological Intervention in Diabetic Wound Healing

Study Design	Methodology	Duration	Drug and Dose	Clinical Outcomes	Reference
Randomized clinical trial	21 patients (divided into 3 groups)	6 weeks	Angipars (100 mg/kg, 3% gel with oral form)	Increased blood flow and oxygenation in tissues, promoted angiogenesis	Bahrami et al., 2008
Randomized, double-blind, placebo-controlled trial	60 patients (aged 40–85)	12 weeks	Zinc (220 mg, 50 mg/kg)	Enhance lipid profile, insulin resistance, wound healing by the regulation of NF–κB, which activates B-cells and enhances immune response	Momen-Heravi et al., 2017
Randomized clinical trial	37 patients (divided into 2 groups)	21 days	*Actinidia deliciosa* (kiwifruit)	Reduce surface area of ulcer, significantly increased the level of collagen and granulation, and promoted angiogenesis	Mohajeri et al., 2014
Double-blind randomized clinical trial	34 patients (divided into 2 groups)	4 weeks	Olive oil (topical)	Reduced the area of the wound	Nasiri et al., 2015
Randomized, double-blind, placebo-controlled trial	216 patients	12 weeks	Polydeoxy ribonucleotide PDRN (5.625 mg in 3 ml)	Improved wound closure, shortened the time needed for healing, and promoted re-epithelialization	Squadrito et al., 2014

metformin was proven effective in confirming its efficacy in patients with nonhealing lower limb traumatic wounds or ulcers (Zhao et al., 2017). It also accelerated wound healing by enhancing epidermis, hair follicles, and collagen deposition when applied topically in patients experiencing an excision wound (Tawfeek et al., 2020). The clinical response observed was related to enhanced synthesis of TGF-β in the wounded area, a growth factor known to impact angiogenesis, inflammatory processes, granulation tissue formation, ECM deposition, re-epithelization, and remodeling, thereby boosting the healing process (El Gazaerly et al., 2013).

2.7.1.4 Dipeptidyl Peptidase 4 (DPP4) Inhibitors

The transmembrane serine aminopeptidase DPP4, often referred to as CD26, cleaves N-terminal dipeptides in the presence of proline, hydroxyproline, and alanine. In actuality, DPP4 is expressed by keratinocytes, some subsets of macrophages, endothelial and epithelial cells, and hematopoietic progenitor cells (Boonacker et al., 2003). GLP1 (or its analogs) exhibits antioxidant and anti-inflammatory characteristics, and enhances NO production and endothelial NO synthase (eNOS) activation in endothelial cells (Erdogdu et al., 2010), and promotes angiogenesis by encouraging cell migration and proliferation. One of the helper T cell's accessory molecules, DPP4, promotes T cell activation (Morimoto et al., 1998). Additionally, it interacts with antigen T cell caveolin-1, which activates NF–κB and the inflammatory response (Ohnuma et al., 2005). Additionally, DPP4 appears to be important at the skin's surface, where dermal fibroblasts and epidermal keratinocytes are the main enzyme producers. Its expression has been seen to fluctuate during various stages of the wound repair process. More precisely, the existence of elevated DPP4 levels in diabetic wounds in mice is a sign of the maintenance of an inflammatory state that hinders the process of healing. Consequently, suppression of DPP4 activity leads to increased re-epithelialization of poor wound healing in diabetic animals, and DPP4 knockout mice exhibit rapid wound closure (Baticic et al., 2017). DPP4 inhibition in keratinocytes has also been linked to improved epithelial to mesenchymal transition, which is important for wound repair, resulting in increased production of SDF1α. It has also been demonstrated that increased SDF1α at the damage site restores normal recruitment of progenitor cells (Gallagher et al., 2007), hence encouraging angiogenesis. Furthermore, GLP1 seems to also play a role in the remodeling stage of the healing process by inducing the production of TGF-β1, which is essential for the healing event through enhanced fibroblast proliferation, collagen synthesis, stimulation of MMP activity, and extracellular matrix remodeling (Roan et al., 2017). The suppression of DPP4 in a mouse excisional wound model resulted in enhanced wound healing that was characterized by both improved skin thickness and increased vascularization (Whittam et al., 2019).

2.7.2 Topical Therapies

2.7.2.1 Growth Factors

Numerous essential cellular processes involved in the healing process, such as cell migration, division, neovascularization, and the creation of extracellular matrix components, are heavily dependent on growth factors (Hong et al., 2006). Thus, topical growth factor formulations have been employed in wound healing throughout the past 20 years (Saad et al., 2011). The only growth factor that has received US Food and Drug Administration (FDA) approval for use in diabetic foot ulcer treatment is PDGF, which is sold under the brand name Regranex Gel (Smith & Nephew Ltd., London, UK). This was first made available for purchase in 1998 (Margolis et al., 2005). A protein called PDGF is typically present in endothelial cells, macrophages, and platelets. In addition to inducing cell migration and mitosis, it promotes the creation of proteins and extracellular matrix elements. It also activates inflammatory cells and is commonly utilized in DFUs. PDGF therapy has been shown to increase 12-week healing rates to 34% (Niezgoda et al., 2005). Growth factors isolated at high concentrations from platelet-rich plasma (PRP) include recombinant human epidermal growth factor (rh-EGF), nerve growth factor (NGF), recombinant human vascular endothelial growth factor (rhVEGF), and basic fibroblast growth factor (bFGF), which have been used as treatments to stimulate the wound healing process in numerous ways, such as regulating cell growth, proliferation, and differentiation or stimulating angiogenesis (Maier et al., 2013).

2.7.2.2 Cytokines

Activating or attracting anti-inflammatory macrophages (M2 macrophages) to the wounded tissue in order to promote angiogenesis and cell proliferation is one therapeutic approach that may enhance wound healing.

In diabetic wounds, activated macrophages (M1 macrophages) generate a range of pro-inflammatory cytokines, including IL-1, which appears to be essential for maintaining a pro-inflammatory macrophage phenotype.

It has been demonstrated that suppressing the IL-1β pathway in diabetic mice facilitates the change of pro-inflammatory macrophages into reparative or anti-inflammatory macrophages, which in turn stimulates the production of growth factors and, ultimately, the healing of wounds (Mirza et al., 2013). More extensive clinical research is required to assess the safety and effectiveness of these novel therapies and determine whether they should be suggested to promote wound healing and lessen lower limb amputations in diabetics.

2.7.2.3 Negative Pressure Wound Therapy

Local subatmospheric pressure is applied to a wound as part of negative pressure wound therapy (NPWT). Several mechanisms could account for the effectiveness of NPWT, including the elimination of excessive third-space fluid from the affected area (Plikaitis et al., 2006), the decrease in bacterial load, a mechanical impact on the wound bed, or the creation of more granulation tissue with a greater number of small blood vessels through the mobilization of EPCs (Seo et al., 2013). Nonetheless, other research has produced contradictory findings. For example, a retrospective analysis involving 25 patients discovered that 43% of wounds treated with NPWT had an increase in the bacterial bioburden. Positive suggestions for DFUs without ischemia were made by an international agreement about the use of negative pressure therapy in wound healing (Vig et al., 2011). However, additional data are required to support the role of NPWT in standard clinical practice.

2.7.2.4 Stem Cells

Adult stem cells are injected into the wound as part of stem cell-based therapies, which have demonstrated encouraging results in both human trials and research on animals. Furthermore, EPCs are known to be a viable target for therapeutic angiogenesis, which is used to treat both macrovascular and microvascular problems associated with diabetes. Numerous clinical trials have looked at the application of stem cells to diabetic lesions, with encouraging outcomes when injected intramuscularly into the ischemic lower limb of patients with diabetes mellitus and critical limb ischemia. Autologous peripheral blood mononuclear cells (PBMNCs), bone marrow mesenchymal stem cells (BM-MSCs), or bone marrow-derived mononuclear cells (BM-MNCs) substantially enhanced blood flow and resulted in full wound healing. Notably, no major adverse events were reported (Huang et al., 2005). Local application of BM-MSCs using a fibrin polymer spray system resulted in augmented wound healing (Falanga et al., 2007), while autologous bone marrow cells have also been used in diabetic patients with chronic venous and neuroischemic wounds and have been shown to increase tissue blood circulation and enhance wound healing with no systemic complications (Humpert et al., 2005).

2.7.3 Systemic Therapies

2.7.3.1 Hyperbaric Oxygen Therapy

Despite its high cost, hyperbaric oxygen therapy has been generally recommended for the treatment of diabetic foot ulcers that are resistant to other treatments.

It is hypothesized that HBOT will raise oxygen levels in hypoxic wounds, improving leukocyte and fibroblast function, reducing inflammatory cytokines, and encouraging angiogenesis (Dinh et al., 2011). The majority of the data supporting the efficacy of HBOT has come from anecdotal sources; however, there is growing evidence in the medical literature as well.

Abidia et al.'s study (2003), which was double-blinded, randomized, and controlled, demonstrated that, following 30 therapy sessions, the healing rates of chronic diabetic foot ulcers treated with hyperbaric oxygen were superior to those treated with air. However, because only Wagner grade 1 and 2 ulcers were included in the small sample size, this study had limitations.

Because of this, the authors advised larger clinical trials to assess the clinical efficacy and cost-effectiveness of HBOT and cautioned that it should only be used as a supplementary treatment. The effectiveness of HBOT in treating 94 diabetic foot ulcers was examined by Löndahl et al. (2010) in a randomized, single-center, double-blinded, placebo-controlled clinical trial. In contrast to 29% in the placebo group, 52% of patients with Wagner grade 2, 3, or 4 ulcers were cured after a year. This is one of the biggest randomized controlled trials to date, and it has the extra advantages of blinding and placebo control, which removes variables that hampered previous research. Before routine HBOT treatment on all diabetic foot ulcers is contemplated, more research to address these challenging problems may be required (Wu et al., 2010).

2.7.4 Emerging Strategies

2.7.4.1 Gene Therapy

It has been suggested that delivering genes that encode growth factors, like VEGF and PDGF, which promote wound healing, could be a viable therapy option in the future for ulcers that are resistant to the present standard of care. The most well-established method for delivering genes is by viral vectors. Replication-defective adenoviruses encoding PDGF or VEGF have been employed in investigations on humans and animals with positive effects on the wound healing process. More precisely, topical application of a bovine collagen gel containing a replication-defective adenovirus expressing PDGF led to a complete healing of ulcers and a quick reduction in wound size in individuals with neuropathic ulcers and diabetes, with no significant safety issues noted (Mulder et al., 2009). In animal experiments, local delivery of VEGF via an adenovirus vector has also been shown to improve wound healing (Brem et al., 2009).

2.7.4.2 Bioengineered Skin Substitutes

The FDA has approved two medications in this group—Apligraf and Dermagraf (Organogenesis, Inc., Canton, Massachusetts, USA)—for the treatment of DFU. Apligraf is a graft made from neonatal foreskin that consists of a cultured live dermis and a progressively cultivated epidermis. Extracellular matrix, viable allogenic dermal fibroblasts, epidermal keratinocytes, and a stratum corneum are its four constituent parts. Bovine type I collagen lattice arranged into fibrils and proteins made by fibroblasts make up the extracellular matrix. Growth factors that promote wound healing, aid in the development of new dermal tissue, or support the maintenance of the overlying epidermis are produced by dermal fibroblasts. The epidermis is made up of epidermal keratinocytes, which also produce growth factors to promote biologic wound healing involving chemotaxis, angiogenesis, fibrosis, and other cellular processes that are frequently disrupted in chronic wounds (Shen et al., 2003). Dermagraft is an allogeneic culture of human dermal fibroblasts taken from neonates, cultured on a scaffold made of bioabsorbable polyglactin mesh. Collagen and glycosaminoglycans are two examples of the extracellular matrix components that are released by fibroblasts in high concentrations (Marston et al., 2004). Growth factors found in Dermagraft include VEGF, insulin-like growth factor (IGF), heparin-binding epidermal growth factor (HBEGF), PDGF-A, KGF, transforming growth factors (TGF-α, TGF-β1, TGF-β3), and mitogen for keratinocytes (KGF).

2.8 CONCLUSION AND FUTURE PERSPECTIVES

The process of healing a wound is dynamic and complex. Still, the majority of diabetic wounds are not easily healed. Diabetes-related wound healing is a complex symphony of factors including hyperglycemia, neuropathy, blood flow, matrix turnover, wound contraction, and the microbiome. Wound healing is regulated by oxidative stress via a number of connected signaling mechanisms. It is clear from the preceding discussion that endothelial dysfunction in diabetes is complex and results in defective angiogenesis. Some of the factors responsible are elevated ROS and AGEs, lowered cytokines and growth factors, and modified immune cell responses. Scientists are attempting to determine several substances that may offer vascular advantages for diabetes.

A significant obstacle to our understanding has been the absence of easily accessible animal models that accurately mimic the healing of diabetes wounds in humans, particularly those that are chronic. Moreover, research on diabetes wound healing has only emphasized the early phases of the healing process rather than any deficiencies that can impede the resolution of diabetic wounds. In order to develop treatments and interventions that can help these nonhealing wounds return to a state of integrity and health, more research is required to understand the variations that occur in the diabetic condition during all stages of wound healing. Current research on treatments targeted at various variables of illness progression could serve as a supplement to the existing conventional medications. Enhancing clinical approaches and procedures may also aid in determining the degree of endothelial injury, potentially reducing the chance of disease progression. Research on combination therapy may be helpful in improving protection against vascular problems associated with diabetes.

With the increasing understanding of the pathophysiology underlying the development and nonhealing of diabetic foot ulcers, technologies designed to address these problems also emerge.

These technologies include, for instance, topical growth factors, HBOT, and NPWT. But how we treat chronic diabetic foot ulcers is also influenced by medical research innovations created for other fields. The utilization of stem cell therapy as a source of regenerative cells and tissues for use in cell-based therapies, as well as a means of testing new medications, is an area of particular interest.

REFERENCES

Abdelkader, D. H., Osman, M. A., Elgizaway, S. A., Faheem, A., & McCarron, P. A. (2016). The role of insulin in wound healing process: mechanism of action and pharmaceutical applications. *Journal of Analytical & Pharmaceutical Research*, *2*(1).

Abidia, A., Laden, G., Kuhan, G., Johnson, B. F., Wilkinson, A. R., Renwick, P. M., ... McCollum, P. T. (2003). The role of hyperbaric oxygen therapy in ischaemic diabetic lower extremity ulcers: a double-blind randomised-controlled trial. *European Journal of Vascular and Endovascular Surgery*, *25*(6), 513–518.

Alexiadou, K., & Doupis, J. (2012). Management of diabetic foot ulcers. *Diabetes Therapy*, *3*, 1–15.

Armstrong, D. G. (2000). The 10-g monofilament: the diagnostic divining rod for the diabetic foot? *Diabetes Care*, *23*(7), 887–887.

Armstrong, D. G., Boulton, A. J., & Bus, S. A. (2017). Diabetic foot ulcers and their recurrence. *New England Journal of Medicine*, *376*(24), 2367–2375.

Armstrong, D. G., & Gurtner, G. C. (2018). A histologically hostile environment made more hospitable? *Nature Reviews Endocrinology*, *14*(9), 511–512.

Armstrong, D. G., Lavery, L. A., & Harkless, L. B. (1998). Who is at risk for diabetic foot ulceration? *Clinics in Podiatric Medicine and Surgery*, *15*(1), 11–19.

Aronson, D. (2008). Hyperglycemia and the pathobiology of diabetic complications. *Cardiovascular Diabetology: Clinical, Metabolic and Inflammatory Facets*, *45*, 1–16.

Aschner, P., Karuranga, S., James, S., Simmons, D., Basit, A., Shaw, J. E., et al. (2021). The International Diabetes Federation's guide for diabetes epidemiological studies. *Diabetes Research and Clinical Practice*, *172*, 108630. doi:10.1016/j.diabres.2020.108630.

Babaei, S., Bayat, M., Nouruzian, M., & Bayat, M. (2013). Pentoxifylline improves cutaneous wound healing in streptozotocin-induced diabetic rats. *European Journal of Pharmacology*, *700*(1–3), 165–172.

Babizhayev, M. A., Strokov, I. A., Nosikov, V. V., Savel'yeva, E. L., Sitnikov, V. F., Yegorov, Y. E., & Lankin, V. Z. (2015). The role of oxidative stress in diabetic neuropathy: generation of free radical species in the glycation reaction and gene polymorphisms encoding antioxidant enzymes to genetic susceptibility to diabetic neuropathy in population of type I diabetic patients. *Cell Biochemistry and Biophysics*, *71*, 1425–1443.

Bagdas, D., Etoz, B. C., Gul, Z., Ziyanok, S., Inan, S., Turacozen, O., ... Gurun, M. S. (2015). In vivo systemic chlorogenic acid therapy under diabetic conditions: wound healing effects and cytotoxicity/genotoxicity profile. *Food and Chemical Toxicology*, *81*, 54–61.

Bagheri, M., Jahromi, B. M., Mirkhani, H., Solhjou, Z., Noorafshan, A., Zamani, A., & Amirghofran, Z. (2011). Azelnidipine, a new calcium channel blocker, promotes skin wound healing in diabetic rats. *Journal of Surgical Research*, *169*(1), e101–e107.

Bahrami, A., Kamali, K., ALI, A. A., Hosseini, P., Heshmat, R., Khoram, K. H., ... Madani, S. H. (2008). Clinical application of oral form of ANGIPARS™ and in combination with topical form as a new treatment for diabetic foot ulcers: A randomized clinical trial. *Daru*, *16*, 41–48.

Balasubramanian, G., Vas, P., Chockalingam, N., & Naemi, R. (2020). A synoptic overview of neurovascular interactions in the foot. *Frontiers in Endocrinology*, *11*, 513867.

Baldini, E., Testa, E., Voellenkle, C., De Domenico, E., Cianfarani, F., Martelli, F., ... Odorisio, T. (2020). Dysregulation of microRNA expression in diabetic skin. *Journal of Dermatological Science*, *98*(3), 186–194.

Barrientos, S., Stojadinovic, O., Golinko, M. S., Brem, H., & Tomic-Canic, M. (2008). Growth factors and cytokines in wound healing. *Wound Repair and Regeneration*, *16*(5), 585–601.

Baticic Pucar, L., Pernjak Pugel, E., Detel, D., & Varljen, J. (2017). Involvement of DPP IV/CD26 in cutaneous wound healing process in mice. *Wound Repair and Regeneration*, *25*(1), 25–40.

Becerra, S. P., & Notario, V. (2013). The effects of PEDF on cancer biology: mechanisms of action and therapeutic potential. *Nature Reviews Cancer*, *13*(4), 258–271.

Bennett, N. T., & Schultz, G. S. (1993). Growth factors and wound healing: biochemical properties of growth factors and their receptors. *The American Journal of Surgery*, *165*(6), 728–737.

Bereda, G. (2021). Brief overview of diabete mellitus. *Diabetes Manag S*, *1*, 21–27.

Bhadri, N., Sanji, T., Guggilla, H. M., & Razdan, R. (2013). Amelioration of behavioural, biochemical, and neurophysiological deficits by combination of monosodium glutamate with resveratrol/alpha-lipoic acid/coenzyme Q10 in rat model of cisplatin-induced peripheral neuropathy. *The Scientific World Journal*, *2013*, *1*: 565813.

Bitto, A., Minutoli, L., Altavilla, D., Polito, F., Fiumara, T., Marini, H., ... Squadrito, F. (2008). Simvastatin enhances VEGF production and ameliorates impaired wound healing in experimental diabetes. *Pharmacological Research*, *57*(2), 159–169.

Boonacker, E., & Van Noorden, C. J. (2003). The multifunctional or moonlighting protein CD26/DPPIV. *European Journal of Cell Biology*, *82*(2), 53–73.

Boulton, A. J., Armstrong, D. G., Hardman, M. J., Malone, M., Embil, J. M., Attinger, C. E., & Kirsner, R. S. (2020). *Diagnosis and management of diabetic foot infections*. Arlington, VA: American Diabetes Association.

Brem, H., Kodra, A., Golinko, M. S., Entero, H., Stojadinovic, O., Wang, V. M., ... Tomic-Canic, M. (2009). Mechanism of sustained release of vascular endothelial growth factor in accelerating experimental diabetic healing. *Journal of Investigative Dermatology*, *129*(9), 2275–2287.

Broadhead, M. L., Becerra, S. P., Choong, P. F., & Dass, C. R. (2010). The applied biochemistry of PEDF and implications for tissue homeostasis. *Growth Factors*, *28*(4), 280–285.

Brownlee, M. (2001). Biochemistry and molecular cell biology of diabetic complications. *Nature*, *414*(6865), 813–820.

Brownlee, M., Vlassara, H., & Cerami, A. (1984). Nonenzymatic glycosylation and the pathogenesis of diabetic complications. *Annals of Internal Medicine*, *101*(4), 527–537.

Brudno, Y., Ennett-Shepard, A. B., Chen, R. R., Aizenberg, M., & Mooney, D. J. (2013). Enhancing microvascular formation and vessel maturation through temporal control over multiple pro-angiogenic and pro-maturation factors. *Biomaterials*, *34*(36), 9201–9209.

Bucala, R., Tracey, K. J., & Cerami, A. (1991). Advanced glycosylation products quench nitric oxide and mediate defective endothelium-dependent vasodilatation in experimental diabetes. *The Journal of Clinical Investigation*, *87*(2), 432–438.

Burgess, J. L., Wyant, W. A., Abdo Abujamra, B., Kirsner, R. S., & Jozic, I. (2021). Diabetic wound-healing science. *Medicina*, *57*(10), 1072.

Cano Sanchez, M., Lancel, S., Boulanger, E., & Neviere, R. (2018). Targeting oxidative stress and mitochondrial dysfunction in the treatment of impaired wound healing: a systematic review. *Antioxidants*, *7*(8), 98.

Cazzell, S., Moyer, P. M., Samsell, B., Dorsch, K., McLean, J., & Moore, M. A. (2019). A prospective, multicenter, single-arm clinical trial for treatment of complex diabetic foot ulcers with deep exposure using acellular dermal matrix. *Advances in Skin & Wound Care*, *32*(9), 409–415.

Ceriello, A., Esposito, K., Piconi, L., Ihnat, M. A., Thorpe, J. E., Testa, R., ... Giugliano, D. (2008). Oscillating glucose is more deleterious to endothelial function and oxidative stress than mean glucose in normal and type 2 diabetic patients. *Diabetes*, *57*(5), 1349–1354.

Chen, C. Y., Abell, A. M., Moon, Y. S., & Kim, K. H. (2012). An advanced glycation end product (AGE)-receptor for AGEs (RAGE) axis restores adipogenic potential of senescent preadipocytes through modulation of p53 protein function. *Journal of Biological Chemistry*, *287*(53), 44498–44507.

Chen, X., Li, C., He, T., Mao, J., Li, C., Lyu, J., & Meng, Q. H. (2016). Metformin inhibits prostate cancer cell proliferation, migration, and tumor growth through upregulation of PEDF expression. *Cancer Biology & Therapy*, *17*(5), 507–514.

Chen, Y. Q., Su, M., Walia, R. R., Hao, Q., Covington, J. W., & Vaughan, D. E. (1998). Spl sites mediate activation of the plasminogen activator inhibitor-1 promoter by glucose in vascular smooth muscle cells. *Journal of Biological Chemistry*, *273*(14), 8225–8231.

Daeschlein, G., Hinz, P., Kiefer, T., & Jünger, M. (2019). Role of the microbiome in chronic wounds. *Der Hautarzt*, *70*, 422–431.

David, J. A., Rifkin, W. J., Rabbani, P. S., & Ceradini, D. J. (2017). The Nrf2/Keapl/ARE pathway and oxidative stress as a therapeutic target in type II diabetes mellitus. *Journal of Diabetes Research*, *2017*, *1*, 4826724.

Demidova-Rice, T. N., Durham, J. T., & Herman, I. M. (2012). Wound healing angiogenesis: innovations and challenges in acute and chronic wound healing. *Advances in Wound Care*, *1*(1), 17–22.

Demyanenko, I. A., Zakharova, V. V., Ilyinskaya, O. P., Vasilieva, T. V., Fedorov, A. V., Manskikh, V. N., ... Popova, E. N. (2017). Mitochondria-targeted antioxidant SkQ1 improves dermal wound healing in genetically diabetic mice. *Oxidative Medicine and Cellular Longevity*, *2017*, *1*, 6408278.

Diller, R. B., & Tabor, A. J. (2022). The role of the extracellular matrix (ECM) in wound healing: a review. *Biomimetics*, *7*(3), 87.

Dimmeler, S., Aicher, A., Vasa, M., Mildner-Rihm, C., Adler, K., Tiemann, M., ... Zeiher, A. M. (2001). HMG-CoA reductase inhibitors (statins) increase endothelial progenitor cells via the PI 3-kinase/Akt pathway. *The Journal of Clinical Investigation*, *108*(3), 391–397.

Dinh, T., Elder, S., & Veves, A. (2011). Delayed wound healing in diabetes: considering future treatments. *Diabetes Management*, *1*(5), 509.

Domingueti, C. P., Dusse, L. M. S. A., das Graças Carvalho, M., de Sousa, L. P., Gomes, K. B., & Fernandes, A. P. (2016). Diabetes mellitus: the linkage between oxidative stress, inflammation, hypercoagulability and vascular complications. *Journal of Diabetes and its Complications*, *30*(4), 738–745.

Du, X. L., Edelstein, D., Rossetti, L., Fantus, I. G., Goldberg, H., Ziyadeh, F., ... Brownlee, M. (2000). Hyperglycemia-induced mitochondrial superoxide overproduction activates the hexosamine pathway and induces plasminogen activator inhibitor-1 expression by increasing Spl glycosylation. *Proceedings of the National Academy of Sciences*, *97*(22), 12222–12226.

Dunnill, C., Patton, T., Brennan, J., Barrett, J., Dryden, M., Cooke, J., ... Georgopoulos, N. T. (2017). Reactive oxygen species (ROS) and wound healing: the functional role of ROS and emerging ROS-modulating technologies for augmentation of the healing process. *International Wound Journal*, *14*(1), 89–96.

Duraisamy, Y., Slevin, M., Smith, N., Bailey, J., Zweit, J., Smith, C., ... Gaffney, J. (2001). Effect of glycation on basic fibroblast growth factor induced angiogenesis and activation of associated signal transduction pathways in vascular endothelial cells: possible relevance to wound healing in diabetes. *Angiogenesis*, *4*, 277–288.

Dworzański, J., Strycharz-Dudziak, M., Kliszczewska, E., Kiełczykowska, M., Dworzańska, A., Drop, B., & Polz-Dacewicz, M. (2020). Glutathione peroxidase (GPx) and superoxide dismutase (SOD) activity in patients with diabetes mellitus type 2 infected with Epstein-Barr virus. *PLos One*, *15*(3), e0230374.

El Gazaerly, H., Elbardisey, D. M., Eltokhy, H. M., & Teaama, D. (2013). Effect of transforming growth factor Beta 1 on wound healing in induced diabetic rats. *International Journal of Health Sciences*, *7*(2), 160.

El-Osta, A., Brasacchio, D., Yao, D., Pocai, A., Jones, P. L., Roeder, R. G., ... Brownlee, M. (2008). Transient high glucose causes persistent epigenetic changes and altered gene expression during subsequent normoglycemia. *The Journal of Experimental Medicine*, *205*(10), 2409–2417.

ElSayed, N. A., Aleppo, G., Aroda, V. R., Bannuru, R. R., Brown, F. M., Bruemmer, D., et al. (2023). Classification and diagnosis of diabetes: Standards of care in diabetes—2023. *Diabetes Care*, *46*, S19–S40. doi:10.2337/dc23–S002

Eming, S. A., Brachvogel, B., Odorisio, T., & Koch, M. (2007). Regulation of angiogenesis: wound healing as a model. *Progress in Histochemistry and Cytochemistry*, *42*(3), 115–170.

Erdogdu, Ö., Nathanson, D., Sjöholm, Å., Nyström, T., & Zhang, Q. (2010). Exendin-4 stimulates proliferation of human coronary artery endothelial cells through eNOS-, PKA-and PI3K/Akt-dependent pathways and requires GLP-1 receptor. *Molecular and Cellular Endocrinology*, *325*(1–2), 26–35.

Fagiani, E., & Christofori, G. (2013). Angiopoietins in angiogenesis. *Cancer Letters*, *328*(1), 18–26.

Falanga, V., Iwamoto, S., Chartier, M., Yufit, T., Butmarc, J., Kouttab, N., ... Carson, P. (2007). Autologous bone marrow–derived cultured mesenchymal stem cells delivered in a fibrin spray accelerate healing in murine and human cutaneous wounds. *Tissue Engineering*, *13*(6), 1299–1312.

Fernyhough, P., & McGavock, J. (2014). Mechanisms of disease: mitochondrial dysfunction in sensory neuropathy and other complications in diabetes. *Handbook of Clinical Neurology*, *126*, 353–377.

Filleur, S., Nelius, T. I. O. V. S., De Riese, W., & Kennedy, R. C. (2009). Characterization of PEDF: a multi-functional serpin family protein. *Journal of Cellular Biochemistry*, *106*(5), 769–775.

Fowler, M. J. (2011). Microvascular and macrovascular complications of diabetes. *Clinical Diabetes*, *29*(3), 116–122.

Frank, G. D., Eguchi, S., & Motley, E. D. (2005). The role of reactive oxygen species in insulin signaling in the vasculature. *Antioxidants & Redox Signaling*, *7*(7–8), 1053–1061.

Freeley, M., Kelleher, D., & Long, A. (2011). Regulation of protein kinase C function by phosphorylation on conserved and non-conserved sites. *Cellular Signalling*, *23*(5), 753–762.

Fukumura, D. A. I., & Jain, R. K. (2008). Imaging angiogenesis and the microenvironment. *Apmis*, *116*(7–8), 695–715.

Galeano, M., Polito, F., Bitto, A., Irrera, N., Campo, G. M., Avenoso, A., ... Altavilla, D. (2011). Systemic administration of high-molecular weight hyaluronan stimulates wound healing in genetically diabetic mice. *Biochimica et Biophysica Acta (BBA)-Molecular Basis of Disease*, *1812*(7), 752–759.

Galeshkalami, N. S., Abdollahi, M., Najafi, R., Baeeri, M., Jamshidzade, A., Falak, R., ... Hosseini, A. (2019). Alpha-lipoic acid and coenzyme Q10 combination ameliorates experimental diabetic neuropathy by modulating oxidative stress and apoptosis. *Life Sciences*, *216*, 101–110.

Gallagher, K. A., Liu, Z. J., Xiao, M., Chen, H., Goldstein, L. J., Buerk, D. G., ... Velazquez, O. C. (2007). Diabetic impairments in NO-mediated endothelial progenitor cell mobilization and homing are reversed by hyperoxia and SDF-1α. *The Journal of Clinical Investigation*, *117*(5), 1249–1259.

Ganesh, S., & Don, K. R. (2020). The role of matrix metalloproteinases in wound healing. *International Journal of Pharmaceutical Research*, 09752366.

Ghaisas, M. M., Kshirsagar, S. B., & Sahane, R. S. (2014). Evaluation of wound healing activity of ferulic acid in diabetic rats. *International Wound Journal*, *11*(5), 523–532.

Giugliano, D., Ceriello, A., & Esposito, K. (2008). Glucose metabolism and hyperglycemia. *The American Journal of Clinical Nutrition*, *87*(1), 217S–222S.

Goldberg, R. B. (2009). Cytokine and cytokine-like inflammation markers, endothelial dysfunction, and imbalanced coagulation in development of diabetes and its complications. *The Journal of Clinical Endocrinology & Metabolism*, *94*(9), 3171–3182.

Goldin, A., Beckman, J. A., Schmidt, A. M., & Creager, M. A. (2006). Advanced glycation end products: sparking the development of diabetic vascular injury. *Circulation*, *114*(6), 597–605.

Grazul-Bilska, A. T., Johnson, M. L., Bilski, J. J., Redmer, D. A., Reynolds, L. P., Abdullah, A., & Abdullah, K. M. (2003). Wound healing: The role of growth factors. *Drugs Today (Barc)*, *39*(10), 787–800.

Guo, Y., Lin, C., Xu, P., Wu, S., Fu, X., Xia, W., & Yao, M. (2016). AGEs induced autophagy impairs cutaneous wound healing via stimulating macrophage polarization to M1 in diabetes. *Scientific Reports*, *6*(1), 36416.

Hanefeld, M., Duetting, E., & Bramlage, P. (2013). Cardiac implications of hypoglycaemia in patients with diabetes–a systematic review. *Cardiovascular Diabetology*, *12*, 1–11.

Hayat, M., Nawaz, A., Chinnam, S., Muzammal, M., Latif, M. S., Yasin, M., Ashique, S., Zengin, G., & Farid, A. (2023, July 1). Formulation development and optimization of herbo synthetic gel: In vitro biological evaluation and in vivo wound healing studies. *Process Biochemistry*, *130,* 116–126.

Hegazi, S., Aly, R., Mesilhy, R., & Aljohary, H. (2024). *Diabetic foot ulcer wound healing and tissue regeneration: Signaling pathways and mechanisms*. IntechOpen. doi: 10.5772/intechopen.1004267.

Ho, E. C., Lam, K. S., Chen, Y. S., Yip, J. C., Arvindakshan, M., Yamagishi, S. I., ... Chung, S. K. (2006). Aldose reductase–deficient mice are protected from delayed motor nerve conduction velocity, increased c-Jun NH2-terminal kinase activation, depletion of reduced glutathione, increased superoxide accumulation, and DNA damage. *Diabetes*, *55*(7), 1946–1953.

Hong, J. P., Jung, H. D., & Kim, Y. W. (2006). Recombinant human epidermal growth factor (EGF) to enhance healing for diabetic foot ulcers. *Annals of Plastic Surgery*, *56*(4), 394–398.

Hosseini, A., & Abdollahi, M. (2013). Diabetic neuropathy and oxidative stress: therapeutic perspectives. *Oxidative Medicine and Cellular Longevity*, *2013*, *1* ,168039.

Hou, Z., Nie, C., Si, Z., & Ma, Y. (2013). Deferoxamine enhances neovascularization and accelerates wound healing in diabetic rats via the accumulation of hypoxia-inducible factor-1α. *Diabetes Research and Clinical Practice*, *101*(1), 62–71.

Hozzein, W. N., Badr, G., Al Ghamdi, A. A., Sayed, A., Al-Waili, N. S., & Garraud, O. (2015). Topical application of propolis enhances cutaneous wound healing by promoting TGF-beta/Smad-mediated collagen production in a streptozotocin-induced type I diabetic mouse model. *Cellular Physiology and Biochemistry*, *37*(3), 940–954.

Hozzein, W. N., Badr, G., Badr, B. M., Allam, A., Al Ghamdi, A., Al-Wadaan, M. A., & Al-Waili, N. S. (2018). Bee venom improves diabetic wound healing by protecting functional macrophages from apoptosis and enhancing Nrf2, Ang-1 and Tie-2 signaling. *Molecular Immunology*, *103*, 322–335.

Hrynyk, M., & Neufeld, R. J. (2014). Insulin and wound healing. *Burns*, *40*(8), 1433–1446.

Huang, P., Li, S., Han, M., Xiao, Z., Yang, R., & Han, Z. C. (2005). Autologous transplantation of granulocyte colony–stimulating factor–mobilized peripheral blood mononuclear cells improves critical limb ischemia in diabetes. *Diabetes Care*, *28*(9), 2155–2160.

Huang, X., Sun, J., Chen, G., Niu, C., Wang, Y., Zhao, C., ... Zhu, Z. (2019). Resveratrol promotes diabetic wound healing via SIRT1-FOXO1-c-Myc signaling pathway-mediated angiogenesis. *Frontiers in Pharmacology*, *10*, 421.

Humpert, P. M., Bärtsch, U., Konrade, I., Hammes, H. P., Morcos, M., Kasper, M., ... Nawroth, P. P. (2005). Locally applied mononuclear bone marrow cells restore angiogenesis and promote wound healing in a type 2 diabetic patient. *Experimental and Clinical Endocrinology & Diabetes*, *113*(09), 538–540.

Jere, S. W., & Houreld, N. N. (2022). Regulatory processes of the canonical Wnt/β-catenin pathway and photobiomodulation in diabetic wound repair. *International Journal of Molecular Sciences*, *23*(8), 4210.

Jere, S. W., Houreld, N. N., & Abrahamse, H. (2019). Role of the PI3K/AKT (mTOR and GSK3β) signalling pathway and photobiomodulation in diabetic wound healing. *Cytokine & Growth Factor Reviews*, *50*, 52–59.

Jhamb, S., Vangaveti, V. N., & Malabu, U. H. (2016). Genetic and molecular basis of diabetic foot ulcers: clinical review. *Journal of Tissue Viability*, *25*(4), 229–236.

Ju, C. C., Liu, X. X., Liu, L. H., Guo, N., Guan, L. W., Wu, J. X., & Liu, D. W. (2024). Epigenetic modification: A novel insight into diabetic wound healing. *Heliyon*, *10*(6), e28086. https://doi.org/10.1016/j.heliyon.2024.e28086

Kandhare, A. D., Ghosh, P., & Bodhankar, S. L. (2014). Naringin, a flavanone glycoside, promotes angiogenesis and inhibits endothelial apoptosis through modulation of inflammatory and growth factor expression in diabetic foot ulcer in rats. *Chemico-Biological Interactions*, *219*, 101–112.

Kang, R., Tang, D., Lotze, M. T., & Zeh, III, H. J. (2011). RAGE regulates autophagy and apoptosis following oxidative injury. *Autophagy*, *7*(4), 442–444.

Kerbel, R. S. (2008). Tumor angiogenesis. *New England Journal of Medicine*, *358*(19), 2039–2049.

Kim, J. H., Yang, B., Tedesco, A., Lebig, E. G. D., Ruegger, P. M., Xu, K., ... Martins-Green, M. (2019). High levels of oxidative stress and skin microbiome are critical for initiation and development of chronic wounds in diabetic mice. *Scientific Reports*, *9*(1), 19318.

Kim, J. H., Yoon, N. Y., Kim, D. H., Jung, M., Jun, M., Park, H. Y., ... Choi, E. H. (2018). Impaired permeability and antimicrobial barriers in type 2 diabetes skin are linked to increased serum levels of advanced glycation end-product. *Experimental Dermatology*, *27*(8), 815–823.

Kolm-Litty, V., Sauer, U., Nerlich, A., Lehmann, R., & Schleicher, E. D. (1998). High glucose-induced transforming growth factor beta1 production is mediated by the hexosamine pathway in porcine glomerular mesangial cells. *The Journal of Clinical Investigation*, *101*(1), 160–169.

Kolte, D., McClung, J. A., & Aronow, W. S. (2016). Vasculogenesis and angiogenesis. In *Translational research in coronary artery disease*, Aronow, W. S. and McClung, J. A., eds. (pp. 49–65). Academic Press.

Komarcević, A. (2000). The modern approach to wound treatment. *Medicinski Pregled*, *53*(7–8), 363–368.

Kunkemoeller, B., Bancroft, T., Xing, H., Morris, A. H., Luciano, A. K., Wu, J., ... Kyriakides, T. R. (2019). Elevated thrombospondin 2 contributes to delayed wound healing in diabetes. *Diabetes*, *68*(10), 2016–2023.

Lan, C. C. E., Wu, C. S., Huang, S. M., Wu, I. H., & Chen, G. S. (2013). High-glucose environment enhanced oxidative stress and increased interleukin-8 secretion from keratinocytes: new insights into impaired diabetic wound healing. *Diabetes*, *62*(7), 2530–2538.

Larjava, H. (Ed.). (2012). *Oral wound healing: cell biology and clinical management*. John Wiley & Sons.

Li, X., Wu, G., Han, F., Wang, K., Bai, X., Jia, Y., ... Hu, D. (2019). SIRT1 activation promotes angiogenesis in diabetic wounds by protecting endothelial cells against oxidative stress. *Archives of Biochemistry and Biophysics*, *661*, 117–124.

Liu, L., Marti, G. P., Wei, X., Zhang, X., Zhang, H., Liu, Y. V., ... Harmon, J. W. (2008). Age-dependent impairment of HIF-1α expression in diabetic mice: correction with electroporation-facilitated gene therapy increases wound healing, angiogenesis, and circulating angiogenic cells. *Journal of Cellular Physiology*, *217*(2), 319–327.

Löndahl, M., Katzman, P., Nilsson, A., & Hammarlund, C. (2010). Hyperbaric oxygen therapy facilitates healing of chronic foot ulcers in patients with diabetes. *Diabetes Care*, *33*(5), 998–1003.

Long, M., Rojo de la Vega, M., Wen, Q., Bharara, M., Jiang, T., Zhang, R., ... Zhang, D. D. (2016). An essential role of NRF2 in diabetic wound healing. *Diabetes*, *65*(3), 780–793.

Maier, H. M., Ilich, J. Z., Kim, J. S., & Spicer, M. T. (2013). Nutrition supplementation for diabetic wound healing: a systematic review of current literature. *Skinmed*, *11*(4), 217–224.

Maisonpierre, P. C., Suri, C., Jones, P. F., Bartunkova, S., Wiegand, S. J., Radziejewski, C., ... Yancopoulos, G. D. (1997). Angiopoietin-2, a natural antagonist for Tie2 that disrupts in vivo angiogenesis. *Science*, *277*(5322), 55–60.

Margolis, D. J., Bartus, C., Hoffstad, O., Malay, S., & Berlin, J. A. (2005). Effectiveness of recombinant human platelet-derived growth factor for the treatment of diabetic neuropathic foot ulcers. *Wound Repair and Regeneration*, *13*(6), 531–536.

Marston, W. A. (2004). Dermagraft®, a bioengineered human dermal equivalent for the treatment of chronic nonhealing diabetic foot ulcer. *Expert Review of Medical Devices*, *1*(1), 21–31.

McIntyre, H. D., Catalano, P., Zhang, C., Desoye, G., Mathiesen, E. R., & Damm, P. (2019). Gestational diabetes mellitus. *Nature Reviews Disease Primers*, *5*(1), 47.

Mieczkowski, M., Mrozikiewicz-Rakowska, B., Kowara, M., Kleibert, M., & Czupryniak, L. (2022). The problem of wound healing in diabetes—from molecular pathways to the design of an animal model. *International Journal of Molecular Sciences*, *23*(14), 7930.

Mirza, R. E., Fang, M. M., Ennis, W. J., & Koh, T. J. (2013). Blocking interleukin-1β induces a healing-associated wound macrophage phenotype and improves healing in type 2 diabetes. *Diabetes*, *62*(7), 2579–2587.

Moeini, S., Gottlieb, H., Jørgensen, T. S., Aagaard, T. V., Kornø, M. R., Larsen, M. R. B., & Brorson, S. (2022). Inforatio technique to promote wound healing of diabetic foot ulcers: study protocol for a parallel-group, evaluator-blinded, randomised clinical trial. *BMJ Open*, *12*(9), e062344.

Moffarah, A. S., Mohajer, M. A., Hurwitz, B. L., & Armstrong, D. G. (2016). Skin and soft tissue infections. *Diagnostic Microbiology of the Immunocompromised Host*, 691–708.

Mohajeri, G., Safaee, M., & Sanei, M. H. (2014). Effects of topical Kiwifruit on healing of neuropathic diabetic foot ulcer. *Journal of Research in Medical Sciences: The Official Journal of Isfahan University of Medical Sciences*, *19*(6), 520.

Momen-Heravi, M., Barahimi, E., Razzaghi, R., Bahmani, F., Gilasi, H. R., & Asemi, Z. (2017). The effects of zinc supplementation on wound healing and metabolic status in patients with diabetic foot ulcer: a randomized, double-blind, placebo-controlled trial. *Wound Repair and Regeneration*, *25*(3), 512–520.

Monnier, L., Mas, E., Ginet, C., Michel, F., Villon, L., Cristol, J. P., & Colette, C. (2006). Activation of oxidative stress by acute glucose fluctuations compared with sustained chronic hyperglycemia in patients with type 2 diabetes. *Jama*, *295*(14), 1681–1687.

Morimoto, C., & Schlossman, S. F. (1998). The structure and function of CD26 in the T-cell immune response. *Immunological Reviews*, *161*(1), 55–70.

Moura, L. I., Dias, A. M., Suesca, E., Casadiegos, S., Leal, E. C., Fontanilla, M. R., ... Carvalho, E. (2014). Neurotensin-loaded collagen dressings reduce inflammation and improve wound healing in diabetic mice. *Biochimica et Biophysica Acta (BBA)-Molecular Basis of Disease*, *1842*(1), 32–43.

Moy, L. S., Tan, E. M., Holness, R., & Uitto, J. (1985). Phenytoin modulates connective tissue metabolism and cell proliferation in human skin fibroblast cultures. *Archives of Dermatology*, *121*(1), 79–83.

Mrozikiewicz-Rakowska, B., Szabłowska-Gadomska, I., Cysewski, D., Rudziński, S., Płoski, R., Gasperowicz, P., ... Lewandowska-Szumiel, M. (2023). Allogenic adipose-derived stem cells in diabetic foot ulcer treatment: clinical effectiveness, safety, survival in the wound site, and proteomic impact. *International Journal of Molecular Sciences*, *24*(2), 1472.

Mulder, G., Tallis, A. J., Marshall, V. T., Mozingo, D., Phillips, L., Pierce, G. F., ... Sosnowski, B. K. (2009). Treatment of nonhealing diabetic foot ulcers with a platelet-derived growth factor gene-activated matrix (GAM501): Results of a Phase 1/2 trial. *Wound Repair and Regeneration*, *17*(6), 772–779.

Nasiri, M., Fayazi, S., Jahani, S., Yazdanpanah, L., & Haghighizadeh, M. H. (2015). The effect of topical olive oil on the healing of foot ulcer in patients with type 2 diabetes: a double-blind randomized clinical trial study in Iran. *Journal of Diabetes & Metabolic Disorders*, *14*, 1–10.

National Diabetes Statistics Report (2021). https://www.cdc.gov/diabetes/php/data-research/index.html

Negre-Salvayre, A., Salvayre, R., Augé, N., Pamplona, R., & Portero-Otin, M. (2009). Hyperglycemia and glycation in diabetic complications. *Antioxidants & Redox Signaling*, *11*(12), 3071–3109.

Ngo, B. T., Hayes, K. D., DiMiao, D. J., Srinivasan, S. K., Huerter, C. J., & Rendell, M. S. (2005). Manifestations of cutaneous diabetic microangiopathy. *American Journal of Clinical Dermatology*, *6*, 225–237.

Niezgoda, J. A., Van Gils, C. C., Frykberg, R. G., Hodde, J. P., & OASIS Diabetic Ulcer Study Group. (2005). Randomized clinical trial comparing OASIS wound matrix to regranex gel for diabetic ulcers. *Advances in Skin & Wound Care*, *18*(5), 258–266.

Nissen, N. N., Polverini, P., Koch, A. E., Volin, M. V., Gamelli, R. L., & DiPietro, L. A. (1998). Vascular endothelial growth factor mediates angiogenic activity during the proliferative phase of wound healing. *The American Journal of Pathology*, *152*(6), 1445.

Niu, Y. W., Miao, M. Y., Dong, W., Dong, J. Y., Cao, X. Z., & Lu, S. L. (2012). Effects of advanced glycation end products and its receptor on oxidative stress in diabetic wounds. *Zhonghua Shao Shang za zhi= Zhonghua Shaoshang Zazhi= Chinese Journal of Burns*, *28*(1), 32–35.

Obrosova, I. G. (2003). Update on the pathogenesis of diabetic neuropathy. *Current Diabetes Reports*, *3*(6), 439–445.

Ohnuma, K., Yamochi, T., Uchiyama, M., Nishibashi, K., Iwata, S., Hosono, O., ... Morimoto, C. (2005). CD26 mediates dissociation of Tollip and IRAK-1 from caveolin-1 and induces upregulation of CD86 on antigen-presenting cells. *Molecular and Cellular Biology*, *25*(17), 7743–7757.

Oryan, A., & Alemzadeh, E. (2017). Effects of insulin on wound healing: a review of animal and human evidences. *Life Sciences*, *174*, 59–67.

Pang, L., Lian, X., Liu, H., Zhang, Y., Li, Q., Cai, Y., ... Yu, X. (2020). Understanding diabetic neuropathy: focus on oxidative stress. *Oxidative Medicine and Cellular Longevity, 2020*, 1–13.

Park, K. G., Kim, M. J., Kim, H. S., Lee, S. J., Song, D. K., & Lee, I. K. (2004). Prevention and treatment of macroangiopathy: focusing on oxidative stress. *Diabetes Research and Clinical Practice, 66*, S57–S62.

Patel, D. K., Kumar, R., Laloo, D., & Hemalatha, S. (2012). Diabetes mellitus: an overview on its pharmacological aspects and reported medicinal plants having antidiabetic activity. *Asian Pacific Journal of Tropical Biomedicine, 2*(5), 411–420.

Patel, S., Srivastava, S., Singh, M. R., & Singh, D. (2019). Mechanistic insight into diabetic wounds: Pathogenesis, molecular targets and treatment strategies to pace wound healing. *Biomedicine & Pharmacotherapy, 112*, 108615.

Patil, M. M., Sahoo, J., Kamalanathan, S., & Pillai, V. (2015). Phenytoin induced osteopathy-Too common to be neglected. *Journal of Clinical and Diagnostic Research: JCDR, 9*(11), OD11.

Pereira, C. A. Z., & de A Alchorne, A. D. O. (2010). Assessment of the effect of phenytoin on cutaneous healing from excision of melanocytic nevi on the face and on the back. *BMC Dermatology, 10*, 1–7.

Plikaitis, C. M., & Molnar, J. A. (2006). Subatmospheric pressure wound therapy and the vacuum-assisted closure device: basic science and current clinical successes. *Expert Review of Medical Devices, 3*(2), 175–184.

Potter, C. F., Kuo, N. T., Farver, C. F., McMahon, J. T., Chang, C. H., Agani, F. H., ... Martin, R. J. (1999). Effects of hyperoxia on nitric oxide synthase expression, nitric oxide activity, and lung injury in rat pups. *Pediatric Research, 45*(1), 8–13.

Pries, A. R., & Secomb, T. W. (2014). Making microvascular networks work: angiogenesis, remodeling, and pruning. *Physiology, 29*(6), 446–455.

Rabbani, P. S., Ellison, T., Waqas, B., Sultan, D., Abdou, S., David, J. A., ... Ceradini, D. J. (2018). Targeted Nrf2 activation therapy with RTA 408 enhances regenerative capacity of diabetic wounds. *Diabetes Research and Clinical Practice, 139*, 11–23.

Ren, J., Yang, M., Xu, F., Chen, J., & Ma, S. (2019). Acceleration of wound healing activity with syringic acid in streptozotocin induced diabetic rats. *Life Sciences, 233*, 116728.

Roan, J. N., Cheng, H. N., Young, C. C., Lee, C. J., Yeh, M. L., Luo, C. Y., ... Lam, C. F. (2017). Exendin-4, a glucagon-like peptide-1 analogue, accelerates diabetic wound healing. *Journal of Surgical Research, 208*, 93–103.

Romana-Souza, B., Nascimento, A. P., & Monte-Alto-Costa, A. (2009). Propranolol improves cutaneous wound healing in streptozotocin-induced diabetic rats. *European Journal of Pharmacology, 611*(1–3), 77–84.

Rosenberger, D. C., Blechschmidt, V., Timmerman, H., Wolff, A., & Treede, R. D. (2020). Challenges of neuropathic pain: focus on diabetic neuropathy. *Journal of Neural Transmission, 127*(4), 589–624.

Saad Setta, H., Elshahat, A., Elsherbiny, K., Massoud, K., & Safe, I. (2011). Platelet-rich plasma versus platelet-poor plasma in the management of chronic diabetic foot ulcers: a comparative study. *International Wound Journal, 8*(3), 307–312.

Salazar, J. J., Ennis, W. J., & Koh, T. J. (2016). Diabetes medications: impact on inflammation and wound healing. *Journal of Diabetes and its Complications, 30*(4), 746–752.

Sayeski, P. P., & Kudlow, J. E. (1996). Glucose metabolism to glucosamine is necessary for glucose stimulation of transforming growth factor-α gene transcription. *Journal of Biological Chemistry, 271*(25), 15237–15243.

Schultz, G. S., & Wysocki, A. (2009). Interactions between extracellular matrix and growth factors in wound healing. *Wound Repair and Regeneration, 17*(2), 153–162.

Seo, S. G., Yeo, J. H., Kim, J. I., Kim, J. B., Cho, T. J., & Lee, D. Y. (2013). Negative-pressure wound therapy induces endothelial progenitor cell mobilization in diabetic patients with foot infection or skin defects. *Experimental & Molecular Medicine, 45*(11), e62–e62.

Sharp, A., & Clark, J. (2011). Diabetes and its impact on wound healing. *Nursing Standard, 25*(45), 41–47.

Shen, J. T., & Falanga, V. (2003). Innovative therapies in wound healing. *Journal of Cutaneous Medicine and Surgery: Incorporating Medical and Surgical Dermatology, 7*(3), 217–224.

Singh, D., Singh, M. R., Saraf, S., & Saraf, S. (2011). Development of delivery cargoes for debriding enzymes effective in wound healing. *Trends in Applied Sciences Research, 6*(8), 863.

Spichler, A., Hurwitz, B. L., Armstrong, D. G., & Lipsky, B. A. (2015). Microbiology of diabetic foot infections: from Louis Pasteur to 'crime scene investigation'. *BMC Medicine, 13*, 1–13.

Squadrito, F., Bitto, A., Altavilla, D., Arcoraci, V., De Caridi, G., De Feo, M. E., ... Cucinotta, D. (2014). The effect of PDRN, an adenosine receptor A2A agonist, on the healing of chronic diabetic foot ulcers: results of a clinical trial. *The Journal of Clinical Endocrinology & Metabolism, 99*(5), E746–E753.

Srivastava, S. K., Yadav, U. C., Reddy, A. B., Saxena, A., Tammali, R., Shoeb, M., ... Ramana, K. V. (2011). Aldose reductase inhibition suppresses oxidative stress-induced inflammatory disorders. *Chemico-Biological Interactions, 191*(1–3), 330–338.

Staton, C. A., Valluru, M., Hoh, L., Reed, M. W. R., & Brown, N. J. (2010). Angiopoietin-1, angiopoietin-2 and Tie-2 receptor expression in human dermal wound repair and scarring. *British Journal of Dermatology, 163*(5), 920–927.

Steed, D. L. (1997). The role of growth factors in wound healing. *Surgical Clinics of North America, 77*(3), 575–586.

Stirban, A., Negrean, M., Stratmann, B., Gawlowski, T., Horstmann, T., Gotting, C. H. R. I. S. T. I. A. N., ... Tschoepe, D. (2006). Benfotiamine prevents macro-and microvascular endothelial dysfunction and oxidative stress following a meal rich in advanced glycation end products in individuals with type 2 diabetes. *Diabetes Care, 29*(9), 2064–2071.

SYDNEY Trial Authors, for the SYDNEY Trial Study Group:, Ametov, A. S., Barinov, A., Dyck, P. J., Hermann, R., Kozlova, N., ... Ziegler, D. (2003). The sensory symptoms of diabetic polyneuropathy are improved with α-lipoic acid: the SYDNEY trial. *Diabetes Care, 26*(3), 770–776.

Tandara, A. A., & Mustoe, T. A. (2004). Oxygen in wound healing—more than a nutrient. *World Journal of Surgery, 28*, 294–300.

Tawfeek, H. M., Abou-Taleb, D. A., Badary, D. M., Ibrahim, M., & Abdellatif, A. A. (2020). Pharmaceutical, clinical, and immunohistochemical studies of metformin hydrochloride topical hydrogel for wound healing application. *Archives of Dermatological Research, 312*, 113–121.

Ursini, F., Russo, E., Pellino, G., D'Angelo, S., Chiaravalloti, A., De Sarro, G., ... De Giorgio, R. (2018). Metformin and autoimmunity: a "new deal" of an old drug. *Frontiers in Immunology, 9*, 1236.

Vaidyanathan, L. (2021). Growth factors in wound healing—a review. *Biomedical and Pharmacology Journal, 14*, 1469–1481.

Valluru, M., Staton, C. A., Reed, M. W., & Brown, N. J. (2011). Transforming growth factor-β and endoglin signaling orchestrate wound healing. *Frontiers in Physiology, 2*, 89.

Vig, S., Dowsett, C., Berg, L., Caravaggi, C., Rome, P., Birke-Sorensen, H., ... International Expert Panel on Negative Pressure Wound Therapy [NPWT-EP. (2011). Evidence-based recommendations for the use of negative pressure wound therapy in chronic wounds: steps towards an international consensus. *Journal of Tissue Viability, 20*, S1–S18.

Vijayasingham, S. M., Dykes, P. J., & Marks, R. (1991). Phenytoin has little effect on in-vitro models of wound healing. *British Journal of Dermatology, 125*(2), 136–139.

Wang, Y., Zhu, J., Chen, J., Xu, R., Groth, T., Wan, H., & Zhou, G. (2022). The signaling pathways induced by exosomes in promoting diabetic wound healing: a mini-review. *Current Issues in Molecular Biology, 44*(10), 4960–4976.

Whittam, A. J., Maan, Z. N., Duscher, D., Barrera, J. A., Hu, M. S., Fischer, L. H., ... Gurtner, G. C. (2019). Small molecule inhibition of dipeptidyl peptidase-4 enhances bone marrow progenitor cell function and angiogenesis in diabetic wounds. *Translational Research, 205*, 51–63.

Wietecha, M. S., Cerny, W. L., & DiPietro, L. A. (2013). Mechanisms of vessel regression: toward an understanding of the resolution of angiogenesis. *New Perspectives in Regeneration*, 3–32.

Wu, S. C., Marston, W., & Armstrong, D. G. (2010). Wound care: the role of advanced wound-healing technologies. *Journal of the American Podiatric Medical Association, 100*(5), 385–394.

Xiang, G. D., Sun, H. L., Zhao, L. S., Hou, J., Yue, L., & Xu, L. (2008). The antioxidant alpha-lipoic acid improves endothelial dysfunction induced by acute hyperglycaemia during OGTT in impaired glucose tolerance. *Clinical Endocrinology, 68*(5), 716–723.

Zayed, M. A., Wei, X., Park, K. M., Belaygorod, L., Naim, U., Harvey, J., ... Semenkovich, C. F. (2017). N-Acetylcysteine accelerates amputation stump healing in the setting of diabetes. *The FASEB Journal, 31*(6), 2686.

Zhang, H., Nie, X., Shi, X., Zhao, J., Chen, Y., Yao, Q., ... Yang, J. (2018). Regulatory mechanisms of the Wnt/β-catenin pathway in diabetic cutaneous ulcers. *Frontiers in Pharmacology, 9*, 1114.

Zhao, H., Huang, J., Li, Y., Lv, X., Zhou, H., Wang, H., ... Liu, Z. (2020). ROS-scavenging hydrogel to promote healing of bacteria infected diabetic wounds. *Biomaterials, 258*, 120286.

Zhao, P., Sui, B. D., Liu, N., Lv, Y. J., Zheng, C. X., Lu, Y. B., ... Jin, Y. (2017). Anti-aging pharmacology in cutaneous wound healing: effects of metformin, resveratrol, and rapamycin by local application. *Aging Cell, 16*(5), 1083–1093.

3 Chronic Wound Healing
Existing Treatment Approaches and Limitations

Aakriti Patel[1], Priyanti Halder[2], Jayshree Manahty[2], Radheshyam Pal[2], Sumel Ashique[3], Suman Ghosh[4], Shakira Ghazanfar[5], and Farzad Taghizadeh-Hesary[6]*

[1] Lovely Professional University, Punjab, India
[2] Bengal College of Pharmaceutical Sciences and Research, West Bengal, India
[3] Bharat Technology, West Bengal, India
[4] Guru Nanak Institute of Pharmaceutical Science and Technology, West Bengal, India
[5] National Agricultural Research Centre (NARC), Islamabad, Pakistan
[6] Iran University of Medical Sciences, Tehran, Iran
* Corresponding Author

3.1 INTRODUCTION

The skin of mammals acts as a protective layer against any cutaneous injury. In response to any wound, the skin itself repairs slowly and becomes scar-free by adopting a few cellular signaling pathways (Greaves S. N. et al., 2013). In the clinical aspect, a chronic wound is a wound that does not progress through all the consequent stages of healing in an orderly and timely manner. In the current scenario, chronic wounds have become a massive concern affecting a larger population. Although the proper definition and etiology of chronic wounds are not fully outlined in the literature, most chronic wounds are associated with subsequent tissue damage in a prolonged inflammatory state. Usually, the wound can be termed as a sore on the skin or mucous membrane, accompanied by the breakdown of the epidermal layer integrity leading to tissue injury. The wound results in complete loss of the epidermis with portions of the dermis and even subcutaneous fat. An inflamed wound appears as the surrounding skin becomes reddish, swollen, and tender to the touch. Sometimes, skin wounds can occur from the immobility of body parts, which creates prolonged pressure on the tissues of that region, reducing blood circulation as well as tissue perfusion (Iqbaal A. et al., 2017). Although broadly conflicted, a wound is referred to as "chronic" when the healing is not visible even after more than three months. A chronic wound is distinguished by persistent inflammation with defective vasculogenesis and angiogenesis; impaired re-epithelialization; dysregulated levels of cytokines, chemokines, and growth factors; and increased protease activity (Las Heras K. et al., 2020). Normal physiological wound repair is completed within four to six weeks of injury. It is a complex cascade of events split into four continuous and overlapping phases: (1) hemostasis (closure of the wound by blood clotting); (2) followed by vasodilation and the immediate inflammatory response, triggering the release of cytokines and other mediators; (3) proliferation leading to the formation of granulation tissue, re-epithelialization, and neovascularization where new epithelium,

DOI: 10.1201/9781003504948-3

blood vessels, and extracellular matrix (ECM) are generated; and (4) remodeling (maturation of granulation tissue into scar and increased tissue tensile strength over weeks to months). The aforementioned highly regulated cellular and molecular mechanisms are the potentially intact interplay for any chronic wound to regenerate resurfaced rejuvenated skin (Zhao R. et al., 2016). Even with proper wound care, chronic wounds are restricted to a self-perpetuating inflammatory stage and remain obdurate and uncontrollable. The nonhealing state of chronic wounds promotes skin ulcers associated with several comorbid complications. In acute wounds, neutrophil infiltration is the first response mechanism enhancing immunity, and the associated reactive oxygen species (ROS) burst acts as the main defense inducing the destruction of pathogens present in the wound (Diegelmann R. F. & Evans M. C. et al., 2004). Enormous factors such as chronic disease, vascular insufficiency, diabetes, malnutrition, aging, massive trauma, and a few local factors like infection and edema are responsible for the collapse of keratinocyte migration around wound edges, which aids in closure and further causes chronic wounds (Lierte S. et al., 2020). By eliminating these primary adjuvant factors, chronic wounds can be effectively managed. However, chronic wounds with mixed etiologies prevalently exist as a silent epidemic throughout the world, affecting around 10.5 million people (based on 2022 data) (Sen C. K. et al., 2023). A chronic wound is a heterogeneous group of disorders, including venous ulcers, pressure sores, and diabetic ulcers (Nwomeh B. C. et al., 1998; Mustoe T. A. et al., 2006).

3.2 TYPES OF CHRONIC WOUNDS

Chronic wounds are broadly classified into three main categories, which are further divided into various subtypes underlying different mechanisms (Goldberg S. R. & Diegelmann R. F., 2020; Zhao R. et al., 2016; Eaglstein W. H. & Falanga V. et al., 1997; Nwomeh B. C. et al., 1998).

3.2.1 Venous Ulcers

Venous ulcers mainly occur in the legs due to irregular circulation in the leg veins. Mostly, they affect women and geriatric people (1%–2% of adults). Around 65% to 70% of leg ulcers are delayed manifestations of chronic venous insufficiency with serious complications. Due to venous thrombosis and valvular incompetence (the damage of valves inside leg veins), the blood pressure in leg veins is increased during walking. Thus, venous hypertension is generated from the ankle and restricts blood flow, causing extreme swelling and a chronic inflammatory response. When this condition is prolonged over time, it promotes venous ulcers on the skin accompanied by lipodermatosclerosis and hemosiderin deposition (Abbade L. P. F. & Lastoria S. et al., 2005; Crawford J. et al., 2017). The sustained pressure in the lower extremity facilitates blood vessel permeability and leads to the leakage of macromolecules such as fibrinogen into the extravascular space where they may act as chemoattractants for leukocyte infiltration (Demidova-Rice T. N. et al., 2012; Word R. et al., 2010). As a result, edema is caused by obstructed diffusion of growth factors, oxygen, and other nutrients in the wound (Trent J. T. et al., 2005). Venous ulcers tend to be large and palpable with irregular margins (Fonder M. A. et al., 2008; Zhao R. et al., 2016; Eaglstein W. & Falanga V., 1997; Nwomeh B. C. et al., 1998; Goldberg R. & Diegelmann F., 2020).

3.2.2 Pressure Sores

A pressure sore is a type of chronic localized wound in the skin and underlying tissue that arises because of constant direct pressure, shear force or friction or a combination of these on one area for a prolonged period (National Pressure Ulcer Advisory Panel, 2014; Alhaug J. et al., 2017). A pressure sore is also called a bedsore, pressure ulcer, or decubitus ulcer. It is quite common in bedridden patients who have gone through spinal injury, hip fracture, or paralyzed or any orthopedic problem, or geriatric patients with poor mobility due to sensory impairment or neuropathy. If they are confined to resting on a bed or chair for more than a week, the unrelenting sustained pressure decreases capillary blood flow, causing occlusion of blood vessels and lymphatic vessels, leading to tissue ischemia, and the epidermal layer of the skin disrupts forming pressure ulcers. Most pressure sores heal with treatment, but some do not recover completely if associated with concomitant arterial and venous insufficiency or osteomyelitis (Grey J. E. et al., 2006). Normally the underlying blood capillaries of the skin can tolerate external pressure from 16 to around 35 mm Hg in different segments. When external pressure rises to more than 50 mm Hg to 200 mm Hg over time between the bony prominence and external surface of surrounding tissue, it occludes the capillaries, causing cell death and leading to soft tissue necrosis (Agrawal K. & Chauhan N., 2012).

Stages of pressure sores:

a) Non-blanchable erythema
b) Partial-thickness skin loss
c) Full-thickness skin loss
d) Extensive destruction, tissue necrosis with or without full-thickness skin loss (Alhaug J. et al., 2017; Grey J. E. et al., 2006; Allman R. M. et al., 1997).

3.2.3 Diabetic Ulcers

Most diabetic ulcers occur on the feet, usually in areas withstanding strain. Persons suffering from diabetes mellitus are more likely to develop diabetic foot ulcers during their lifespan. The pathophysiology of diabetic ulcers includes neuropathic, vascular, and immunological factors associated with elevated glucose levels giving rise to high levels of reactive oxygen species, protein kinase C, nitric oxide blockade, DNA alteration, ECM stiffness, ischemia, and inflammation (Goldberg S. R. & Diegelmann R., 2020; Zhao R. et al., 2016). Diabetic neuropathy and peripheral arterial disease (PAD) are the major risk factors involved in diabetic foot ulcers. These two factors either act individually or in combination with other lesions such as limited joint mobility and microvascular disease to increase the susceptibility of diabetic ulcers. Diabetic neuropathy is a common complication in patients with diabetes mellitus where a progressive loss of peripheral nerve fibers occurs due to elevated glycemic levels. The intensity and duration of hyperglycemia potentially present the severity of neuropathy. Neuropathy sometimes results in an insentient, deformed ulcerative foot with an abnormal walking pattern. Moreover, diabetic hyperglycemia along with oxidative stress tends to cause cellular dysfunction, modulating different metabolic cycles of the body (Rathur H. M. & Boulton A. J. M., 2005; Bašić-Kes V. et al., 2011; Andrews K. L. et al., 2015). PAD is characterized by partial or complete occlusion of noncardiac, nonintracranial, peripheral arteries of the upper and lower limbs that may give rise to atherosclerosis from reduced blood flow or tissue loss. In diabetes mellitus, the arteries of the lower limbs get affected by PAD, which results in diabetic foot ulcers (Kullo I. J. & Rooke T. W., 2016; Soyoye D. et al., 2021).

3.3 ADJUNCTIVE FACTORS IN CHRONIC WOUNDS

3.3.1 TGF-β Signaling

Transforming growth factor-β (TGF-β) has a crucial role in re-epithelialization, angiogenesis, and granulation tissue formation during wound healing. Keratinocyte migration and proliferation throughout re-epithelialization are modulated by multiple growth factors, including TGF-β. The TGF-β family consists of three isoforms of pluripotent cytokines: TGFβ-1, TGFβ-2, and TGFβ-3, where TGF-β1 represents the most influential factor in cutaneous wound healing (Barrientos S. et al., 2008). TGF-β1 lies in the stratum granulosum and stratum corneum, while TGF-β2 and a limited amount of TGF-β3 are found in suprabasal layers with their different functions (Cho H. R. et al., 2004). After an injury, TGF-β1 is released by keratinocytes, platelets, macrophages, monocytes, and fibroblasts and quickly upregulated to initiate inflammation, myofibroblast differentiation, keratinocyte migration, angiogenesis, and granulation tissue formation by inducing vascular endothelial growth factor. TGF-β2 facilitates re-epithelialization, whereas TGF-β3 improves scarless wound healing through collagen binding and fibroblast deposition at the wound site. These three cytokines are secreted primarily in inactive forms and further activated by proteases, integrins, thrombospondin-1, reactive oxygen species, low pH, heat, and shear forces (Worthington J. J. et al., 2012). After activation, TGF-β mediates its signaling by binding to transmembrane TGF-β receptor II (TGF-βRII), followed by its heterodimerization and phosphorylation via serine/threonine kinases of transmembrane TGF-β receptor I (TGF-βRI). TGF-βRI and TGF-βRII show similar affinity for isoforms TGF-β1 and TGF-β3, whereas only TGF-βRII binds with TGF-β2 with less intensity. TGF-β facilitates keratinocyte migration through integrin induction and Smad-dependent signaling (Buschke S. et al., 2011). The main intracellular mediator of TGF-β signaling is the Smad protein. In chronic wounds, the TGF-β signaling cascade is suppressed, leading to loss of tissue homeostasis and the inability of keratinocyte migration and tissue regeneration to close wounds. Attenuated TGF-β signaling in chronic wounds results in multiple consequences, increasing nitric oxide synthase (NOS) activity and elevated NO synthesis, as TGF-β1 is considered to downregulate NOS activity in epithelial cells and macrophages. Though NO can stimulate keratinocyte migration and angiogenesis, an excessive amount shows an inhibitory effect on the wound healing process. The absence of TGF-βRII induces the chronicity of wounds or nonhealing ulcers, as TGF-βRII is associated with potential wound healing. Suppression of TGF-β signaling in the chronic wound cannot promote keratinocyte activation or differentiation and forms a thick, hyperproliferative, hyperkeratotic, as well as parakeratotic epidermis due to c-Myc overexpression. As TGF-β activity is correlated with the suppression of growth-promoting transcription factors like c-Myc, the vitiated TGF-β signaling cascade in the hyperproliferative epidermis accelerates c-Myc stimulation. Functional suppression of all three TGF-β receptors, dysregulation of TGF-β target genes, and decrease of Smad2 and Smad7 protein implies the chronicity of nonhealing wounds (Lierte S. et al., 2020; Ramirez H. et al., 2014).

3.3.2 Matrix Metalloproteinases (MMPs)

Matrix metalloproteinases (MMPs) play a vital role in regulating extracellular matrix degradation and wound re-epithelialization. Neutrophils from the wound site release MMPs, which further consume the surrounding necrotic tissue. Limited expression of activated MMPs is considered for wound healing. The MMP family comprises a series of seven calcium-dependent zinc-containing enzymes: (1) collagenases, (2) gelatinases, (3) stromelysins, (4) matrilysins, (5) metalloelastases, (6) membrane-type MMPs, and (7) other MMPs. Overproduction of MMPs damages vital growth factors such as platelet-derived growth factor and transforming growth factor (TGF-β), interleukins, and interferons, which are necessary for wound healing, thus delaying re-epithelialization. In acute injury, a lesser extent of MMPs is produced from normal tissue along with fibroblasts, keratinocytes, endothelial cells, cytokines, and other hormones. Overexpression of MMP-1 delays re-epithelialization and induces chronic wounds associated with diabetic foot ulcers. Extracellular release of reactive oxygen species and proteases like MMPs accelerates tissue damage by impairing collagen deposition. Protease activity is controlled by multiple protease inhibitors. Elevated levels of elastase, α1-antitrypsin, and α2-macroglobulin are responsible for fibronectin degradation in chronic wounds (Caley M. et al., 2015; Chen P. & Parks C., 2009; Steffensen B. et al., 2001; Goldberg R. & Diegelmann F., 2020).

3.3.3 Pro-Inflammatory Mediators (IL-8, IL-6, and TNF-α)

Different proinflammatory mediators are responsible for causing chronic and prolonged inflammation. Interleukin (IL-6) is a multifunctional potent cytokine that is a key modulator in wound repair. It is involved in the differentiation and proliferation of keratinocytes, fibroblasts, and chemotaxis of leukocytes. As inflammation becomes protracted, IL-6 signaling becomes responsive toward wound healing (Lin Z. et al., 2003; Johnson B. et al., 2020). Another major angiogenic factor, interleukin 8 (IL-8), profoundly contributes to wound healing through the migration of keratinocytes. IL-8 is a chemoattractant neutrophil-activating peptide belonging to the CXC chemokine family. IL-8 mediates its action through its receptor IL-8R, which has two forms: IL-8RA and IL-8RB. IL-8 is abundantly present in keratinocytes, and its response to inflammatory stimuli is expressed by tissue plasminogen activator (tPA). In chronic injury, IL-8 is released in remarkably low amounts, which leads to impaired wound healing along with the formation of chronic ulcers (Jiang W. G. et al., 2012). Tissue necrosis factor (TNF-α) is also a pro-inflammatory cytokine that plays a crucial role in the chronicity of nonhealing ulcers. During the proliferation phase of normal wound healing, TNF-α is secreted from macrophages, which facilitates the re-formation of the ECM at the wound, followed by the synthesis of proteoglycan and fibronectin by fibroblasts. In chronic wounds, a higher concentration of TNF-α is released to establish a highly inflammatory environment. The overexpression of TNF-α and the reduced secretion of other pro-healing mediators in prolonged inflammation lead to nonhealing chronic diabetic ulcers (Raziyeva K. et al., 2021).

3.3.4 Intrinsic Factors

3.3.4.1 Tissue Ischemia and Hypoxia

The transport of oxygen to the inflamed tissue is highly required in wound healing. A group of systemic and vascular disorders, including arterial occlusive disease and chronic venous insufficiency, affects the transfer of oxygen and other macronutrients to the injured tissue.

Mainly, arterial occlusive disease is characterized by atherosclerosis and thrombosis of blood vessels, associated with the disruption of endothelial integrity by autoimmune disorders and damage to blood vessels by trauma. Difficulty in blood flow through damaged blood vessels to the wound site decreases tissue oxygenation, thus inducing chronic hypoxia, which further results in a nonhealing wound (Gordillo G. M. & Sen C. K., 2003). Hypoxia-inducible factor (HIF-1α) is the key modulator of oxygen homeostasis, which effectively influences wound healing by releasing growth factors, promoting cell migration and differentiation, and facilitating matrix synthesis. HIF-1α activation stimulates re-epithelialization and angiogenesis from disrupted vasculature surrounding the wound. HIF-1α deficiency gives rise to an immediate inflammatory influx at the wound associated with high metabolic oxygen demand, thus producing nonhealing ulcers in response to chronic hypoxia (Hong W. X. et al., 2014).

3.3.4.2 Biofilm Formation and Bacterial Infection

Open wounds are highly prone to contamination by bacterial colonization, which plays a vital role in wound healing. Biofilm potentially contributes to bacterial attachment at the wound site. Biofilm is a slime layer comprising multiple species of microbes with genetic diversity embedded in exopolysaccharide (EPS) to produce chronic infections. Biofilm interferes with the natural wound healing process by allowing bacteria to impair skin barrier function and thus prolong the inflammatory state. Biofilm makes the wound worse and chronic with delayed epithelialization and disrupted granulation tissue formation, causing antimicrobial recalcitrance. The microenvironment of chronic wounds facilitates bacterial growth involving necrotic debris and less oxygenation along with inducing elevated protease and pro-inflammatory cytokine levels, reducing growth factor (epidermal growth factor, platelet-derived growth factor, vascular endothelial growth factor) levels. Biofilm formation is classified into four stages; (i) surface attachment, (ii) microcolony formation, (iii) biofilm maturation, (iv) dispersal and detachment (Wu Y. K. et al., 2019; Clinton A. & Carter T., 2015).

3.3.4.3 Radiation

Radiation therapy shows a detrimental outcome toward chronic wounds. The severe impact of radiation depends upon the location of the wound and the dose of therapy. Late progressive effects of radiation become chronic over time, causing lymphedema, osteoradionecrosis, fibrosis, etc. Radiation injury disrupts the vasculature of the wound, producing endarteritis that impairs angiogenesis and also induces hypoxia and blockage at the cellular level, thus inhibiting wound healing (Granick M. et al., 1993).

3.3.4.4 Aging and Stress

Chronic wounds are often seen in elderly people. As people get older, they are more likely to develop different disease states and stress; hence, the rate of wound healing slows. It is hypothetically considered that the fibroblasts that appear from chronic wounds in geriatric patients are unable to upregulate for surviving through normal healing phases under hypoxic conditions (Mogford J. et al., 2004). As people age, the outer skin layer gets thinner, and blood vessels become more fragile, leading to bruising and bleeding under the skin. The hypodermis also becomes lean, reducing its normal padding, which protects against injury (Blair J. et al., 2020).

3.3.5 Extrinsic Factors

3.3.5.1 Diabetes Mellitus

Diabetes mellitus is a complex metabolic disorder that affects more than 340 million people, and around 20% of them are likely to develop diabetic wounds worldwide (Nicolucci A. et al., 2014; Patel S. et al., 2019). Foot ulcer is the most commonly occurring wound in patients affected by diabetes mellitus. Diabetic patients have a diminished ability to metabolize glucose under hyperglycaemic conditions, which further delays the wound healing process, getting associated with several complications, psychiatric stress, and depression; hence, chronic wounds are generated.

3.3.5.2 Steroids

Generally, glucocorticoids are abundantly used in the treatment of different dermatological disorders. Steroids can alter the normal stages of wound healing after long-term daily use with higher doses. Prolonged systemic steroid administration reduces the inflammatory response, which is essential for wound healing, by inhibiting macrophage activity, the release of lysosomal enzymes, wound contraction, fibrogenesis, and angiogenesis. However, the consumption of vitamin A counteracts the unfavorable effects of glucocorticoids allowing the release of lysosomal enzymes (Hunt K. et al., 1969; Burns L. et al., 2003).

3.3.5.3 Malnutrition

Nutrition and wound healing are highly correlated. Malnutrition affects the normal progression of wound healing by prolonging the inflammatory phase, reducing the proliferation of fibroblasts, and impairing collagen synthesis. Additionally, nutrient deficiency decreases wound tensile strength and increases the rate of infection. Malnourished patients are more susceptible to easily developing nonhealing chronic ulcers. The metabolic rate increases when a wound occurs in the skin, resulting in dehydration and protein breakdown. Loss of protein facilitates energy demand and shows a negative impact on wound healing, involving continuous wound bed drainage (Stechmiller J. K., 2010).

3.3.5.4 Smoking

Smoking significantly damages the immune system of patients affected by any injury. It delays the healing process and increases the risk of infection at the wound site. Even a small amount of tobacco or nicotine reduces the transport of oxygen and other essential nutrients to the wound site for healing because tobacco is a potent vasoconstrictor. Nicotine present in tobacco causes systemic vasoconstriction by acting through the sympathetic nervous system. Moreover, tobacco smoke contains carbon monoxide, which binds to hemoglobin with higher affinity and reduces the oxygen-carrying capacity of hemoglobin, thereby causing cellular hypoxia (Jensen J. et al., 1991).

3.3.5.5 Psychological Stress

Intense anxiety and depression are catastrophically associated with wound healing. More psychological stress delays wound healing, resulting in the activation of the hypothalamic–pituitary–adrenal axis (HPA axis) and the sympathetic–adrenal–medullary axis (SAM axis). Excessive production of glucocorticoids and catecholamines directly affects the wound healing process (Gouin J. P. & Kiecolt-Glaser J., 2012).

3.4 PATHOPHYSIOLOGY OF CHRONIC WOUNDS

The pathophysiology of chronic wounds is elucidated by multiple cellular and molecular mechanisms. Chronic wound healing is a dynamic, integrated cascade of interrelated events involving numerous components, such as the ECM, epidermal and dermal cells, lysosomal enzymes, proteases, reactive oxygen species, pro-inflammatory mediators, tissue necrosis factors, and various growth factors, among others. The complex process of wound healing is regulated by homeostasis and healthy granulation tissue formation, with the final aim of restoring injured tissue. A chronic wound appears as a nonhealing wound when it fails to progress through the well-defined cascade of wound healing within a limited period. Unlike acute wounds, the underlying mitogenic activities of the surrounding wound tissue are obstructed in chronic wounds. The uncontrolled release of different mediators and their impaired functions delay wound healing and cause scarring, which often leads to neoplastic progression. Neutrophils initially infiltrate in huge amounts from the site of inflammation, acting as a biological marker and stimulus of the innate immune response. Excessive infiltration of neutrophils affects the chronic wound healing process, causing overproduction of ROS. Elevated levels of ROS directly damage cell membranes and the ECM. Additionally, neutrophils secrete MMPs, elastases, and collagenases, which destroy essential growth factors like platelet-derived growth factor (PDGF) and transforming growth factor (TGF), and inactivate the components of the ECM. Also, the bioavailability of growth factors also decreases. Infiltrated neutrophils and activated macrophages release pro-inflammatory cytokines such as interleukin-1, interleukin-8, interleukin-13, and TNF-α. ECM organization, collagen binding, and keratinocyte migration are hampered by the imbalanced distribution of mediators. This imbalance results in fibroblast proliferation and degradation of the ECM, which further impairs re-epithelialization and prolongs the healing process. Along with the dysregulation of these systemic factors, a few intrinsic factors such as ischemia–reperfusion injury followed by local hypoxia, bacterial colonization, foreign bodies, and aging coexist to alter the stages of wound healing, thereby producing chronicity (Zhao R. et al., 2016; Mustoe T. et al., 2006; Medina A. et al., 2005). Figure 3.1 shows the pathophysiology of a chronic wound.

3.5 PROCESS OF WOUND HEALING

3.5.1 Stages of Wound Healing

The process of wound healing is rather slow and regulated. It involves a series of steps that help to restore the integrity of the disrupted part (Eming S. A. et al., 2007). After the damage of the affected area, the tedious reaction between the cellular components such as keratinocytes and fibroblasts occurs. The blockage of the skin wound happens with the continuous regeneration and restoration of the affected area (Reinke J. M. et al., 2012). The complex process of wound healing can be broken down into three simpler phases or stages (Takeo M. et al., 2015; Tyavambiza C. et al., 2022):

1. Inflammatory phase
2. Proliferative phase
3. Remodeling phase

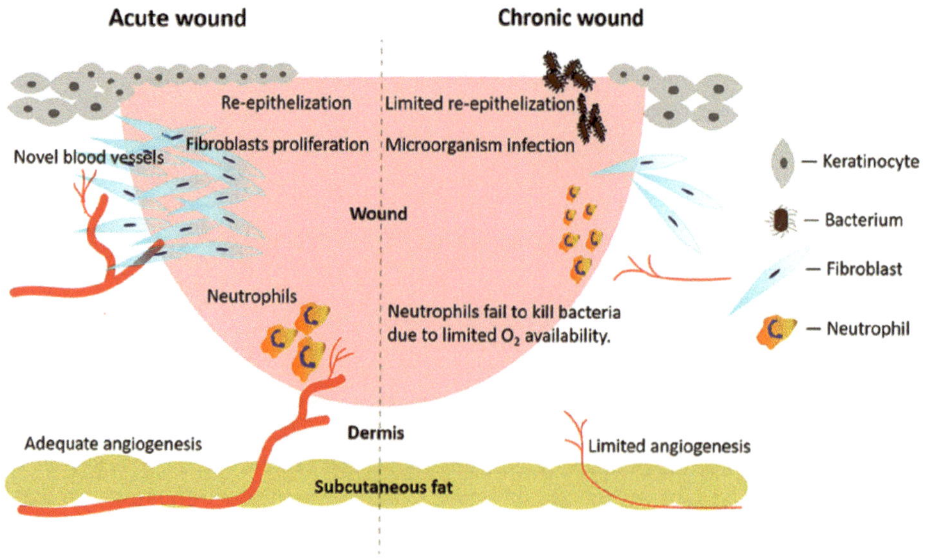

FIGURE 3.1 Pathophysiology of chronic wounds. (Adapted from Wang Y, Armato U, Wu J, Targeting tunable physical properties of materials for chronic wound care, *Frontiers in Bioengineering and Biotechnology,* 2020 June 11; 8:584. This article is an open access article distributed under the terms and conditions of the Creative Commons Attribution (CC BY) license.)

The Von Willebrand factor is first secreted during hemostasis, which encourages platelet adhesion and the subsequent release of the mediator. This leads to the production of fibrin clots, and the bleeding is stopped. The blood flow decreases as the vessels continue to narrow. The phenomena happen in a few minutes (Cañedo-Dorantes L. & Cañedo-Ayala M., 2019).Figure 3.2 is a schematic representation of the different phases of wound healing.

3.5.1.1 Inflammatory Phase

In the inflammatory phase, the blood vessels contract in response to the stimulation produced by the wound. The thrombocytes and platelets start accumulating at the site and a network of fibrin is made (Sun B. K. et al., 2014). This series of processes leads to the reestablishment of homeostasis,

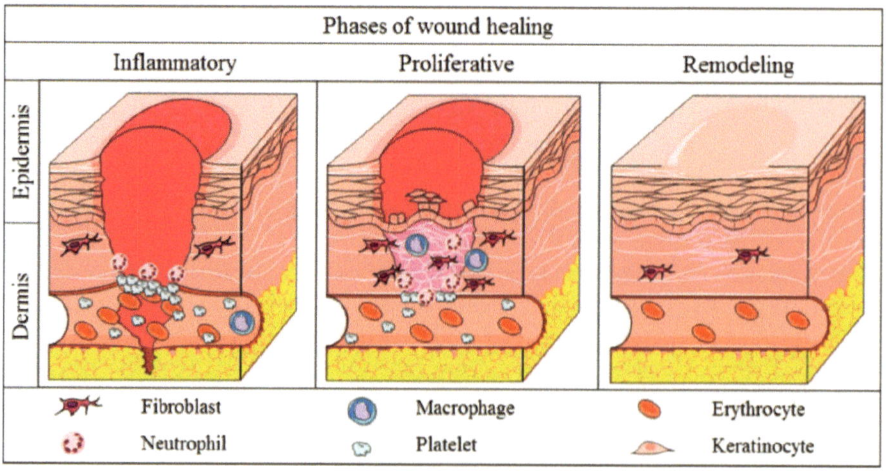

FIGURE 3.2 Schematic representation of different phases of wound healing (Gushiken L. F. et al., 2021).

the formation of a protective barrier against infections, and the laying down of a temporary matrix for cell migration, restoring the integrity of the skin (Shaw T. J. et al., 2016). The cell response happens in the first 24 hours and might extend for two days (Gonzalez et al., 2016).

3.5.1.2 Proliferative Phase

This phase lasts for 3–12 days (Cañedo-Dorantes L. & Cañedo-Ayala M., 2019). During this period, granulation tissues are formed from fibroblasts along with the regulation, proliferation, and migration of keratinocytes. In addition to this, macrophages simultaneously secrete growth factors. The maturation phase involves collagen restoration and wound contraction due to myofibroblasts. Growth factors regulate transitions between mesenchymal–mesenchymal and endothelial–mesenchymal phenotypes (Yoshida M. et al., 2012).

3.5.1.3 Remodeling Phase

During this phase, all other processes happening previously are paused; myofibroblasts, isolated endothelial cells, and macrophages undergo apoptosis or are removed from the wound (Enoch S. et al., 2008). The primary objective of this phase is to increase the tensile strength by replacing collagen III with collagen I (Gonzalez A. et al., 2016) along with reorganizing and resynthesizing the extracellular matrix. Collagen I differs from the regular basket-weave collagen found in the healthy dermis; these are arranged in small parallel bundles (Gurtner G. C. et al., 2000). Subsequently, the myofibroblasts reduce the surface of the forming scar and induce wound contractions through their multiple attachments to collagen (Marshall C. D. et al., 2018; Profyris C. et al., 2012). Moreover, the process of angiogenesis stops, the blood flow diminishes, and the overall metabolic activity retards and finally ceases (Reinke J. M. et al., 2012).

3.5.2 Cellular and Molecular Mechanisms

Based on their fundamental mechanisms of action, which involve a complex interplay of cellular processes and molecular signaling pathways, the stages of wound healing can be thoroughly categorized. This complex process requires the coordinated involvement of various cell types and the secretion of growth factors and other bioactive molecules. Through examining the stages from the perspective of molecular and cellular mechanisms, a deeper understanding of the complex biological mechanisms underlying wound repair becomes clear. During hemostasis, at the cellular level, platelets are the principal contributors; they are crucial in the recruitment of immune cells to the injury site along with the secretion of release factors like PDGF and TGF-β to initiate inflammation (Velnar T. et al., 2009). At the molecular level, the wound triggers vasoconstriction and clotting of the wound (Guo S A. et al., 2010). Similarly, in the inflammation stage, neutrophils and macrophages play an important role in clearing the debris and pathogens at the cellular level (Wilkinson H. N. et al., 2020). While initiating the inflammatory response, various cytokines such as TNF-α and growth factors like PDGF and TGF-β are released (Schultz G. S. et al., 2011). The proliferative phase of wound healing is characterized by extensive stimulation of keratinocytes, fibroblasts, macrophages, and endothelial cells, which form the structural framework for tissue repair, while endothelial cells proliferate to form new blood vessels (angiogenesis) (Shaw T. J. et al., 2016). At the molecular level, the phase is marked by the release of fibroblast growth factor, insulin-like growth factor, and certain chemokines like monocyte chemoattractant proteins that help with proliferation and migration (Falanga V. et al., 2022; Kadam S. et al., 2019). Finally, in the maturation or remodeling phase, myofibroblasts emerge as the predominant cell type tasked with the remodeling of the wound ECM. They replace the initial fibrin clot with hyaluronan, fibronectin, and proteoglycans, subsequently forming mature collagen fibrils as the repair progresses (Darby I. A. et al., 2014). Molecularly, chemokine (chemokine [C-C motif] ligand), cytokines (interleukin 6, interleukin 8), and growth factors (insulin-like growth factor) are involved in the breakdown and remodeling of the extracellular matrix (Tyavambiza C. et al., 2022; Schultz G. S. et al., 2011).

3.5.3 Factors Affecting Wound Healing

Several reasons can hamper wound healing. In broad terms, two types of elements affect repair: systemic and local. Both of these factors are interrelated and can affect or influence the flow of wound repair.

3.5.3.1 Local Factors Affecting Wound Repair

3.5.3.1.1 Oxygenation

Oxygen plays a major role in the process of wound healing. It is the driving factor of cell metabolism and energy production. It prevents any infection and induces angiogenesis, increases keratinocyte production and migration as well as enhances the proliferation of fibroblasts and collagen synthesis. Insufficient oxygenation leads to impaired healing. Temporary hypoxia triggers healing, while prolonged or chronic hypoxia delays it (Bishop A, 2008).

3.5.3.1.2 Infections

An open wound can easily get infected by the microorganisms that are present on the skin surface (Edwards R. & Harding K. G., 2004). Inflammation is a normal part of wound healing, crucial for removing contaminating microorganisms. However, without effective decontamination, inflammation may persist due to incomplete microbial clearance. Bacteria and endotoxins can elevate pro-inflammatory cytokines, leading to chronic inflammation and failure to heal.

3.5.3.2 Systemic Factors Affecting Wound Repair

3.5.3.2.1 Diabetes

Diabetic individuals experience impaired wound healing and are at risk of developing chronic non-healing diabetic foot ulcers (DFUs). These wounds are influenced by complex pathophysiological mechanisms, including hypoxia, dysregulated cellular functions, and defects in T-cell immunity, leukocyte chemotaxis, phagocytosis, and bactericidal capacity (Mathieu D., 2006; Tandara A. A. & Mustoe T. A., 2004).

3.5.3.2.2 Age

It has been observed that in the elderly population (people over 60 years of age) the process of wound healing is relatively slower although it's not impaired. The delayed wound healing is mostly associated with an altered inflammatory response (Bonifant H. & Holloway S. 2019; Pence B. D. & Woods J. A. 2014). Recent research has revealed that nearly all gene expression variations between human wounds from young and elderly males are estrogen-regulated. The effects of estrogen on wound healing are shown via controlling several genes linked to inflammation, regeneration, matrix formation, protease inhibition, and epidermal function (Levine M. E. et al., 2018).

3.5.3.2.3 Drugs and Radiation

Several drugs hinder or affect the process of blood clotting, the function of platelets, etc., or in general terms, the overall wound healing process. Drugs like glucocorticoid steroids, NSAIDs, and chemotherapeutic agents have shown probable inhibition of wound repair (Almadani Y. et al., 2021). Ionizing radiation can disrupt single or double strands of DNA or cause the double helix to become cross-linked. This is caused by excited subatomic particles. Additionally, free radicals produced by radiation harm cell membranes and proteins (Gieringer M. et al., 2011).

3.6 OVERVIEW OF EXISTING TREATMENTS

3.6.1 Conventional Wound Dressing

The process of wound healing is one of the most essential and basic practices in human civilization (Powers J. G. et al., 2016). Accounts of wound treatment, from preventing infection to making bandages and homemade dressings with honey, grease, and lint, can be found in Egyptian papyri and Crimean battlefields (Broughton G. et al., 2006). Table 3.1 lists traditional wound dressings.

3.6.2 Advanced Wound Dressing Techniques

Advanced wound dressing techniques are made of several components that help speed up the healing process. As these novel treatments can typically stay active on the site of wounds for several days, fewer procedures and drugs are required to replace them (Dreifke M. B. et al., 2015). Drug advancements play a major role in the healing of complicated wounds, including chronic wounds, which lowers medical expenses (Tottoli E. M. et al., 2020). Traditional dressings are applied to mild lesions, such as minor surgical wounds, erythema, or nonsevere ulcers, which tend to heal quickly and exhibit little exudate. Although the dressing is placed on the skin to shield the wound, it has no active part in the healing process (Stashak T. S., 2007). To overcome such obstacles, various studies have been done to create advanced dressing techniques. Some of the recently developed techniques are discussed next.

TABLE 3.1
Traditional Wound Dressings

Type of Dressing	Description	Marketed Formulations	Reference
Gauze	Used in deep wounds, absorbs exudates, inexpensive	Curity, Vaseline Gauze, Xeroform	Lawrence J. C., 1994
Impregnated gauze	Impregnated with petroleum, iodine, zinc, etc.; adds moisture to the wound bed; used in skin grafting as a skin layer	DermaGauze, Kendall, Dynagel	Ubbink D. T. et al., 2008
Transparent film dressings	Closed surgical incision sites or superficial wounds, retain moisture	Polyskin, Suresite	Rubio P. A., 1991
Foam dressings	Higher absorptive properties, insulating properties	Lyofoam, Allevyn, Optifoam	Seaman S., 2002
Hydrogels	Moisturize dry wounds, softening and loosening necrotic wounds, easy removal	Carrasyn, Curagel, Nu-Gel	Lopes et al., 2024
Hydrocolloids	Self-adhesive and gel-forming, absorbs exudates, thermal insulation, prevent contamination	DuoDERM, Granuflex	Varghese M. C. et al., 1986
Hydrofibers	High adsorbent properties, easy removal	Aquacel Hydrofiber	Hurlow., 2012
Silicone dressings	Pressure garments, relax or soften scar	Dimora, Carbou	Perkins K. et al., 1983
Silver dressings	Topical antimicrobials, reduce inflammation	Acticoat 7, Allevyn AG,	Warriner R. & Burrell R., 2005
Manuka honey	Antibacterial agent, anti-inflammatory, immune modifying effect	Melcare's Manuka Wound Gel™, Medi Honey ointment	Blair S. E. et al., 2009
Iodine dressings	Antimicrobial, Antiseptic	Iodoflex, Iodosorb, Inadine	Burks R. I., 1998
Charcoal dressings	Reduces wound odor	Actisorb™, Clinisorb	Parry A. D. et al., 1995

3.6.2.1 Nanofiber-Based Wound Dressing

A significant amount of curiosity has been spent on bioactive dressings that can be used to treat wounds that are acute as well as chronic. Electrospun nanofibers are considered advanced dressing options because of their adept resemblance to the extracellular matrix for wound healing and skin regeneration, high porosity, permeability to air and water, and effective barrier properties against external pathogens (Zhang X. et al., 2024). Electrospinning nanofibers have a larger specific surface area, which can absorb exudate, control bleeding, and aid in drug loading and release (Ding Y. et al., 2019). Their random mesh structure and high porosity minimize excessive water loss, keep the environment moist, and inhibit microbial invasion, making them perhaps the best wound dressing (Abrigo et al., 2014).

3.6.2.2 Antimicrobial Peptide Dressings

Chronic wound infections frequently produce multicellular biofilms that may defend bacteria and develop antibiotic resistance owing to their impermeability and poor drug penetration (Miao F. et al., 2021). Antimicrobial peptides are tiny proteins with cationic charges and amphipathic structures that interact with the anionic phospholipids on the microbial cell membrane. They have both destructive and inhibitory effects on biofilms, making them an effective weapon against persistent skin wound infections (Koo H. et al., 2017). They contribute to immunomodulation and wound healing. Antimicrobial peptide delivery techniques such as nanoparticles, PLGA electrospun fibrous scaffolds, and hydrogels have been created to prevent infection and control wound healing rates while retaining antimicrobial characteristics.

3.6.2.3 Electroceutical Dressings for Wound Healing

Bioelectrochemical fields direct the movement of cells and affect wound healing. Electroceuticals, such as ionic, wireless, and battery-powered devices, are being tested on critically wounded patients. This equipment can alter electrochemical signals, influencing molecular charge distributions and biological activity. The majority of clinical investigations suggest that cell migration, wound healing rates, and local perfusion are enhanced in contrast to traditional open-coil therapy (Evans J. et al., 2023).

3.6.2.4 3D-Printed Dressings for Wound Healing

3D printing methods enable high-resolution, precise, and spatially accurate production of biomaterials. They have been utilized successfully for wound healing in vitro and in vivo. 3D printing is computerized, thus each level of the printed part may be fully manipulated in situ to generate the desired dressings. It has the potential to be used for personalized patient therapy with higher clinical efficacy and lower costs. It contains a range of contents loaded in it, such as antibiotics and antibacterial nanoparticles, which helps expedite the process of wound healing (Tabriz A. G. et al., 2022; Tsegay F. et al., 2022).

3.6.2.5 Stem Cell-Derived Dressings

In the treatment of chronic wounds, the application of cells—adult and/or mesenchymal stromal cells in dressings—is quickly growing and redefining it. Cell-based dressings arose in response to the demand for more complicated and biomimetic therapies that provide moisture retention, protection, and structural or pharmacological qualities (Las Heras K. et al., 2020). Adipose stem cells have recently received interest as a potential therapy for persistent wounds. Some preclinical and clinical trials have already produced encouraging outcomes for adipose stem cell therapy in the treatment of secondary lymphedema. Previous studies have indicated the role and usefulness of adipose stem cells in improving tissue regeneration, particularly in soft tissue reconstruction, such as breast reconstruction, wound healing, and the treatment of scars and soft tissue abnormalities (Lu J. H. et al., 2023).

3.7 ASSESSMENT OF CHRONIC WOUNDS

Understanding the biology of a chronic wound is crucial for treating it effectively. Chronic wounds, whatever their etiology, share some traits, including high levels of proinflammatory cytokines, persistent infections, the occurrence of drug-resistant microbial biofilms, and senescent cells that fail to react to reparative stimuli (Frykberg R. G., 2015). As a result, clinical examination is a critical stage in determining a final diagnosis and developing a specific therapeutic strategy. Despite the type of wound and its root cause or etiology, a full examination of the patient's health, involving the medical and pharmaceutical history, is required (Janowska A. et al., 2019).

3.7.1 Etiology of Chronic Wounds

Clinicians analyze wounds to find the best treatment for various disease processes. To properly categorize and treat a wound, the interprofessional team should ask basic questions during the initial and subsequent examinations. Understanding these characteristics has a tremendous impact on the treatment strategy, since without good understanding, the odds of recovery decrease dramatically (Nagle S. M. et al., 2018). They are divided by etiology into four types (Hoversten K. P. et al., 2020), each having its characteristic location, depth, and physical features.

3.7.1.1 Arterial Ulcers

Arterial ulcers are deep and frequently reveal tendons and bones. The initial assessment comprises ankle–brachial index (ABI) measurement and pulse palpation. ABIs less than 0.8 suggest arterial disease, while those larger than 1.2 indicate noncompressible arteries. Ultrasonography, pulse volume recording, skin perfusion pressure, and transcutaneous oximetry can be employed to conduct further vascular assessments. If test findings show inadequate perfusion, a referral to a specialist is recommended (Whitney J. D. et al., 2005).

3.7.1.2 Diabetic Ulcers

Diabetic ulcers demand good blood glucose control and a check for underlying peripheral vascular disease. The dense callus around the ulcers demands surgical debridement. The ideal therapy is to unload the foot and alleviate pressure with a whole contact cast. Hyperbaric oxygen therapy may be useful for certain ulcers (Lalieu R. C. et al., 2020).

3.7.1.3 Pressure Ulcers

Pressure injuries are tissue traumas induced by increased mechanical force that frequently occur on bony prominences. They can be divided into four phases according to the National Pressure Ulcer Advisory Panel (NPUAP) staging scheme. Treatment includes identifying at-risk patients and evaluating their mobility, tissue health, and tissue perfusion. Nutrition, hydration, diabetes, and a low albumin level could have an influence. Risk assessment is advisable, and a variety of measures, including frequent mobilization, relocation, pressure offloading, specialized mattresses, and minimizing excessive moisture or dryness, can help prevent pressure injuries (Tottoli E. M. et al., 2020).

3.7.1.4 Venous Ulcers

Venous ulcers are the most prevalent type of chronic wound. They are usually shallow and situated on the medial supramalleolar aspect of the lower limb. Typical venous symptoms of hypertension may include edema, hemosiderin staining, and lipodermatosclerosis. The initial evaluation should include an examination of the arterial state using the ABI and pulse palpation to exclude mixed arterial and venous disease and ensure adequate perfusion prior to compression. If required, additional testing with venous duplex ultrasonography with reflux and/or arterial duplex ultrasonography will aid in the diagnosis (Raffetto J. D. et al., 2020).

3.7.2 Protocol for Wound Care Management

Wound management is difficult because of numerous reasons, wound types, and patient characteristics. Objective treatment should be determined by wound type, healing phase, and comorbidities (Robson et al., 2006). Anamnesis should cover wound appearance, therapies, progression, illnesses, and patient status. Impaired wounds can lead to systemic disorders, resulting in a healing cycle (Tatarusanu et al., 2022). Table 3.2 displays the methods for monitoring chronic wounds. Table 3.3 lists imaging modalities for chronic wounds.

3.8 CHALLENGES OF EXISTING TREATMENT APPROACHES

Chronic wounds burden healthcare with financial costs and have a destructive influence on morbidity. In India, financing in healthcare has been recorded at 5% of the gross revenue of the domestic product, with more than 80% of healthcare expenses met from out-of-pocket expenses (Prinja S. et al., 2012; Frykberg et al., 2015). These not only indicate financial loss but also reflect frustration and impaired quality of life experienced by chronic wound patients. Hence, they remain a major clinical challenge in long-term care, affecting the quality of life for patients and healthcare expenses (Ruedrich E. et al., 2013). Chronic wounds are frequently depicted by pathological reactions resulting in fibrosis and nonhealing chronic ulcers. This clinical plight is a result of undiagnosed or incurable wounds. The major clinical symptoms of a patient with a chronic wound include agony, redness of the skin, dropsy, heat, and pus with high wound bioburden (Gardner S. et al., 2001; Menke et al., 2007). Moreover, signs and symptoms are particular to secondary wounds frequently noticed in the proliferative phase comprising: (1) wound breakdown, (2) drainage of serosal fluid with simultaneous swelling, (3) malodorous and (4) friable granulation tissue, and (5) delayed healing (Martin G. M., 2001). However, these signs and symptoms need to be further verified for efficient chronic wound management and treatment. A study performed by Sue et al. to verify the clinical signs of chronic wounds proved that the symptoms specific to secondary wounds were better biomarkers of chronic wound infection than the classical symptoms, with a mean sensitivity of 0.62 and 0.38. Other than that, increasing pain and wound breakdown were both adequate indicators with a

TABLE 3.2
Methods for Monitoring Chronic Wounds

Monitoring Approach	Indications	Assessment Tools
Physical assessment	• Visual observation is the primary tool for the preliminary diagnosis of chronic wounds. • Wound assessment charts like the Bates–Jensen Wound Assessment Chart (BJWAT), Pressure Ulcer Grade Recording Charts, and Wagner Grading Scale are used. • These charts do not predict the regenerative phase of the wound, which is crucial as wound healing usually stalls at the inflammatory phase.	• Wound measurement tools like wound photography and digital tracings are efficient in two-dimensional wound assessment. • Other noninvasive tools include the oximeter and subepidermal moisture scanner.
Microbiological and biochemical assessment of wounds	• Biochemical analysis of wound exudate, collected from capillaries, can determine healing potential. • These tests take 3–5 days to yield conclusive results, necessitating a multifaceted tool for multiple bacteria types.	• Swab tests and biopsies are commonly used to assess microbes on wounds.

Source: Li S. et al. (2020).

TABLE 3.3
Imaging Modalities for Chronic Wounds

Imaging Tool	Applications
Computed tomography	• An invasive imaging • Assess diabetic foot ulcers • High anatomic detail and resolution
Magnetic resonance imaging	• Soft tissue and bone pathology evaluation • Enhanced resolution and higher sensitivity to infection
Single-photon emission computed tomography	• Radiotracer-based imaging • High-sensitivity 3D images
Ultrasound imaging	• Safety, low cost, functionality • High resolution • Sends acoustic pulse to wound region
Terahertz spectroscopy	• Analyzes changes in water content • Detects interstitial edema and density of skin
Digital camera imaging	• Cost-effective, noninvasive • Accurate high-resolution wound recording and size measurement
Hyperspectral imaging	• Noninvasive • Generates a three-dimensional data cube • Identifies microvasculature abnormality and tissue oxygenation
Optical coherence tomography	• High-resolution cross-sectional images • Noninvasive diagnostic • Assesses microvascular structure in burn wounds
Laser Doppler imaging	• Blood flow quantification • Burn wound depth measurement
Spatial frequency domain imaging	• Optical properties of wounded tissue • Measures wound depth and severity, and identifies infection • Quantifies tissue oxygen saturation and blood volume

Sources: Li S. et al. (2020) and Mukherjee R. et al. (2017).

specificity of 100%. These signs and symptoms can, in turn, be compared to conditions that show their physiological environment. Hence, it has been proven that chronic wounds face a challenging environment that triggers the wound to a nonhealing condition. It is most important to deeply understand the fundamental mechanisms included in its pathological conditions. This is feasible only by identifying the clinically specific biological markers that are the important factors for the chronicity of a wound. Hence, identifying a biomarker would possibly help in reversing the condition of chronic wounds (Ferreira D. et al., 2020). Chronic wounds are frequently characterized by infection, which boosts the production of inflammatory and pro-inflammatory cytokines, resulting in a rise in the production of matrix metalloproteases and a decrease in their inhibitors. The delay in inflammation results in a reduction of extracellular matrix and fibroblast production. Due to the inhibition of fibroblast production, the collagen fibers necessary for wound remodeling are not produced. Chronic wound-associated fibroblasts show fibroblast senescence. The chronic wound environment damages angiogenesis and stalls epithelialization. These characteristics represent a chronic wound environment that causes delayed healing (Strausberg J. et al., 2007; Shukla V. et al., 2005). It has been predicted that approximately 1% of the population will suffer from leg ulceration during their lifetime (Falanga V. et al., 2005; Afraz E. et al., 2022). Based on the epidemiology of chronic wounds, it has been predicted that the prevalence at the rate of 4.5/1,000 population (Robson M. et al., 2006) partially treated acute traumatic wounds are often a cause of these chronic wounds. Wounds associated with chronic diseases, including diabetes and cancer, pose a major challenge to treat and manage, which is mainly due to molecular complexity.

It is found that diabetes affects wound healing through decreased inflammatory responses, loss of protective sensations due to neuropathy, and an increased risk of infection (Han J. et al., 2020). Chronic nonhealing pressure ulcers are serious and frequent occurrences among immobile and physically weak patients (Diegelmann et al., 2004). Based on this information, it is important to identify the key factors responsible for delayed wound healing and also understand chronic wound healing at the molecular level. Hence, the current research for the treatment of chronic wounds focuses on understanding the fundamental molecular mechanisms in clinically different wounds and at various stages of their advancement and healing (Sen C. K., 2019; Monika P. et al., 2022).

3.9 ADVANCED THERAPEUTIC MODALITIES FOR CHRONIC WOUND HEALING

3.9.1 Hyperbaric Oxygen

Hyperbaric oxygen therapy (HBO2T) (O'Reilly et al., 2011) is a method of administering oxygen at pressures higher than sea level and is incorporated by placing the patient inside a pressure vessel and allowing them to breathe oxygen. It has been believed to be beneficial for an extensive variety of chronic wounds for over two decades (Wu S. et al., 2010). It is based on the principle that it can enhance fibroblast proliferation, improve immune function, and boost the process of angiogenesis, among other functions. Despite this, these ideals have led to some extent of controversy in its use. When this therapy is tested on a patient in a hyperbaric oxygen chamber, localized delivery of oxygen is not efficient, which could lead to various side effects such as myopia and oxygen toxicity in the brain leading to seizures (Wu S. et al., 2010). In another study, only Cochrane was able to exemplify a modest rise in the possibility of diabetic foot ulcer healing in one year (Kranke et al., 2015). At this point, it looks like hyperbaric oxygen can only be used in wounds where an ischemic diabetic ulcer has shown hypoxia (Fonder M. et al., 2008). In some ways, this therapy was emphasized in a recent randomized controlled trial where hyperbaric oxygen was used in wound healing, but even the authors discovered that this therapy is only for "selected patients with diabetes" (Löndahl M. et al., 2010; O'Reilly et al., 2011).

3.9.2 Negative Pressure Wound Therapy

Negative pressure wound therapy (NPWT) is an effective wound management system that uses controlled negative (subatmospheric) pressure applied uniformly to the wound through an open cell foam or other interface dressing, either continuously or intermittently. NPWT is used for the removal of exudate from chronic, acute, traumatic, subacute, and dehisced wounds; partial-thickness burns; ulcers (such as diabetic, venous, or pressure); and surgically closed incisions, flaps, and grafts. It starts healing immediately and also diminishes numerous complications. NPWT is contraindicated or should be used with caution with inadequately drained wounds, the presence of necrotic tissue such as eschar or adherent slough, exposed blood vessels, anastomotic sites, organs, tendons or nerves, wounds containing malignancy, fistulas, untreated osteomyelitis, and actively bleeding wounds (Stretubel P. et al., 2012). NPWT has been shown to rapidly improve granulation tissue formation, increase the rate of wound contraction, improve perfusion, remove toxic exudates, and likely impact surface biofilm. The devices available allow for optimum patient management with hospital devices used for larger wounds, dressings, and devices allowing for instillation and cleansing of wounds, as well as the management of larger volumes of exudate. Available outpatient devices are smaller, more portable, and available in single-use delivery systems that employ mechanical or battery-powered collection devices and dressings. One important aspect is that retained foam or gauze after NPWT is a relatively common complication and is reportable as an adverse event (Venturi M. et al., 2005).

3.9.3 Layering Materials for Dressing

Another new method of wound healing is the use of different properties of the material to make layered dressings. Clinically, absorbing dressings used to consist of a non-adhesive layer or semi-adhesive layer and a highly absorbent layer of fibers such as rayon fabrics, cotton, and other compounds. In this regard, Gunawati et al. created a three-layer dressing (Falabella A., 2006). In this dressing, the layer in contact with the wound is made of nonwoven polyester fibers, the absorbent layer is made of nonwoven bamboo fibers, and the top layer is made of 6/PCL nylon (Simoes D. et al., 2018). Sussman et al. combined a layer of alginate with a layer of hydrofiber to form a sheet of fibers. These layers attach to a layer of activated carbon and combine with an additional viscous outer layer (Dhivya S. et al., 2015). Another example of a multilayer dressing is a combination of hydrocolloids and alginates, which are especially suitable for the treatment of superficial foot wounds, burns, and pressure sores. Lopez et al. developed a multilayer dressing of a combination of hydrogel foam and polyurethane layers for the treatment of chronic wounds in which the physiological environment of the wound should be further preserved during the granulation and epithelialization stages (Zahedi P. et al., 2010). Recently, a study has been conducted on how to combine the materials used in multilayer wound dressings with combination drugs to optimize the treatment of chronic wounds. This study aimed to prepare a multilayer wound dressing with a combination of drugs from two different pharmacodynamic groups. PTE (first layer) includes lidocaine, which was effective for immediate release and rapid pain relief, and diclofenac in alginate (second layer) and viscose (third layer), which were effective for long-term pain relief (Thu H. et al., 2012).

3.9.4 Debridement

A completely different method of wound care is the use of larvae. Larvae from greenflies (*Lucilia sericata*) have beneficial effects on chronic incurable wounds. These properties include the removal of necrotic tissue (debridement), wound disinfection, and active strengthening of granulation tissue formation. The interactions of the two have a significant effect on the formation of new tissue. Sterile larvae are placed in a cage in a wound dressing. The larvae usually remain in the dressing area for 24–60 hours, then are washed with saline solution. If necessary, this is repeated by new sterile larvae (Varaprasad K. et al., 2020). Collagenase ointment (250 units per gram), derived from the bacterium *Clostridium histolyticum*, is very effective for dry wounds with fibrin debris and no granulation tissue, especially when surgical procedures are not possible. Collagenase increases endothelial cells and keratinocyte migration (Balakrishnan B. et al., 2005). Enzymatic destroyers are an effective option for removing necrotic material from pressure ulcers, foot ulcers, and partial-thickness wounds. Biological debridement that uses medical maggots is a rapid and efficient debridement method commonly used for fibrinous wounds. This method is often used less due to high pain and intolerance of the patient and the provider. A recent randomized controlled trial showed that people treated with larvae experienced more discomfort than those treated with hydrogel dressings (Jin S. et al., 2016).

3.9.5 Nanomedicine in Wound Healing

Nanomedicine has made advances in diagnosing and treating different types of diseases, including cancer, cardiovascular disease, tissue engineering, diabetes, regenerative medicine, inflammatory diseases, and wound healing (Riehemann K. et al., 2009). Due to their unique physical, chemical, and biological properties, nanoparticles are used to suppress inflammation, control microbial infections, and accelerate wound healing (Sharifi S. et al., 2020). Dressings made of nanoparticles, due to greater porosity, the ability to absorb wound secretions, increased wound respiration, and mimicking the structure of the ECM, and producing ROS or releasing ions by affecting several cellular and molecular pathways, accelerate wound healing (Berthet M. et al., 2017).

3.9.6 Bioengineered Cellular Therapies

Human fibroblast-derived dermal substitute (HFDS), i.e., Dermagraft (Organogenesis, Canton, MA), is made of three-dimensional, absorbable, cryopreserved polyglactin mesh substrate seeded with living neonatal dermal fibroblasts. These cells release a variety of growth factors, cytokines, matrix proteins, and glycosaminoglycans that promote tissue regeneration by encouraging the creation of granulation tissue and the ingrowth of host fibroblasts and keratinocytes. This process is similar to the replacement of bilayered skin (Edmonds N. et al., 2009). In the crucial study conducted by Marston et al. in 2003, the efficacy of this dermal replacement in treating chronic DFUs was demonstrated. After 12 weeks, participants in this multicenter clinical trial who were allocated to the HFDS group healed far more quickly than those in the routine wound care group (30% vs. 18.3%, $p = 0.023$). The study group had a 64% improvement in healing despite the seemingly low overall healing rate compared with the standard care group (Harding K. et al., 2013).

There was a bigger margin of effect than any other trial's results. A major VLU trial did not show superiority in its primary outcome of complete healing; however, for a subgroup of ulcers with a duration of ≤12 months ($p = 0.029$), significant differences in complete healing were observed (Angirasa A. et al., 2006). According to other reports, dermal replacement has been successfully employed along with different wound healing techniques in a range of lower extremity wounds (Frykberg R. et al., 2011). In the two arms of the DFU pivotal trial, the incidence of amputations and bone resections was examined in a recent study. The results showed that the HFDS group experienced fewer complications from these procedures, which were probably caused by a lower rate of infection and quicker healing in the experimental treatment group (Kashefsky et al., 2012).

3.9.7 Electrical Stimulation in the Treatment of Wounds

There are naturally occurring electrical fields in plants, amphibians, and mammals (Tai G. et al., 2018). Certain charged particles, such as ions, move around and produce these fields. These electrical fields are crucial for healthy tissue growth, regeneration, and injury repair (Levin M. et al., 2007). Over the skin, there are naturally existing electrical potentials that are helpful for promoting cell proliferation. As a result, electrical stimulation has been suggested as a practical technique that can be applied to encourage tissue regeneration and healing (Foulds I. et al., 1983). There are two different kinds of stimulation techniques. The first kind makes use of electrodes to supply monopolar energy, such as those that produce low-intensity direct current (LIDC) and high-voltage pulsed current (HVPC) (Finger S. et al., 2013). This approach is more efficient than the second kind of stimulation, which includes placing bipolar electrodes. Although the bipolar approach reduces the risk of infection because the electrodes are not transferred between the patients, the bipolar treatment acts more slowly than monopolar stimulation (Kloth L. C. et al., 2010). The electrical stimulation electrodes are applied to the skin next to the wound site or in the wound bed. To transfer electricity to the skin or wound bed, a wet conductive medium, such as regular saline-soaked gauze or amorphous hydrogel-impregnated gauze, is typically employed beneath the electrodes.

3.9.8 Low-Frequency Ultrasound

Chronic wounds have been treated with low-frequency ultrasound as an adjuvant. By eliminating necrotic tissue and debris, it causes a debriding effect (mostly through cavitational and acoustic streaming events). Additionally, it has been observed that ultrasound breaks down the biofilm in vitro, making bacteria more sensitive to antibiotics (Chen H. et al., 2023). It is suggested that it works well to promote cellular proliferation, increase angiogenesis, decrease the inflammatory phase, and stimulate collagen formation (Rastogi A. et al., 2019). When low-frequency ultrasound therapy is used to heal wounds, multiple clinical investigations have demonstrated a decrease in the size of the wound (Federman D. et al., 2016).

3.9.9 Extracorporeal Shock Wave Therapy

By providing acoustic pulsed energy to tissues, extracorporeal shock wave therapy, or ESWT, has been proposed as a means of promoting wound healing. ESWT appears to upregulate growth factor and cytokine responses, decrease the anti-inflammatory response, accelerate circulation, and stimulate angiogenesis (Mittermayr et al., 2012). The efficacy and acceptability of ESWT in wounds with various etiologies were shown in a clinical trial (Schaden W. et al., 2007). Additionally, based on enhanced angiogenesis, tissue perfusion, and cellular responses with decreased cell apoptosis, and a greater ulcer healing rate, a review found that ESWT is more beneficial for patients with diabetic foot ulcers than hyperbaric oxygen therapy (Cho Y. et al., 2016).

3.10 FUTURE DIRECTIONS

It will come as no surprise that there is an abundance of research being done on wound healing given its effects on patient care and the economy. Molecular biology and materials science, two seemingly separate domains, are the sources of new research on wound healing. Several of these have concentrated on cutting-edge compounds with potential for use in wound healing, like nitric oxide, and have employed a variety of creative material design strategies to achieve this goal (Han G. et al., 2012). Although these methods may be beneficial for chronic wound healing and may also, in a sense, get around some of the restrictions placed on individual growth factors or cytokines (particularly as more widely acting, gregarious molecules involved in the wound healing cascade are identified), they will not come close to the "holy grail" of wound healing.

Although it has long been known that embryonic wounds do not scar, it has been more difficult to figure out (Mowbray M. et al., 2008). Even though this is a different issue in terms of wound healing—the majority of previous discussions focused on chronic, nonhealing wounds—the concept that a simple wound can heal with little to no scarring has been alluring. Wounds in fetuses heal via regeneration instead of systematic healing (Lo D. et al., 2012). This could be related to the comparatively pathogen-free environment seen in utero and may explain why the wound-healing "switch" activates in the final trimester of pregnancy (Larson B. et al., 2010). The animal kingdom provides a corollary to this, namely, the regeneration of limbs in reptiles and the African spiny mouse (*Acomys*), which was documented more recently (Wilgus T., 2007). While some studies have worked to identify specific proteins, including TGF and matrix metalloproteinases (Seifert A. et al., 2012) that are either up- or downregulated in the fetal wound, stem cell therapy may hold the greatest promise. More recent studies have concentrated on using stem cells as scar-attenuating adjuncts to wound healing, leading to tissue regeneration rather than repair with the formation of fibrotic scar tissue (Degen K. & Gourdie R., 2012). These efforts run parallel to numerous successful research efforts in chronic wounds that focus on injecting stem cells from bone marrow into nonhealing wounds (Badiavas E. et al., 2003).

3.11 CONCLUSION

The physiological process of wound healing is complex, involving the growth of new cells, re-epithelialization, collagen fiber deposition, and tissue regeneration. Consequently, it is essential to choose appropriate polymers, bioactive substances, and wound dressings that can promote and enhance the healing process of wounds. A single-wound dressing will not be effective for every kind of wound since different wounds produce exudate at distinct rates and have different surface shapes. A major concern for the clinician and wound care specialist is managing chronic wounds. Nothing can replace the requirement of adhering to the fundamental principles of wound care, which include thorough assessment with medical and nutritional optimization, debridement (including surgery), management of ischemia, and management of infection, even with all the recent advancements in wound care therapies and our understanding of the pathophysiology underlying chronic wounds.

When these guidelines and accepted wound care practices are followed, positive results should frequently occur. Furthermore, it is advised to utilize advanced wound care therapies rationally when wounds do not adequately respond to good routine care after four weeks or sooner, depending on the situation.

REFERENCES

Abbade, L. P. F., & Lastória, S. (2005). Venous ulcer: Epidemiology, physiopathology, diagnosis and treatment. *International Journal of Dermatology*, *44*(6), 449–456.

Abrigo, M., McArthur, S. L., & Kingshott, P. (2014). Electrospun nanofibers as dressings for chronic wound care: Advances, challenges, and future prospects. *Macromolecular Bioscience*, *14*(6), 772–792.

Afraz, E. S., Sahraee, S., Varnamkhasti, A. G., & Mani, H. (2022). Healing effect of decellularized human placenta and gelatin hydrogel on diabetic wound in animal model. *Journal of Biological Studies*, *5*(3), 637–647.

Agrawal, K., & Chauhan, N. (2012). Pressure ulcers: Back to the basics. *Indian Journal of Plastic Surgery*, *45*(2), 244–254.

Alhaug, J., Gay, C. L., Henriksen, C., & Lerdal, A. (2017). Pressure ulcer is associated with malnutrition as assessed by Nutritional Risk Screening (NRS 2002) in a mixed hospital population. *Food & Nutrition Research*.

Allman, R. M. (1997). Pressure ulcer prevalence, incidence, risk factors, and impact. *Clinics in Geriatric Medicine*, *13*(3), 421–436.

Almadani, Y. H., Vorstenbosch, J., Davison, P. G., & Murphy, A. M. (2021, August). Wound healing: A comprehensive review. In *Seminars in plastic surgery* (Vol. 35, No. 03, pp. 141–144). Thieme Medical Publishers, Inc.

Andrews, K. L., Houdek, M. T., & Kiemele, L. J. (2015). Wound management of chronic diabetic foot ulcers: From the basics to regenerative medicine. *Prosthetics and Orthotics International*, *39*(1), 29–39.

Angirasa, A. K., Willrich, A., Cooper, B., & Stuck, R. (2006). Combining bioengineered human dermal replacement and multilayered compression dressings to manage ulcers in a person with diabetes mellitus: A case study. *Ostomy/Wound Management*, *52*(5), 60–64.

Badiavas, E. V., & Falanga, V. (2003). Treatment of chronic wounds with bone marrow–derived cells. *Archives of Dermatology*, *139*(4), 510–516.

Balakrishnan, B., Mohanty, M., Umashankar, P. R., & Jayakrishnan, A. (2005). Evaluation of an in situ forming hydrogel wound dressing based on oxidized alginate and gelatin. *Biomaterials*, *26*(32), 6335–6342.

Barrientos, S., Stojadinovic, O., Golinko, M. S., Brem, H., & Tomic-Canic, M. (2008). Growth factors and cytokines in wound healing. *Wound Repair and Regeneration*, *16*(5), 585–601.

Bašić-Kes, V., Zavoreo, I., Rotim, K., Bornstein, N., Rundek, T., & Demarin, V. (2011). Recommendations for diabetic polyneuropathy treatment. *Acta Clinica Croatica*, *50*(2), 289–302.

Berthet, M., Gauthier, Y., Lacroix, C., Verrier, B., & Monge, C. (2017). Nanoparticle-based dressing: The future of wound treatment. *Trends in Biotechnology*, *35*(8), 770–784.

Bishop, A. (2008). Role of oxygen in wound healing. *J Wound Care*, *17*(9), 399–402.

Blair, M. J., Jones, J. D., Woessner, A. E., & Quinn, K. P. (2020). Skin structure–function relationships and the wound healing response to intrinsic aging. *Advances in Wound Care*, *9*(3), 127–143.

Blair, S. E., Cokcetin, N. N., Harry, E. J., & Carter, D. A. (2009). The unusual antibacterial activity of medical-grade Leptospermum honey: Antibacterial spectrum, resistance and transcriptome analysis. *European Journal of Clinical Microbiology & Infectious Diseases*, *28*, 1199–1208.

Bonifant, H., & Holloway, S. (2019). A review of the effects of ageing on skin integrity and wound healing. *British Journal of Community Nursing*, *24*(Sup3), S28–S33.

Burks, R. I. (1998). Povidone-iodine solution in wound treatment. *Physical Therapy*, *78*(2), 212–218.

Burns, J. L., Mancoll, J. S., & Phillips, L. G. (2003). Impairments to wound healing. *Clinics in Plastic Surgery*, *30*(1), 47–56.

Buschke, S., Stark, H. J., Cerezo, A., Prätzel-Wunder, S., Boehnke, K., Kollar, J., ... Boukamp, P. (2011). A decisive function of transforming growth factor-β/Smad signaling in tissue morphogenesis and differentiation of human HaCaT keratinocytes. *Molecular Biology of the Cell*, *22*(6), 782–794.

Caley, M. P., Martins, V. L., & O'Toole, E. A. (2015). Metalloproteinases and wound healing. *Advances in Wound Care*, *4*(4), 225–234.

Cañedo-Dorantes, L., & Cañedo-Ayala, M. (2019). Skin acute wound healing: A comprehensive review. *International Journal of Inflammation*, *1*, 3706315

Chen, H., Yu, Z., Liu, N., Huang, J., Liang, X., Liang, X., ... Ni, J. (2023). The efficacy of low-frequency ultrasound as an added treatment for chronic wounds: A meta-analysis. *International Wound Journal, 20*(2), 448–457.

Chen, P., & Parks, W. C. (2009). Role of matrix metalloproteinases in epithelial migration. *Journal of Cellular Biochemistry, 108*(6), 1233–1243.

Cho, H. R., Hong, S. B., Kim, Y. I., Lee, J. W., & Kim, N. I. (2004). Differential expression of TGF-β isoforms during differentiation of HaCaT human keratinocyte cells: Implication for the separate role in epidermal differentiation. *Journal of Korean Medical Science, 19*(6), 853.

Cho, Y. S., Joo, S. Y., Cui, H., Cho, S. R., Yim, H., & Seo, C. H. (2016). Effect of extracorporeal shock wave therapy on scar pain in burn patients: A prospective, randomized, single-blind, placebo-controlled study. *Medicine, 95*(32), e4575.

Clinton, A., & Carter, T. (2015). Chronic wound biofilms: Pathogenesis and potential therapies. *Laboratory Medicine, 46*(4), 277–284.

Crawford, J. M., Lal, B. K., Durán, W. N., & Pappas, P. J. (2017). Pathophysiology of venous ulceration. *Journal of Vascular Surgery: Venous and Lymphatic Disorders, 5*(4), 596–605.

Darby, I. A., Laverdet, B., Bonté, F., & Desmoulière, A. (2014). Fibroblasts and myofibroblasts in wound healing. *Clinical, Cosmetic and Investigational Dermatology, 7*, 301–311.

Degen, K. E., & Gourdie, R. G. (2012). Embryonic wound healing: A primer for engineering novel therapies for tissue repair. *Birth Defects Research Part C: Embryo Today: Reviews, 96*(3), 258–270.

Demidova-Rice, T. N., Hamblin, M. R., & Herman, I. M. (2012). Acute and impaired wound healing: Pathophysiology and current methods for drug delivery, part 1: Normal and chronic wounds: Biology, causes, and approaches to care. *Advances in Skin & Wound Care, 25*(7), 304–314.

Dhivya, S., Padma, V. V., & Santhini, E. (2015). Wound dressings–a review. *BioMedicine, 5*(4), 22.

Diegelmann, R. F., & Evans, M. C. (2004). Wound healing: An overview of acute, fibrotic and delayed healing. *Frontiers in Bioscience, 9*(1), 283–289.

Ding, Y., Li, W., Zhang, F., Liu, Z., Zanjanizadeh Ezazi, N., Liu, D., & Santos, H. A. (2019). Electrospun fibrous architectures for drug delivery, tissue engineering and cancer therapy. *Advanced Functional Materials, 29*(2), 1802852.

Dreifke, M. B., Jayasuriya, A. A., & Jayasuriya, A. C. (2015). Current wound healing procedures and potential care. *Materials Science and Engineering: C, 48*, 651–662

Eaglstein, W. H., & Falanga, V. (1997). Chronic wounds. *Surgical Clinics, 77*(3), 689–700.

Edmonds, M., & European and Australian Apligraf Diabetic Foot Ulcer Study Group. (2009). Apligraf in the treatment of neuropathic diabetic foot ulcers. *The International Journal of Lower Extremity Wounds, 8*(1), 11–18.

Edwards, R., & Harding, K. G. (2004). Bacteria and wound healing. *Current Opinion in Infectious Diseases, 17*(2), 91–96.

Eming, S. A., Krieg, T., & Davidson, J. M. (2007). Inflammation in wound repair: Molecular and cellular mechanisms. *Journal of Investigative Dermatology, 127*(3), 514–525.

Enoch, S., & Leaper, D. J. (2008). Basic science of wound healing. *Surgery (Oxford), 26*(2), 31–37.

Evans, J. P., & Sen, C. K. (2023). Electrochemical devices in cutaneous wound healing. *Bioengineering, 10*(6), 711.

Falabella, A. F. (2006). Debridement and wound bed preparation. *Dermatologic Therapy, 19*(6), 317–325.

Falanga, V. (2005). Wound healing and its impairment in the diabetic foot. *The Lancet, 366*(9498), 1736–1743.

Falanga, V., Isseroff, R. R., Soulika, A. M., Romanelli, M., Margolis, D., Kapp, S., ... Harding, K. (2022). Chronic wounds. *Nature Reviews Disease Primers, 8*(1), 50.

Federman, D. G., Ladiiznski, B., Dardik, A., Kelly, M., Shapshak, D., Ueno, C. M., ... Hopf, H. W. (2016). Wound healing society 2014 update on guidelines for arterial ulcers. *Wound Repair & Regeneration, 24*(1).

Ferreira, D. W., Ulecia-Morón, C., Alvarado-Vázquez, P. A., Cunnane, K., Moracho-Vilriales, C., Grosick, R. L., ... Romero-Sandoval, E. A. (2020). CD163 overexpression using a macrophage-directed gene therapy approach improves wound healing in ex vivo and in vivo human skin models. *Immunobiology, 225*(1), 151862.

Finger, S., Piccolino, M., & Stahnisch, F. W. (2013). Alexander von Humboldt: Galvanism, animal electricity, and self-experimentation part 2: The electric eel, animal electricity, and later years. *Journal of the History of the Neurosciences, 22*(4), 327–352.

Fonder, M. A., Lazarus, G. S., Cowan, D. A., Aronson-Cook, B., Kohli, A. R., & Mamelak, A. J. (2008). Treating the chronic wound: A practical approach to the care of nonhealing wounds and wound care dressings. *Journal of the American Academy of Dermatology, 58*(2), 185–206.

Foulds, I. S., & Barker, A. T. (1983). Human skin battery potentials and their possible role in wound healing. *British Journal of Dermatology*, *109*(5), 515–522.

Frykberg, R., Martin, E., Tallis, A., & Tierney, E. (2011). A case history of multimodal therapy in healing a complicated diabetic foot wound: Negative pressure, dermal replacement and pulsed radio frequency energy therapies. *International Wound Journal*, *8*(2), 132–139.

Frykberg, Robert G. (2015). Challenges in the treatment of chronic wounds. *Advances in Wound Care*, *4*, 560–582.

Gardner, S. E., Frantz, R. A., & Doebbeling, B. N. (2001). The validity of the clinical signs and symptoms used to identify localized chronic wound infection. *Wound Repair and Regeneration*, *9*(3), 178–186.

George Broughton, I. I., Janis, J. E., & Attinger, C. E. (2006). A brief history of wound care. *Plastic and Reconstructive Surgery*, *117*(7S), 6S–11S.

Gieringer, M., Gosepath, J., & Naim, R. (2011). Radiotherapy and wound healing: Principles, management and prospects. *Oncology Reports*, *26*(2), 299–307.

Goldberg, S. R., & Diegelmann, R. F. (2020). What makes wounds chronic. *Surgical Clinics*, *100*(4), 681–693.

Gonzalez, A. C. D. O., Costa, T. F., Andrade, Z. D. A., & Medrado, A. R. A. P. (2016). Wound healing-A literature review. *Anais brasileiros de dermatologia*, *91*, 614–620.

Gordillo, G. M., & Sen, C. K. (2003). Revisiting the essential role of oxygen in wound healing. *The American Journal of Surgery*, *186*(3), 259–263.

Gouin, J. P., & Kiecolt-Glaser, J. K. (2012). The impact of psychological stress on wound healing: Methods and mechanisms. *Critical Care Nursing Clinics*, *24*(2), 201–213.

Granick, M., Solomon, M., & Larson, D. (1993). Management of chronic radiation wounds-preface. *Clinics in Plastic Surgery*, *20*(3), R11–R11.

Greaves, N. S., Iqbal, S. A., Baguneid, M., & Bayat, A. (2013). The role of skin substitutes in the management of chronic cutaneous wounds. *Wound Repair and Regeneration*, *21*(2), 194–210.

Grey, J. E., Enoch, S., & Harding, K. G. (2006). ABC of wound healing: Venous and arterial leg ulcers. *BMJ*, *332*(Suppl S4), 301–311.

Guo, S. A., & DiPietro, L. A. (2010). Factors affecting wound healing. *Journal of Dental Research*, *89*(3), 219–229

Gurtner, G. C., & Evans, G. R. (2000). Advances in head and neck reconstruction. *Plastic and Reconstructive Surgery*, *106*(3), 672–682.

Gushiken, L. F. S., Beserra, F. P., Bastos, J. K., Jackson, C. J., & Pellizzon, C. H. (2021). Cutaneous wound healing: An update from physiopathology to current therapies. *Life*, *11*(7), 665. https://creativecommons.org/licenses/by/4.0/

Han, G., Nguyen, L. N., Macherla, C., Chi, Y., Friedman, J. M., Nosanchuk, J. D., & Martinez, L. R. (2012). Nitric oxide–releasing nanoparticles accelerate wound healing by promoting fibroblast migration and collagen deposition. *The American Journal of Pathology*, *180*(4), 1465–1473.

Han, J. H., Han, S., Jeong, I. S., Cheon, S. H., & Kim, S. W. (2020). Minicircle-based GCP-2 ex vivo gene therapy enhanced the reepithelialization and angiogenic capacity. *Journal of Tissue Engineering and Regenerative Medicine*, *14*(6), 829–839.

Harding, K., Sumner, M., & Cardinal, M. (2013). A prospective, multicentre, randomised controlled study of human fibroblast-derived dermal substitute (Dermagraft) in patients with venous leg ulcers. *International Wound Journal*, *10*(2), 132–137.

Hong, W. X., Hu, M. S., Esquivel, M., Liang, G. Y., Rennert, R. C., McArdle, A., ... Longaker, M. T. (2014). The role of hypoxia-inducible factor in wound healing. *Advances in Wound Care*, *3*(5), 390–399.

Hoversten, K. P., Kiemele, L. J., Stolp, A. M., Takahashi, P. Y., & Verdoorn, B. P. (2020, September). Prevention, diagnosis, and management of chronic wounds in older adults. In *Mayo clinic proceedings* (Vol. 95, No. 9, pp. 2021–2034). Elsevier.

Hunt, T. K., Ehrlich, H. P., Garcia, J. A., & Dunphy, J. E. (1969). Effect of vitamin A on reversing the inhibitory effect of cortisone on healing of open wounds in animals and man. *Annals of Surgery*, *170*(4), 633.

Hurlow, J. (2012). AQUACEL® Ag dressing with hydrofiber® technology. *Advances in Wound Care*, *1*(2), 104–107.

Iqbal, A., Jan, A., Wajid, M. A., & Tariq, S. (2017). Management of chronic non-healing wounds by hirudotherapy. *World Journal of Plastic Surgery*, *6*(1), 9.

Janowska, A., Dini, V., Oranges, T., Iannone, M., Loggini, B., & Romanelli, M. (2019). Atypical ulcers: Diagnosis and management. *Clinical Interventions in Aging*, *14*, 2137–2143.

Jensen, J. A., Goodson, W. H., Hopf, H. W., & Hunt, T. K. (1991). Cigarette smoking decreases tissue oxygen. *Archives of Surgery*, *126*(9), 1131–1134.

Jiang, W. G., Sanders, A. J., Ruge, F., & Harding, K. G. (2012). Influence of interleukin-8 (IL-8) and IL-8 receptors on the migration of human keratinocytes, the role of PLC-γ and potential clinical implications. *Experimental and Therapeutic Medicine*, *3*(2), 231–236.

Jin, S. G., Yousaf, A. M., Kim, K. S., Kim, D. W., Kim, D. S., Kim, J. K., ... Choi, H. G. (2016). Influence of hydrophilic polymers on functional properties and wound healing efficacy of hydrocolloid based wound dressings. *International Journal of Pharmaceutics*, *501*(1–2), 160–166.

Johnson, B. Z., Stevenson, A. W., Prêle, C. M., Fear, M. W., & Wood, F. M. (2020). The role of IL-6 in skin fibrosis and cutaneous wound healing. *Biomedicines*, *8*(5), 101.

Kadam, S., Nadkarni, S., Lele, J., Sakhalkar, S., Mokashi, P., & Kaushik, K. S. (2019). Bioengineered platforms for chronic wound infection studies: How can we make them more human-relevant? *Frontiers in Bioengineering and Biotechnology*, *7*, 418.

Kashefsky, H., & Marston, W. (2012). Total contact casting combined with human fibroblast-derived dermal tissue in 15 DFU patients. *Journal of Wound Care*, *21*(5), 236–243.

Kloth, L. C., & Zhao, M. (2010). Endogenous and exogenous electrical fields for wound healing. In *Wound healing: Evidence based management*. FA Davis.

Koo, H., Allan, R. N., Howlin, R. P., Stoodley, P., & Hall-Stoodley, L. (2017). Targeting microbial biofilms: Current and prospective therapeutic strategies. *Nature Reviews Microbiology*, *15*(12), 740–755.

Kranke, P., Bennett, M. H., Martyn-St James, M., Schnabel, A., Debus, S. E., & Weibel, S. (2015). Hyperbaric oxygen therapy for chronic wounds. *Cochrane Database of Systematic Reviews*, *24*(6), CD004123.

Kullo, I. J., & Rooke, T. W. (2016). Peripheral artery disease. *New England Journal of Medicine*, *374*(9), 861–871.

Lalieu, R. C., Brouwer, R. J., Ubbink, D. T., Hoencamp, R., Bol Raap, R., & van Hulst, R. A. (2020). Hyperbaric oxygen therapy for nonischemic diabetic ulcers: A systematic review. *Wound Repair and Regeneration*, *28*(2), 266–275.

Larson, B. J., Longaker, M. T., & Lorenz, H. P. (2010). Scarless fetal wound healing: A basic science review. *Plastic and Reconstructive Surgery*, *126*(4), 1172–1180.

Las Heras, K., Igartua, M., Santos-Vizcaino, E., & Hernandez, R. M. (2020). Chronic wounds: Current status, available strategies and emerging therapeutic solutions. *Journal of Controlled Release*, *328*, 532–550.

Lawrence, J. C. (1994). Dressings and wound infection. *The American Journal of Surgery*, *167*(1), S21–S24.

Levin, M. (2007). Large-scale biophysics: Ion flows and regeneration. *Trends in Cell Biology*, *17*(6), 261–270.

Levine, M. E., Lu, A. T., Quach, A., Chen, B. H., Assimes, T. L., Bandinelli, S., ... Horvath, S. (2018). An epigenetic biomarker of aging for lifespan and healthspan. *Aging (Albany NY)*, *10*(4), 573.

Li, S., Mohamedi, A. H., Senkowsky, J., Nair, A., & Tang, L. (2020). Imaging in chronic wound diagnostics. *Advances in Wound Care*, *9*(5), 245–263

Liarte, S., Bernabé-García, Á., & Nicolás, F. J. (2020). Role of TGF-β in skin chronic wounds: A keratinocyte perspective. *Cells*, *9*(2), 306.

Lin, Z. Q., Kondo, T., Ishida, Y., Takayasu, T., & Mukaida, N. (2003). Essential involvement of IL-6 in the skin wound-healing process as evidenced by delayed wound healing in IL-6-deficient mice. *Journal of Leucocyte Biology*, *73*(6), 713–721.

Lo, D. D., Zimmermann, A. S., Nauta, A., Longaker, M. T., & Lorenz, H. P. (2012). Scarless fetal skin wound healing update. *Birth Defects Research Part C: Embryo Today: Reviews*, *96*(3), 237–247.

Löndahl, M., Katzman, P., Nilsson, A., & Hammarlund, C. (2010). Hyperbaric oxygen therapy facilitates healing of chronic foot ulcers in patients with diabetes. *Diabetes Care*, *33*(5), 998–1003.

Lopes, A. I., Pintado, M. M., & Tavaria, F. K. (2024). Plant-based films and hydrogels for wound healing. *Microorganisms*, *12*(3), 438.

Lu, J. H., Hsia, K., Su, C. K., Pan, Y. H., Ma, H., Chiou, S. H., & Lin, C. H. (2023). A novel dressing composed of adipose stem cells and decellularized Wharton's jelly facilitated wound healing and relieved lymphedema by enhancing angiogenesis and lymphangiogenesis in a rat model. *Journal of Functional Biomaterials*, *14*(2), 104.

Marshall, C. D., Hu, M. S., Leavitt, T., Barnes, L. A., Lorenz, H. P., & Longaker, M. T. (2018). Cutaneous scarring: Basic science, current treatments, and future directions. *Advances in Wound Care*, *7*(2), 29–45.

Martin, G. M. (2001). Frontiers of aging. *Science*, *294*(5540), 13.

Mathieu, D. (Ed.). (2006). *Handbook on hyperbaric medicine* (Vol. 27, p. 812). Springer.

Medina, A., Scott, P. G., Ghahary, A., & Tredget, E. E. (2005). Pathophysiology of chronic nonhealing wounds. *Journal of Burn Care & Rehabilitation*, *26*(4), 306–319.

Menke, N. B., Ward, K. R., Witten, T. M., Bonchev, D. G., & Diegelmann, R. F. (2007). Impaired wound healing. *Clinics in Dermatology*, *25*(1), 19–25.

Miao, F., Li, Y., Tai, Z., Zhang, Y., Gao, Y., Hu, M., & Zhu, Q. (2021). Antimicrobial peptides: The promising therapeutics for cutaneous wound healing. *Macromolecular Bioscience*, *21*(10), 2100103.

Mittermayr, R., Antonic, V., Hartinger, J., Kaufmann, H., Redl, H., Téot, L., ... Schaden, W. (2012). Extracorporeal shock wave therapy (ESWT) for wound healing: Technology, mechanisms, and clinical efficacy. *Wound Repair and Regeneration*, *20*(4), 456–465.

Mogford, J. E., Sisco, M., Bonomo, S. R., Robinson, A. M., & Mustoe, T. A. (2004). Impact of aging on gene expression in a rat model of ischemic cutaneous wound healing. *Journal of Surgical Research*, *118*(2), 190–196.

Monika, P., Chandraprabha, M. N., Rangarajan, A., Waiker, P. V., & Chidambara Murthy, K. N. (2022). Challenges in healing wound: Role of complementary and alternative medicine. *Frontiers in Nutrition*, 8, 791899.

Mowbray, M., Tan, X., Wheatley, P. S., Morris, R. E., & Weller, R. B. (2008). Topically applied nitric oxide induces T-lymphocyte infiltration in human skin, but minimal inflammation. *Journal of Investigative Dermatology*, *128*(2), 352–360.

Mukherjee, R., Tewary, S., & Routray, A. (2017). Diagnostic and prognostic utility of non-invasive multimodal imaging in chronic wound monitoring: A systematic review. *Journal of Medical Systems*, *41*, 1–17.

Mustoe, T. A., O'shaughnessy, K., & Kloeters, O. (2006). Chronic wound pathogenesis and current treatment strategies: A unifying hypothesis. *Plastic and Reconstructive Surgery*, *117*(7S), 35S–41S.

Nagle, S. M., Stevens, K. A., & Wilbraham, S. C. (2018). Wound assessment. In *StatPearls*. StatPearls Publishing.

Nicolucci, A. C. D. A., Rossi, C., & Lucisano, G. (2014). Facts and figures about diabetes in Italy. *Italian Diabetes Monitor: Fondazione Mario Negri Sud*, *30*, 100–106.

Nwomeh, B. C., Yager, D. R., & Cohen, I. K. (1998). Physiology of the chronic wound. *Clinics in Plastic Surgery*, *25*(3), 341–356.

O'Reilly, D., Linden, R., Fedorko, L., Tarride, J. E., Jones, W. G., Bowen, J. M., & Goeree, R. (2011). A prospective, double-blind, randomized, controlled clinical trial comparing standard wound care with adjunctive hyperbaric oxygen therapy (HBOT) to standard wound care only for the treatment of chronic, non-healing ulcers of the lower limb in patients with diabetes mellitus: A study protocol. *Trials*, *12*, 1–10.

Pacific, P. A. N. (2014). Prevention and treatment of pressure ulcers: Quick reference guide. *National Pressure Ulcer Advisory Panel*, *75*, 11–13.

Parry, A. D., Chadwick, P. R., Simon, D., Oppenheim, B., & McCollum, C. N. (1995). Leg ulcer odour detection identifies β-haemolytic streptococcal infection. *Journal of Wound Care*, *4*(9), 404–406.

Patel, S., Srivastava, S., Singh, M. R., & Singh, D. (2019). Mechanistic insight into diabetic wounds: Pathogenesis, molecular targets and treatment strategies to pace wound healing. *Biomedicine & Pharmacotherapy*, *112*, 108615.

Pence, B. D., & Woods, J. A. (2014). Exercise, obesity, and cutaneous wound healing: Evidence from rodent and human studies. *Advances in Wound Care*, *3*(1), 71–79.

Perkins, K., Davey, R. B., & Wallis, K. A. (1983). Silicone gel: A new treatment for burn scars and contractures. *Burns*, *9*(3), 201–204.

Powers, J. G., Higham, C., Broussard, K., & Phillips, T. J. (2016). Wound healing and treating wounds: Chronic wound care and management. *Journal of the American Academy of Dermatology*, *74*(4), 607–625.

Prinja, S., Bahuguna, P., Pinto, A. D., Sharma, A., Bharaj, G., Kumar, V., ... Kumar, R. (2012). The cost of universal health care in India: A model based estimate. *PLoS One*, *7*(1), e30362.

Profyris, C., Tziotzios, C., & Do Vale, I. (2012). Cutaneous scarring: Pathophysiology, molecular mechanisms, and scar reduction therapeutics: Part I. The molecular basis of scar formation. *Journal of the American Academy of Dermatology*, *66*(1), 1–10.

Raffetto, J. D., Ligi, D., Maniscalco, R., Khalil, R. A., & Mannello, F. (2020). Why venous leg ulcers have difficulty healing: Overview on pathophysiology, clinical consequences, and treatment. *Journal of Clinical Medicine*, *10*(1), 29.

Ramirez, H., Patel, S. B., & Pastar, I. (2014). The role of TGFβ signaling in wound epithelialization. *Advances in Wound Care*, *3*(7), 482–491.

Rastogi, A., Bhansali, A., & Ramachandran, S. (2019). Efficacy and safety of low-frequency, noncontact airborne ultrasound therapy (Glybetac) for neuropathic diabetic foot ulcers: A randomized, double-blind, sham-control study. *The International Journal of Lower Extremity Wounds*, *18*(1), 81–88.

Rathur, H. M., & Boulton, A. J. M. (2005). Recent advances in the diagnosis and management of diabetic neuropathy. *The Journal of Bone & Joint Surgery British Volume*, *87*(12), 1605–1610.

Raziyeva, K., Kim, Y., Zharkinbekov, Z., Kassymbek, K., Jimi, S., & Saparov, A. (2021). Immunology of acute and chronic wound healing. *Biomolecules*, *11*(5), 700.

Reinke, J. M., & Sorg, H. (2012). Wound repair and regeneration. *European Surgical Research*, *49*(1), 35–43

Riehemann, K., Schneider, S. W., Luger, T. A., Godin, B., Ferrari, M., & Fuchs, H. (2009). Nanomedicine—challenge and perspectives. *Angewandte Chemie International Edition*, *48*(5), 872–897.

Robson, M. C., & Barbul, A. (2006). Guidelines for the best care of chronic wounds. *Wound Repair and Regeneration*, *14*(6), 647–648.

Rubio, P. A. (1991). Use of semiocclusive, transparent film dressings for surgical wound protection: Experience in 3637 cases. *International surgery*, *76*(4), 253–254.

Ruedrich, E. D., Henzel, M. K., Hausman, B. S., & Bogie, K. M. (2013). Reference gene identification for reverse transcription-quantitative polymerase chain reaction analysis in an ischemic wound-healing model. *Journal of Biomolecular Techniques: JBT*, *24*(4), 181.

Schaden, W., Thiele, R., Kölpl, C., Pusch, M., Nissan, A., Attinger, C. E., ... Stojadinovic, A. (2007). Shock wave therapy for acute and chronic soft tissue wounds: A feasibility study. *Journal of Surgical Research*, *143*(1), 1–12.

Schultz, G. S., Chin, G. A., Moldawer, L., & Diegelmann, R. F. (2011). 23 principles of wound healing. In *Mechanisms of vascular disease: A reference book for vascular specialists* (p. 423). University of Adelaide Press.

Seaman, S. (2002). Dressing selection in chronic wound management. *Journal of the American Podiatric Medical Association*, *92*(1), 24–33.

Seifert, A. W., Kiama, S. G., Seifert, M. G., Goheen, J. R., Palmer, T. M., & Maden, M. (2012). Skin shedding and tissue regeneration in African spiny mice (Acomys). *Nature*, *489*(7417), 561–565.

Sen, C. K. (2019). Human wounds and its burden: An updated compendium of estimates. *Advances in Wound Care*, *8*(2), 39–48.

Sen, C. K. (2023). Human wound and its burden: Updated 2022 compendium of estimates. *Advances in Wound Care*, *12*(12), 657–670.

Sharifi, S., Hajipour, M. J., Gould, L., & Mahmoudi, M. (2020). Nanomedicine in healing chronic wounds: Opportunities and challenges. *Molecular Pharmaceutics*, *18*(2), 550–575.

Shaw, T. J., & Martin, P. (2016). Wound repair: A showcase for cell plasticity and migration. *Current Opinion in Cell Biology*, *42*, 29–37

Shukla, V. K., Ansari, M. A., & Gupta, S. K. (2005). Wound healing research: A perspective from India. *The International Journal of Lower Extremity Wounds*, *4*(1), 7–8.

Simões, D., Miguel, S. P., Ribeiro, M. P., Coutinho, P., Mendonça, A. G., & Correia, I. J. (2018). Recent advances on antimicrobial wound dressing: A review. *European Journal of Pharmaceutics and Biopharmaceutics*, *127*, 130–141.

Soyoye, D. O., Abiodun, O. O., Ikem, R. T., Kolawole, B. A., & Akintomide, A. O. (2021). Diabetes and peripheral artery disease: A review. *World Journal of Diabetes*, *12*(6), 827.

Stashak, T. S. (2007). Wound dressings: Indications and best use. *Clinical Techniques in Equine Practice*, *3*, 148–163.

Stechmiller, J. K. (2010). Understanding the role of nutrition and wound healing. *Nutrition in Clinical Practice*, *25*(1), 61–68.

Steffensen, B., Häkkinen, L., & Larjava, H. (2001). Proteolytic events of wound-healing—coordinated interactions among matrix metalloproteinases (MMPs), integrins, and extracellular matrix molecules. *Critical Reviews in Oral Biology & Medicine*, *12*(5), 373–398.

Strausberg, J., Lehmann, N., Kröger, K., Maier, I., Schneider, H., & Niebel, W. (2007). Changes in secondary care may explain increasing pressure ulcer rates in an University Clinic in Germany. *Wound Manag*, 5, 194–198.

Streubel, P. N., Stinner, D. J., & Obremskey, W. T. (2012). Use of negative-pressure wound therapy in orthopaedic trauma. *JAAOS-Journal of the American Academy of Orthopaedic Surgeons*, *20*(9), 564–574.

Sun, B. K., Siprashvili, Z., & Khavari, P. A. (2014). Advances in skin grafting and treatment of cutaneous wounds. *Science*, *346*(6212), 941–945.

Tabriz, A. G., & Douroumis, D. (2022). Recent advances in 3D printing for wound healing: A systematic review. *Journal of Drug Delivery Science and Technology*, *74*, 103564.

Tai, G., Tai, M., & Zhao, M. (2018). Electrically stimulated cell migration and its contribution to wound healing. *Burns & Trauma*, *6*.

Takeo, M., Lee, W., & Ito, M. (2015). Wound healing and skin regeneration. *Cold Spring Harbor Perspectives in Medicine*, *5*(1), a023267.

Tandara, A. A., & Mustoe, T. A. (2004). Oxygen in wound healing—more than a nutrient. *World Journal of Surgery*, *28*, 294–300.

Tatarusanu, S., Lupascu, F., Profire, B., Szilagyi, A., Gardikiotis, I., Iacob, A., Caluian, I. et al. (2022). Modern approaches in wounds management. *Polymers*, *15*, (17), 3648.

Thu, H. E., Zulfakar, M. H., & Ng, S. F. (2012). Alginate based bilayer hydrocolloid films as potential slow-release modern wound dressing. *International Journal of Pharmaceutics*, *434*(1–2), 375–383.

Tottoli, E. M., Dorati, R., Genta, I., Chiesa, E., Pisani, S., & Conti, B. (2020). Skin wound healing process and new emerging technologies for skin wound care and regeneration. *Pharmaceutics*, *12*(8), 735.

Trent, J. T., Falabella, A., Eaglstein, W. H., & Kirsner, R. S. (2005). Venous ulcers: Pathophysiology and treatment options. *Ostomy/Wound Management*, *51*(5), 38–54.

Tsegay, F., Elsherif, M., & Butt, H. (2022). Smart 3D printed hydrogel skin wound bandages: A review. *Polymers*, *14*(5), 1012.

Tyavambiza, C., Meyer, M., & Meyer, S. (2022). Cellular and molecular events of wound healing and the potential of silver based nanoformulations as wound healing agents. *Bioengineering*, *9*(11), 712.

Ubbink, D. T., Vermeulen, H., & Van Hattem, J. (2008). Comparison of homecare costs of local wound care in surgical patients randomized between occlusive and gauze dressings. *Journal of Clinical Nursing*, *17*(5), 593–601.

Varaprasad, K., Jayaramudu, T., Kanikireddy, V., Toro, C., & Sadiku, E. R. (2020). Alginate-based composite materials for wound dressing application: A mini review. *Carbohydrate Polymers*, *236*, 116025.

Varghese, M. C., Balin, A. K., Carter, D. M., & Caldwell, D. (1986). Local environment of chronic wounds under synthetic dressings. *Archives of Dermatology*, *122*(1), 52–57.

Velnar, T., Bailey, T., & Smrkolj, V. (2009). The wound healing process: An overview of the cellular and molecular mechanisms. *Journal of International Medical Research*, *37*(5), 1528–1542.

Venturi, M. L., Attinger, C. E., Mesbahi, A. N., Hess, C. L., & Graw, K. S. (2005). Mechanisms and clinical applications of the vacuum-assisted closure (VAC) device: A review. *American Journal of Clinical Dermatology*, *6*, 185–194.

Warriner, R., & Burrell, R. (2005). Infection and the chronic wound: A focus on silver. *Advances in Skin & Wound Care*, *18*(8), 2–12.

Whitney, J. D. (2005). Overview: Acute and chronic wounds. *Nursing Clinics*, *40*(2), 191–205.

Wilgus, T. A. (2007). Regenerative healing in fetal skin: A review of the literature. *Ostomy/Wound Management*, *53*(6), 16–31.

Wilkinson, H. N., & Hardman, M. J. (2020). Wound healing: Cellular mechanisms and pathological outcomes. *Open Biology*, *10*(9), 200223

Word, R. (2010). Medical and surgical therapy for advanced chronic venous insufficiency. *Surgical Clinics*, *90*(6), 1195–1214.

Worthington, J. J., Fenton, T. M., Czajkowska, B. I., Klementowicz, J. E., & Travis, M. A. (2012). Regulation of TGFβ in the immune system: An emerging role for integrins and dendritic cells. *Immunobiology*, *217*(12), 1259–1265.

Wu, S. C., Marston, W., & Armstrong, D. G. (2010). Wound care: The role of advanced wound-healing technologies. *Journal of the American Podiatric Medical Association*, *100*(5), 385–394.

Wu, Y. K., Cheng, N. C., & Cheng, C. M. (2019). Biofilms in chronic wounds: Pathogenesis and diagnosis. *Trends in Biotechnology*, *37*(5), 505–517.

Yoshida, M., Okubo, N., Chosa, N., Hasegawa, T., Ibi, M., Kamo, M., & Ishisaki, A. (2012). TGF-β-operated growth inhibition and translineage commitment into smooth muscle cells of periodontal ligament-derived endothelial progenitor cells through Smad-and p38 MAPK-dependent signals. *International Journal of Biological Sciences*, *8*(7), 1062.

Zahedi, P., Rezaeian, I., Ranaei-Siadat, S. O., Jafari, S. H., & Supaphol, P. (2010). A review on wound dressings with an emphasis on electrospun nanofibrous polymeric bandages. *Polymers for Advanced Technologies*, *21*(2), 77–95.

Zhang, X., Wang, Y., Gao, Z., Mao, X., Cheng, J., Huang, L., & Tang, J. (2024). Advances in wound dressing based on electrospinning nanofibers. *Journal of Applied Polymer Science*, *141*(1), e54746.

Zhao, R., Liang, H., Clarke, E., Jackson, C., & Xue, M. (2016). Inflammation in chronic wounds. *International Journal of Molecular Sciences*, *17*(12), 2085.

4 Current Status and Future Aspects of Nanomaterials for Wound Healing

Shriyansh Srivastava[1,2] *Nandani Jayaswal,*[3] *Sharib Raza Khan,*[4] *Sachin Kumar,*[2] *and Sathvik Belagodu Sridhar*[5]*

[1] Delhi Pharmaceutical Sciences and Research University, New Delhi, India
[2] Galgotias University, Uttar Pradesh, India
[3] Mahayogi Gorakhnath University, Uttar Pradesh, India
[4] CSIR-Central Drug Research Institute, Uttar Pradesh, India
[5] RAK Medical and Health Sciences University, Ras al Khaimah, UAE
*Corresponding Author

4.1 INTRODUCTION

Nanomaterials offer promising opportunities in wound healing by enhancing cellular interactions, biochemical signaling pathways, and the extracellular matrix (Liu et al., 2023; Napagoda et al., 2022). Their unique physicochemical properties have revolutionized wound care, overcoming traditional limitations and accelerating healing (Z. Xu et al., 2023). Despite involving diverse biological processes, the primary goal remains tissue integration and functional restoration (Yang et al., 2021). Challenges such as infections, delayed healing, and tissue regeneration issues hinder this process (Huan & Tay, 2020; Naskar & Kim, 2020), necessitating novel treatment strategies (Pakpahan et al., 2023). Nanotechnology has emerged as a pivotal tool in medical applications, particularly in wound healing, leveraging nanoscale properties to control cellular and molecular processes effectively (Baig et al., 2021; Bhardwaj et al., 2017; Sharma et al., 2022). Nanomaterials facilitate targeted therapy, infection control (Wang et al., 2019), and improved wound care through features like extended drug release (Jana et al., 2011) and modulation of tissue regeneration pathways (Bombin et al., 2020; Pal et al., 2021; Zhang et al., 2021). Metallic nanomaterials such as gold, silver, and zinc oxide exhibit potent antibacterial properties, crucial for combating infections and promoting healing (Ali et al., 2020; Thakkar et al., 2010). Polymeric nanomaterials like nanofibers and hydrogels serve as effective scaffolds for cell adhesion and proliferation, enhancing tissue regeneration (Bellu et al., 2021). Lipid-based nanomaterials enable localized drug delivery, ensuring sustained release at wound sites (Nandhini et al., 2024). Inorganic nanomaterials like silica and calcium promote angiogenesis and cell proliferation, aiding in tissue regeneration (Degli Esposti et al., 2018). Despite their potential, challenges such as toxicity, stability, regulatory concerns, and scalability for clinical applications need to be addressed for widespread adoption (Liu et al., 2022; Sharifi et al., 2012). This overview aims to detail nanomaterial-based approaches in wound healing, highlighting their mechanisms,

DOI: 10.1201/9781003504948-4

types, and current challenges, while also pointing toward future advancements for safer and more efficient treatments (Cruz et al., 2020; Dzulkharnien & Rohani, 2022; Eke et al., 2020).

4.2 NAVIGATING THE COMPLEXITIES: CHALLENGES IN WOUND REPAIR

In healthy people, the epidermal barrier layer is highly effective as compared to that in infected people. When an injury occurs to the skin, the body initiates a compensatory mechanism with a controlled biochemical process for wound healing. First in this process is the stoppage of bleeding; and after that, the stage of inflammation; then the proliferative phase that includes the stages of re-epithelialization, tissue granulation, and neovascularization; and the last is the skin repair remodeling phase. Apart from that, certain conditions can augment the time for delayed wound healing (Sinno & Prakash, 2013). The chronic wounds have distinctive characteristics due to the presence of senescent fibroblasts, less activity of small growth factors, decreased mitotic activity, and the presence of inflammatory mediators like cytokines, proteases, and elastase (McCarty & Percival, 2013). The intensity of inflammation is directly proportional to wound fibrosis, which have been explained by researchers. This is pointed out in fetal scarless wounds, which is supported or repaired by the absence of intrauterine inflammation absence (Martin, 2017). Similarly, in chronic wounds, the abundant infiltration of neutrophils and reactive oxygen species causes persistent inflammation due to macrophages' phagocytic activity. Simultaneously, reactive oxygen species (ROS) play a significant role in protection against harmful pathogens; however, an excessive level of ROS in the case of wound healing can cause an increased level of oxidative stress, leading to tissue damage and hyper production of matrix metalloproteinases (MMPs). The clear relationship between persistent inflammation and the infection-prone nature of open wounds is still unclear (Martin & Nunan, 2015). The results obtained from chronic wounds show that the increased levels of metalloproteinases cause significant destruction of extracellular matrix components and impede cell migration. The inadequate response of growth factors toward neutrophils can cause the degradation of the extracellular matrix, leading to granulation tissue dysfunction (Krzyszczyk et al., 2018). Other challenges in wound healing include single or multiple biofilms, in which protease production stops the migration of keratinocytes and re-epithelialization of skin, causing microorganisms to stick to the microbial surface. After attachment to the surface, the microorganisms can be removed and treatment can continue with antibiotics (Pastar et al., 2014). The infection grows at the biofilm area not controlled by antibiotics or the immune system due to surface adhesion, which initiates the release of a mucopolysaccharide matrix made up of microbial and host parts. At the chronic margins of wounds, keratinocytes are hyperdifferentiative and unable to move because they are not fully differentiated and activated (Mihai et al., 2018).

4.3 TYPES OF NANOMATERIALS FOR WOUND HEALING

Nanomaterials have revolutionized wound healing by offering antibacterial properties, biocompatibility, and enhanced mechanical strength, significantly reducing healthcare costs (Dam et al., 2023). Researchers have utilized various nanomaterials, such as metal, ceramic, and polymer nanoparticles, and nanoparticle-loaded hydrogels, to mimic the extracellular matrix in wounds (Prabhu et al., 2023). Additionally, collagen nanoparticles from natural extracts have shown minimal cytotoxicity and accelerated wound repair, demonstrating their therapeutic potential. Incorporating antibacterial nanomaterials into wound dressings has improved biosafety and treatment outcomes, emphasizing their role in infection control and wound management (Pang et al., 2023). Fabrication methods are crucial in developing wound-healing materials and scaffolds. Three-dimensional (3D) printing allows for the creation of customized scaffolds with controlled porosity and antimicrobial properties, suitable for external wound healing (Kumar et al., 2022). Electrospinning produces nanofibers that promote cell attachment and facilitate the delivery of therapeutic agents (Lotfi et al., 2023). Hydrogels create three-dimensional networks that support cell growth and can deliver therapeutic agents and growth factors (R. Xu et al., 2023). Bioprinting uses living cells and biomaterials to form complex structures with controlled cell distribution (Ho et al., 2024). Hybrid materials combine polymers, ceramics, and biological molecules to offer biocompatibility and mechanical strength (Khan et al., 2019; Zahid et al., 2022). Suture coatings enhance the biocompatibility and antimicrobial activity of sutures, releasing antibacterial agents to prevent infection (Y.-G. Chen et al., 2022). Cell-laden hydrogels incorporate living cells to promote growth and differentiation, while bilayer nanofibrous-hydrogel scaffolds merge nanofibers and hydrogels for antimicrobial activity and cell promotion (Alamouty et al., 2023). Each method provides unique advantages, catering to specific applications and desired scaffold properties (Khan et al., 2024).

4.3.1 Nanofibers

Nanofibers have emerged as a promising approach for wound healing due to their unique properties, such as high surface area, porosity, and structural similarities to the extracellular matrix, allowing for effective drug delivery (Jiang et al., 2023; Kumar et al., 2023). Electrospun nanofibers, particularly those made from chitosan, exhibit biocompatibility, antimicrobial properties, and efficiency in wound regeneration (Ajith et al., 2023). Incorporating plant extracts and phytoconstituents into nanofibers enhances their antibacterial and therapeutic properties, making them ideal for wound dressings (Elbhnsawi et al., 2023). Recent advancements focus on loading nanofibers with drugs to achieve controlled release for improved wound healing, although challenges remain in maintaining drug activity and controlling release rates. Encapsulating eucalyptus oil in nano-drug carriers within electrospun nanofibers has shown enhanced antimicrobial activity and accelerated wound healing in vitro and in vivo studies, highlighting the potential of nanofibers in future wound care applications (M. Hayat et al., 2023).

4.3.2 Nanostructured Surfaces

Nanostructured surfaces play a crucial role in wound healing by providing advanced functionalities such as enhanced mechanical properties, controlled drug release, antibacterial activity, and accelerated wound closure. Research has shown the development of innovative wound dressings with nanostructured surfaces that mimic human skin properties, offering excellent physical characteristics, biocompatibility, and protection (Shi et al., 2023). These nanostructural formulations utilize nanomaterials like nanoparticles, hydrogels, and scaffolds to promote the healing of infectious wounds and prevent complications (Lukhey & Shende, 2023). Hierarchical structural wound dressings with hydrophobic and hydrophilic layers loaded with drugs and nanoparticles have demonstrated outstanding surface wettability, mechanical strength, rapid drug release, and antibacterial properties, leading to accelerated wound healing and complete recovery within a short period (Xu et al., 2022). Nanostructured antimicrobial-loaded dressings and hydrogels have shown promise in reducing pathogenic bioburden and restoring physiological balance in chronic non-healing wounds, highlighting the potential of nanotechnology in wound care (Arruebo, 2022). Additionally, nanomaterials from natural and synthetic sources have emerged as effective therapies for wound healing, with advancements in split and complete thickness skin alternatives and various nanocarrier-based drug delivery systems contributing to improved treatment outcomes (Rajput et al., 2022). Table 4.1 presents the application of nanoparticles in enhancing wound healing.

4.4 ANTIMICROBIAL, ANTI-INFLAMMATORY, AND ANGIOGENIC EFFECTS OF NANOMATERIALS IN WOUND HEALING

Nanomaterials have emerged as a promising approach to enhancing wound healing by creating a suitable environment for cell migration, controlling microbial infections, and managing inflammation. They exhibit antimicrobial effects through silver nanoparticles (AgNPs) that release silver ions ($Ag+$) to destroy bacterial cell membranes and honey-based nanoparticles that utilize hydrogen peroxide, acidity, and nonperoxide components (Bahari et al., 2022; Wong, 2007; Xu et al.,

TABLE 4.1
Applications of Nanoparticles in Enhancing Wound Healing

Application	Description	References
Antimicrobial properties	Nanoparticles are made of metals like silver, gold, and zinc oxide demonstrate potent antimicrobial activity against bacteria, fungi, and pathogens. When incorporated into wound dressings, these nanoparticles prevent or reduce infection risk, crucial for proper wound healing.	(Dam et al., 2023; Kwiatkowska et al., 2022)
Controlled drug delivery	Nanoparticles are carriers that deliver therapeutic agents, such as antibiotics, growth factors, and other bioactive compounds, in a controlled and sustained manner to the wound site. This targeted delivery maximizes treatment efficacy and promotes faster healing.	(Ahmadi et al., 2024; Dam et al., 2023)
Structural support and wound dressings	Nanoparticles are added to hydrogel-based wound dressings, providing structural reinforcement and improved mechanical properties. These nanocomposite dressings better mimic the extracellular matrix of the wound, creating a favorable microenvironment for cell growth and tissue regeneration.	(Ahmadi et al., 2024; Dam et al., 2023)
Stimulation of wound healing processes	Nanoparticles, such as polyphosphate nanoparticles, induce gene activity, promote bone mineralization, and serve as energy suppliers through ATP formation, enhancing various stages of the wound healing process.	(Müller et al., 2024)
Biocompatibility and wound compatibility	Nanoparticles derived from natural materials like alginate, gelatin, and chitosan exhibit excellent biocompatibility and low toxicity, making them suitable for wound healing applications.	(Froelich et al., 2023; Rezaei Kolarijani et al., 2023; Sridhar et al., 2015)

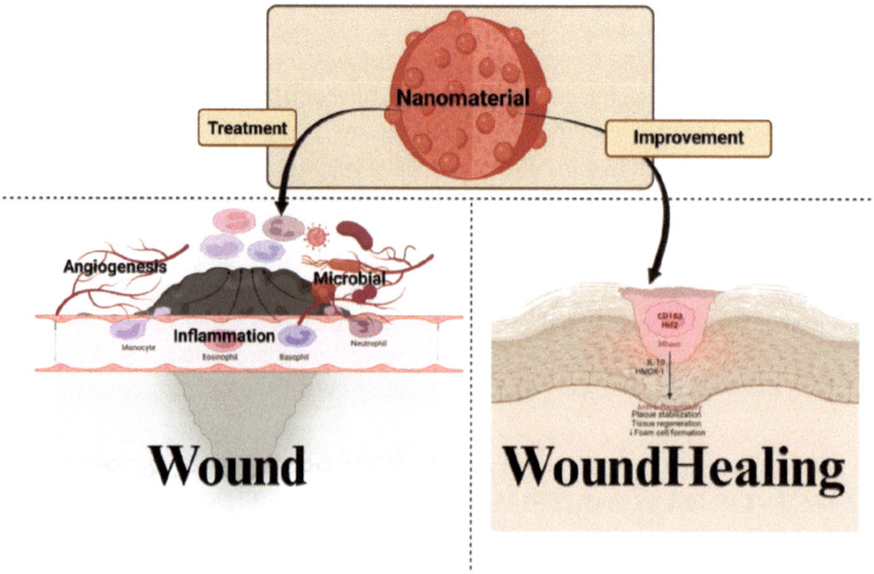

FIGURE 4.1 Molecular mechanisms underlying the therapeutic effects of nanomaterials in wound healing.

2020). Nanomaterials reduce inflammation by scavenging ROS and inhibiting pro-inflammatory cytokines, with honey-based nanoparticles also contributing to anti-inflammatory effects (Bahari et al., 2022; Joorabloo & Liu, 2024) (Figure 4.1). They promote angiogenesis by encouraging new blood vessel growth, essential for tissue oxygenation and nutrient delivery. A benefit is also seen with honey-based nanoparticles (Bahari et al., 2022; Wang et al., 2021). Additionally, nanomaterials stimulate cellular and molecular processes like hemostasis, immune regulation, and tissue proliferation, thus accelerating wound closure and tissue regeneration, aided by their ROS-scavenging properties (Joorabloo & Liu, 2024; Wang et al., 2021). They can be engineered for the targeted delivery of therapeutic agents such as antibiotics and growth factors, enhancing treatment efficacy (Bahari et al., 2022; Wang et al., 2021). Their mechanical properties, including stiffness and elasticity, mimic the natural skin environment, and their biocompatibility reduces the risk of adverse reactions, improving the safety of wound healing treatments. Overall, nanomaterials offer significant promise in wound healing through antimicrobial, anti-inflammatory, and angiogenic effects tailored to specific applications (Bahari et al., 2022; Wang et al., 2021).

4.5 OVERCOMING BARRIERS TO CLINICAL ADOPTION OF NANOMATERIAL INNOVATIONS

Nanomaterials promise to enhance wound healing, but their clinical application faces several challenges. One significant hurdle involves ensuring the biocompatibility of these materials, as they can interact complexly with biological systems, potentially causing toxicity and triggering adverse immune responses. Rigorous safety testing is necessary to validate nanomaterial-based wound dressings before clinical approval (Kushwaha et al., 2022; Zeng et al., 2022). Another critical issue is navigating regulatory requirements. These dressings, classified as combination products, necessitate coordinated review by multiple regulatory agencies. This complexity can extend approval timelines and escalate development costs. Establishing clear regulatory guidelines and streamlining reviews are essential to expedite the clinical adoption of these technologies (Freedman et al., 2023; Kushwaha et al., 2022; Zeng et al., 2022). Furthermore, scaling up the manufacturing of nanomaterial-based wound dressings consistently and cost-effectively poses ongoing challenges. Maintaining batch-to-batch reproducibility and ensuring product quality during scale-up is vital for successful commercialization (Freedman et al., 2023). While nanomaterials offer significant potential for improving wound care, addressing challenges related to biocompatibility, regulatory approval, and manufacturing is crucial for their clinical translation. Continued collaboration among researchers, clinicians, and regulators will be pivotal in overcoming these obstacles and advancing these innovative therapies for patient benefit (Freedman et al., 2023; Zeng et al., 2022).

4.6 CHALLENGES

Nanomaterials are increasingly studied for their potential in wound healing, aiming to enhance the healing process and tackle challenges in wound treatment. They can create a conducive environment for cell migration, control infections, and manage inflammation, thereby improving overall healing. Despite progress, challenges like ensuring biocompatibility (Zou et al., 2023), achieving targeted delivery (Zhu et al., 2024), combating antimicrobial resistance (Yuan et al., 2022), and ensuring scalability and cost-effectiveness remain. Regulatory frameworks are crucial for their safe clinical adoption. Emerging trends include MXene-based materials for antibacterial properties (Li et al., 2024), photothermal therapy for infected wounds (Lu et al., 2023), hydrogel systems for tissue regeneration, and integrating nanomaterials with stem cells to optimize therapeutic outcomes. Nanomaterials hold significant promise for advancing wound healing, contingent upon addressing current challenges and optimizing their design and application.

4.7 FUTURE PERSPECTIVES

Nanomaterials are actively explored to enhance wound healing applications through various formulations such as hydrogels and scaffolds. Electrospun nanofiber scaffolds, mimicking the skin's extracellular matrix, have emerged as particularly promising (Sharma et al., 2023). These multifunctional electrospun nanofiber (MEN) scaffolds and stem cells aim to optimize wound healing (Behere & Ingavle, 2022). Innovative electrospinning techniques enable tailored geometries and properties of MEN scaffolds, enhancing their effectiveness (Behere & Ingavle, 2022). Additionally, nanomaterials integrated into hydrogel dressings improve mechanical strength and antibacterial properties and enable controlled drug delivery for sustained therapeutic effects (Liu et al., 2022). Future research directions include exploring nanomaterials in novel wound healing applications, such as oxygen-carrying nanoparticles, to improve tissue oxygenation (Zhu et al., 2024). Continued advancements in nanomaterial design and their integration into wound care products are anticipated to enhance outcomes in wound healing significantly (Shalaby et al., 2022; Yaşayan et al., 2023). Incorporating nanomaterials with technologies such as artificial intelligence (AI) and 3D bioprinting holds promise for advancing wound healing therapies. AI-assisted high-throughput screening optimizes 3D bioprinting conditions, improving hydrogel scaffolds' mechanical properties and biological performance, thereby enhancing wound treatment efficacy (B. Chen et al., 2022). Bioinspired hydrogels, formed through DNA-induced biomineralization, offer functional wound dressings with enhanced porosity and mechanical flexibility, accelerating healing through reactive oxygen species scavenging and angiogenesis promotion (Kim et al., 2023). AI-guided customization of wound dressings tailors treatments to individual patient needs, which is crucial for managing diabetic chronic wounds (Kim et al., 2023). Nanomaterial-infused bioinks enrich scaffold properties, facilitating accelerated healing through antibacterial drug delivery and tissue-specific emulation (Choudary et al., 2023; Tanfani et al., 2023). Advanced AI-driven technologies also advance functional hydrogel development, supporting effective wound care and improved patient outcomes (Kim et al., 2023). Furthermore, 3D bioprinting enables the creation of skin models that mimic dermal environments, aiding in disease modeling and treatment development (Zhang et al., 2023). Together, these innovations highlight the potential of integrating nanomaterials with AI and 3D bioprinting to revolutionize wound healing therapies, promising more effective treatments across various wound types and skin disorders.

4.8 CONCLUSION

Nanomaterials significantly enhance wound healing processes. Studies demonstrate that various nanomaterials promote healing by creating a favorable wound microenvironment with antibacterial, anti-inflammatory, and angiogenic properties. Nanomaterials' unique physicochemical characteristics, including high surface area and tunable properties, enable the effective delivery of bioactive agents such as antibiotics, growth factors, and natural compounds to the wound site, thereby enhancing wound healing treatments. However, attention to regulatory and quality control aspects of nanomaterial-based medical devices is needed to ensure their safe and effective clinical translation. Continued research and development are essential to transform these promising findings into user-friendly solutions for wound management.

REFERENCES

Ahmadi, M., Sabzini, M., Rastgordani, S., & Farazin, A. (2024). Optimizing wound healing: Examining the influence of biopolymers through a comprehensive review of nanohydrogel-embedded nanoparticles in advancing regenerative medicine. *The International Journal of Lower Extremity Wounds*. doi: 10.1177/15347346241244890.

Ajith, G., Tamilarasi, G. P., Sabarees, G., Gouthaman, S., Manikandan, K., Velmurugan, V., Alagarsamy, V., & Solomon, V. R. (2023). Recent developments in electrospun nanofibers as delivery of phytoconstituents for wound healing. *Drugs and Drug Candidates*, *2*(1), 148–171.

Alamouty, M. A., Asl, N. S., Safari, A., Ebrahimi, M., & Daemi, H. (2023). Fabrication of cell-laden AME-loaded collagen-based hydrogel promotes fibroblast proliferation and wound healing in vitro. *Cell Journal (Yakhteh)*, *25*(4), 255.

Ali, M. A., Ahmed, T., Wu, W., Hossain, A., Hafeez, R., Islam Masum, M. M., Wang, Y., An, Q., Sun, G., & Li, B. (2020). Advancements in plant and microbe-based synthesis of metallic nanoparticles and their antimicrobial activity against plant pathogens. *Nanomaterials*, *10*(6), 1146.

Arruebo, M. (2022). Opportunities and challenges for antimicrobial nanostructured materials in the management of skin infections. *Nano Futures*, *6*(4), 042501.

Bahari, N., Hashim, N., Md Akim, A., & Maringgal, B. (2022). Recent advances in honey-based nanoparticles for wound dressing: A review. *Nanomaterials*, *12*(15), 2560.

Baig, N., Kammakakam, I., & Falath, W. (2021). Nanomaterials: A review of synthesis methods, properties, recent progress, and challenges. *Materials Advances*, *2*(6), 1821–1871.

Behere, I., & Ingavle, G. (2022). In vitro and in vivo advancement of multifunctional electrospun nanofiber scaffolds in wound healing applications: Innovative nanofiber designs, stem cell approaches, and future perspectives. *Journal of Biomedical Materials Research Part A*, *110*(2), 443–461.

Bellu, E., Medici, S., Coradduzza, D., Cruciani, S., Amler, E., & Maioli, M. (2021). Nanomaterials in skin regeneration and rejuvenation. *International Journal of Molecular Sciences*, *22*(13), 7095.

Bhardwaj, N., Chouhan, D., & Mandal, B. (2017). Tissue engineered skin and wound healing: Current strategies and future directions. *Current Pharmaceutical Design*, *23*(24), 3455–3482.

Bombin, A. D. J., Dunne, N. J., & McCarthy, H. O. (2020). Electrospinning of natural polymers for the production of nanofibres for wound healing applications. *Materials Science and Engineering: C*, *114*, 110994.

Chen, B., Dong, J., Ruelas, M., Ye, X., He, J., Yao, R., Fu, Y., Liu, Y., Hu, J., & Wu, T. (2022). Artificial intelligence-assisted high-throughput screening of printing conditions of hydrogel architectures for accelerated diabetic wound healing. *Advanced Functional Materials*, *32*(38), 2201843.

Chen, Y.-G., Li, C.-X., Zhang, Y., Qi, Y.-D., Liu, X.-H., Feng, J., & Zhang, X.-Z. (2022). Hybrid suture coating for dual-staged control over antibacterial actions to match well wound healing progression. *Materials Horizons*, *9*(11), 2824–2834.

Choudary, R., Saini, N., Chopra, D. S., Singh, D., & Singh, N. (2023). A comprehensive review of 3D bioprinting biomaterials: Properties, strategies and wound healing application. *Journal of Materials Research*, *38*(13), 3264–3300.

Cruz, D. M., Mostafavi, E., Vernet-Crua, A., Barabadi, H., Shah, V., Cholula-Díaz, J. L., Guisbiers, G., & Webster, T. J. (2020). Green nanotechnology-based zinc oxide (ZnO) nanomaterials for biomedical applications: A review. *Journal of Physics: Materials*, *3*(3), 034005.

Dam, P., Celik, M., Ustun, M., Saha, S., Saha, C., Kacar, E. A., Kugu, S., Karagulle, E. N., Tasoglu, S., & Buyukserin, F. (2023). Wound healing strategies based on nanoparticles incorporated in hydrogel wound patches. *RSC Advances*, *13*(31), 21345–21364.

Degli Esposti, L., Carella, F., & Iafisco, M. (2018). Inorganic nanoparticles for theranostic use. In *Electrofluidodynamic technologies (EFDTs) for biomaterials and medical devices* (pp. 351–376). Elsevier.

Dzulkharnien, N. S. F., & Rohani, R. (2022). A review on current designation of metallic nanocomposite hydrogel in biomedical applications. *Nanomaterials*, *12*(10), 1629.

Eke, J., Banks, L., Mottaleb, M. A., Morris, A. J., Tsyusko, O. V., & Escobar, I. C. (2020). Dual-functional phosphorene nanocomposite membranes for the treatment of perfluorinated water: An investigation of perfluorooctanoic acid removal via filtration combined with ultraviolet irradiation or oxygenation. *Membranes*, *11*(1), 18.

Elbhnsawi, N. A., Elwakil, B. H., Hassanin, A. H., Shehata, N., Elshewemi, S. S., Hagar, M., & Olama, Z. A. (2023). Nano-chitosan/eucalyptus oil/cellulose acetate nanofibers: Manufacturing, antibacterial and wound healing activities. *Membranes*, *13*(6), 604.

Freedman, B. R., Hwang, C., Talbot, S., Hibler, B., Matoori, S., & Mooney, D. J. (2023). Breakthrough treatments for accelerated wound healing. *Science Advances*, *9*(20), eade7007.

Froelich, A., Jakubowska, E., Wojtyłko, M., Jadach, B., Gackowski, M., Gadziński, P., Napierała, O., Ravliv, Y., & Osmałek, T. (2023). Alginate-based materials loaded with nanoparticles in wound healing. *Pharmaceutics*, *15*(4), 1142.

Hayat, M., Nawaz, A., Chinnam, S., Muzammal, M., Latif, M. S., Yasin, M., Ashique, S., Zengin, G., & Farid, A. (2023, July 1). Formulation development and optimization of herbo synthetic gel: In vitro biological evaluation and in vivo wound healing studies. *Process Biochemistry, 130,* 116–126.

Ho, T. T. P., Tran, H. A., Doan, V. K., Maitz, J., Li, Z., Wise, S. G., Lim, K. S., & Rnjak-Kovacina, J. (2024). Natural polymer-based materials for wound healing applications. *Advanced NanoBiomed Research*, 2300131.

Huan, S., & Tay, L. (2020). Four curious cases of closed-incision negative pressure therapy (ciNPT). *Cureus, 12*(5).

Jana, U., Pal, S., Mohanta, G., Manna, P., & Manavalan, R. (2011). Nanoparticles: A potential approach for drug delivery. *Research Journal of Pharmacy and Technology, 4*(7), 1016–1019.

Jiang, Z., Zheng, Z., Yu, S., Gao, Y., Ma, J., Huang, L., & Yang, L. (2023). Nanofiber scaffolds as drug delivery systems promoting wound healing. *Pharmaceutics, 15*(7), 1829.

Joorabloo, A., & Liu, T. (2024). Recent advances in reactive oxygen species scavenging nanomaterials for wound healing. *Exploration, 4*(3), 20230066.

Khan, R., Aslam Khan, M. U., Stojanović, G. M., Javed, A., Haider, S., & Abd Razak, S. I. (2024). Fabrication of bilayer nanofibrous-hydrogel scaffold from bacterial cellulose, PVA, and gelatin as advanced dressing for wound healing and soft tissue engineering. *ACS Omega, 9*(6), 6527–6536.

Khan, Z. A., Jamil, S., Akhtar, A., Bashir, M. M., & Yar, M. (2019). Chitosan based hybrid materials used for wound healing applications-A short review. *International Journal of Polymeric Materials and Polymeric Biomaterials.*

Kim, N., Lee, H., Han, G., Kang, M., Park, S., Kim, D. E., Lee, M., Kim, M. J., Na, Y., & Oh, S. (2023). 3D-Printed functional hydrogel by DNA-induced biomineralization for accelerated diabetic wound healing. *Advanced Science, 10*(17), 2300816.

Krzyszczyk, P., Schloss, R., Palmer, A., & Berthiaume, F. (2018). The role of macrophages in acute and chronic wound healing and interventions to promote pro-wound healing phenotypes. *Frontiers in Physiology, 9*, 419.

Kumar, M., Keshwania, P., Chopra, S., Mahmood, S., & Bhatia, A. (2023). Correction: Therapeutic potential of nanocarrier-mediated delivery of phytoconstituents for wound healing: Their current status and future perspective. *AAPS PharmSciTech, 24*(7), 206.

Kumar, S. P., Asokan, Y., Balamurugan, K., & Harsha, B. (2022). A review of wound dressing materials and its fabrication methods: Emphasis on three-dimensional printed dressings. *Journal of Medical Engineering & Technology, 46*(4), 318–334.

Kushwaha, A., Goswami, L., & Kim, B. S. (2022). Nanomaterial-based therapy for wound healing. *Nanomaterials, 12*(4), 618.

Kwiatkowska, A., Drabik, M., Lipko, A., Grzeczkowicz, A., Stachowiak, R., Marszalik, A., & Granicka, L. H. (2022). Composite membrane dressings system with metallic nanoparticles as an antibacterial factor in wound healing. *Membranes, 12*(2), 215.

Li, H., Mu, M., Chen, B., Zhou, L., Han, B., & Guo, G. (2024). MXene-based nanomaterials for antibacterial and wound healing. *Materials Research Letters, 12*(2), 67–87.

Liu, T., Lu, Y., Zhan, R., Qian, W., & Luo, G. (2023). Nanomaterials and nanomaterials-based drug delivery to promote cutaneous wound healing. *Advanced Drug Delivery Reviews, 193,* 114670.

Liu, Y., Song, S., Liu, S., Zhu, X., & Wang, P. (2022). Application of nanomaterial in hydrogels related to wound healing. *Journal of Nanomaterials, 2022*(1), 4656037.

Lotfi, Z., Khakbiz, M., Davari, N., Bonakdar, S., Mohammadi, J., Shokrgozar, M. A., & Derhambakhsh, S. (2023). Fabrication and multiscale modeling of polycaprolactone/amniotic membrane electrospun nanofiber scaffolds for wound healing. *Artificial Organs, 47*(8), 1267–1284.

Lu, Y., Jia, C., Gong, C., Wang, H., Xiao, Q., Guo, J., Ni, D., & Xu, N. (2023). A hydrogel system containing molybdenum-based nanomaterials for wound healing. *Nano Research, 16*(4), 5368–5375.

Lukhey, M. S., & Shende, P. (2023). Advancement in wound healing treatment using functional nanocarriers. *International Journal of Polymeric Materials and Polymeric Biomaterials, 72*(17), 1406–1421.

Martin, P., & Nunan, R. (2015). Cellular and molecular mechanisms of repair in acute and chronic wound healing. *British Journal of Dermatology, 173*(2), 370–378.

Martin, W. J. (2017). Tissue regeneration without scarring achieved by enhancing the alternative cellular energy (ACE) pathway. *Journal of Cosmetics, Dermatological Sciences and Applications, 7*(1), 82–98.

McCarty, S. M., & Percival, S. L. (2013). Proteases and delayed wound healing. *Advances in Wound Care, 2*(8), 438–447.

Mihai, M. M., Preda, M., Lungu, I., Gestal, M. C., Popa, M. I., & Holban, A. M. (2018). Nanocoatings for chronic wound repair—modulation of microbial colonization and biofilm formation. *International Journal of Molecular Sciences*, *19*(4), 1179.

Müller, W. E., Neufurth, M., Wang, S., Schröder, H. C., & Wang, X. (2024). Polyphosphate nanoparticles: Balancing energy requirements in tissue regeneration processes. *Small*, 2309528.

Nandhini, J., Karthikeyan, E., & Rajeshkumar, S. (2024). Nanomaterials for wound healing: Current status and futuristic frontier. *Biomedical Technology*, *6*, 26–45.

Napagoda, M., Madhushanthi, P., & Witharana, S. (2022). Nanomaterials for wound healing and tissue regeneration. In *Nanotechnology in modern medicine* (pp. 109–134). Springer.

Naskar, A., & Kim, K.-s. (2020). Recent advances in nanomaterial-based wound-healing therapeutics. *Pharmaceutics*, *12*(6), 499.

Pakpahan, F. D., Rahmiyani, I., & Sukmawan, Y. P. (2023). Wound healing activity of the clitoria ternatea L. flower ethanolic extract gel preparation in diabetic animal model. *Research Journal of Pharmacy and Technology*, *16*(1), 140–144.

Pal, S., Mehta, D., Dasgupta, U., & Bajaj, A. (2021). Advances in engineering of low molecular weight hydrogels for chemotherapeutic applications. *Biomedical Materials*, *16*(2), 024102.

Pang, Q., Jiang, Z., Wu, K., Hou, R., & Zhu, Y. (2023). Nanomaterials-based wound dressing for advanced management of infected wound. *Antibiotics*, *12*(2), 351.

Pastar, I., Stojadinovic, O., Yin, N. C., Ramirez, H., Nusbaum, A. G., Sawaya, A., Patel, S. B., Khalid, L., Isseroff, R. R., & Tomic-Canic, M. (2014). Epithelialization in wound healing: A comprehensive review. *Advances in Wound Care*, *3*(7), 445–464.

Prabhu, A., Baliga, V., Bhat, S., Thenkondar, S. T., Nayak, Y., & Nayak, U. Y. (2023). Transforming wound management: nanomaterials and their clinical impact. *Pharmaceutics*, *15*(5), 1560.

Rajput, A., Keshari, R., Patil, R., & Banerjee, R. (2022). Nanobiomaterials for wound healing. In *Nanotechnology in medicine and biology* (pp. 109–139). Elsevier.

Rezaei Kolarijani, N., Cheraghali, D., Khastar, H., Ehterami, A., Alizade, M., Vaez, A., Amini, S. M., & Salehi, M. (2023). Nanofibrous polycaprolactone/gelatin scaffold containing gold nanoparticles: Physicochemical and biological characterization for wound healing. *Wound Repair and Regeneration*, *31*(6), 804–815.

Shalaby, M. A., Anwar, M. M., & Saeed, H. (2022). Nanomaterials for application in wound healing: Current state-of-the-art and future perspectives. *Journal of Polymer Research*, *29*(3), 91.

Sharifi, S., Behzadi, S., Laurent, S., Forrest, M. L., Stroeve, P., & Mahmoudi, M. (2012). Toxicity of nanomaterials. *Chemical Society Reviews*, *41*(6), 2323–2343.

Sharma, D., Srivastava, S., Kumar, S., Sharma, P. K., Hassani, R., Dailah, H. G., Khalid, A., & Mohan, S. (2023). Biodegradable electrospun scaffolds as an emerging tool for skin wound regeneration: A comprehensive review. *Pharmaceuticals*, *16*(2), 325.

Sharma, M., Chauhan, P., Sharma, R., & Kumar, D. (2022). Application of nanotechnology in clinical research: Present and future prospects. *Nanomaterials in Clinical Therapeutics: Synthesis and Applications*, 75–113.

Shi, S., Wu, H., Zhi, C., Yang, J., Si, Y., Ming, Y., Fei, B., & Hu, J. (2023). A skin-like nanostructured membrane for advanced wound dressing. *Composites Part B: Engineering*, *250*, 110438.

Sinno, H., & Prakash, S. (2013). Complements and the wound healing cascade: An updated review. *Plastic Surgery International*, *2013*(1), 146764.

Sridhar, R., Lakshminarayanan, R., Madhaiyan, K., Barathi, V. A., Lim, K. H. C., & Ramakrishna, S. (2015). Electrosprayed nanoparticles and electrospun nanofibers based on natural materials: Applications in tissue regeneration, drug delivery and pharmaceuticals. *Chemical Society Reviews*, *44*(3), 790–814.

Tanfani, J. D., Monpara, J. D., & Jonnalagadda, S. (2023). 3D bioprinting and its role in a wound healing renaissance. *Advanced Materials Technologies*, *8*(17), 2300411.

Thakkar, K. N., Mhatre, S. S., & Parikh, R. Y. (2010). Biological synthesis of metallic nanoparticles. *Nanomedicine: Nanotechnology, Biology and Medicine*, *6*(2), 257–262.

Wang, M., Huang, X., Zheng, H., Tang, Y., Zeng, K., Shao, L., & Li, L. (2021). Nanomaterials applied in wound healing: Mechanisms, limitations and perspectives. *Journal of Controlled Release*, *337*, 236–247.

Wang, W., Lu, K.-j., Yu, C.-h., Huang, Q.-l., & Du, Y.-Z. (2019). Nano-drug delivery systems in wound treatment and skin regeneration. *Journal of Nanobiotechnology*, *17*(1), 82.

Wong, K. K. (2007). Spherical and anisotropic silver nanomaterials in medical therapy. *Nanotechnologies for the Life Sciences: Online*.

Xu, H., Zhang, F., Wang, M., Lv, H., Yu, D.-G., Liu, X., & Shen, H. (2022). Electrospun hierarchical structural films for effective wound healing. *Biomaterials Advances*, *136*, 212795.

Xu, L., Wang, Y.-Y., Huang, J., Chen, C.-Y., Wang, Z.-X., & Xie, H. (2020). Silver nanoparticles: Synthesis, medical applications and biosafety. *Theranostics, 10*(20), 8996.

Xu, R., Fang, Y., Zhang, Z., Cao, Y., Yan, Y., Gan, L., Xu, J., & Zhou, G. (2023). Recent advances in biodegradable and biocompatible synthetic polymers used in skin wound healing. *Materials, 16*(15), 5459.

Xu, Z., Dong, M., Yin, S., Dong, J., Zhang, M., Tian, R., Min, W., Zeng, L., Qiao, H., & Chen, J. (2023). Why traditional herbal medicine promotes wound healing: Research from immune response, wound microbiome to controlled delivery. *Advanced Drug Delivery Reviews, 195*, 114764.

Yang, F., Bai, X., Dai, X., & Li, Y. (2021). The biological processes during wound healing. *Regenerative Medicine, 16*(04), 373–390.

Yaşayan, G., Nejati, O., Ceylan, A. F., Karasu, Ç., Ugur, P. K., Bal-Öztürk, A., Zarepour, A., Zarrabi, A., & Mostafavi, E. (2023). Tackling chronic wound healing using nanomaterials: advancements, challenges, and future perspectives. *Applied Materials Today, 32*, 101829.

Yuan, X., Tan, S., Gao, J., & Wang, L. (2022). Research progress about photothermal nanomaterials with targeted antibacterial properties and their applications in wound healing. *Sheng wu yi xue gong cheng xue za zhi= Journal of biomedical engineering= Shengwu yixue gongchengxue zazhi, 39*(1), 207–216.

Zahid, A. A., Chakraborty, A., Shamiya, Y., Ravi, S. P., & Paul, A. (2022). Leveraging the advancements in functional biomaterials and scaffold fabrication technologies for chronic wound healing applications. *Materials Horizons, 9*(7), 1850–1865.

Zeng, Q., Qi, X., Shi, G., Zhang, M., & Haick, H. (2022). Wound dressing: from nanomaterials to diagnostic dressings and healing evaluations. *ACS Nano, 16*(2), 1708–1733.

Zhang, M., Zhang, C., Li, Z., Fu, X., & Huang, S. (2023). Advances in 3D skin bioprinting for wound healing and disease modeling. *Regenerative Biomaterials, 10*, rbac105.

Zhang, T., Gao, Y., Cui, W., Li, Y., Xiao, D., & Zhou, R. (2021). Nanomaterials-based cell osteogenic differentiation and bone regeneration. *Current Stem Cell Research & Therapy, 16*(1), 36–47.

Zhu, K., Wang, L., Xiao, Y., Zhang, X., You, G., Chen, Y., Wang, Q., Zhao, L., Zhou, H., & Chen, G. (2024). Nanomaterial-related hemoglobin-based oxygen carriers, with emphasis on liposome and nano-capsules, for biomedical applications: current status and future perspectives. *Journal of Nanobiotechnology, 22*(1), 336.

Zou, W., Zhang, L., Lu, J., & Sun, D. (2023). Recent development of metal–organic frameworks in wound healing: Current status and applications. *Chemical Engineering Journal*, 148220.

5 Smart 3D-Printed Hydrogels for Wound Healing

Md Jasim Uddin[1*] *Samiha Binte Sadiq,*[2] *Rehnuma Binta Hafiz,*[2] *and Prati Shahanaz*[2]

[1] University of Malaya, Kuala Lumpur, Malaysia
[2] Drug Delivery and Therapeutics Lab, Dhaka, Bangladesh
*Corresponding Author

5.1 INTRODUCTION

It is commonly known that human skin can heal cuts and bruises on its own. Chronic wounds affect 20 million persons worldwide and cost healthcare systems more than USD 20 billion [1]. Wounds to the skin are the most prevalent type of wound in the United States. Between USD 28.1 billion and USD 96.8 billion was anticipated to be required in 2018 to treat both chronic and acute wounds [2]. It has been established that wound care represents a substantial financial burden in the UK. Studies show that the annual public spending on wound care is approximately 4% [3]. Treatment expenses for diabetic foot ulcers and outpatient wound care are higher than those for hospital stays [2]. Diabetes and obesity continue to be major global health, social, and economic hazards, making chronic wounds a major issue [2].

The gold standard for wound care practices at this time include dermal replacements, skin flaps, skin expansion techniques, and split and full-thickness skin grafts (Figure 5.1) [4–6]. The lack of donor sites and the formation of keloids or hypertrophic scars, that can result in functional or psychological issues, are the primary drawbacks of these procedures [7]. The issue of worldwide technique availability needs to be addressed in addition to the high costs. Consequently, the development of innovative solutions to these challenges is imperative.

Biocompatible hydrogels and promising 3D printing technology could provide solutions for the challenges related to wound healing that have been brought to light. Complex microcomponent design is made possible by 3D printing technology, which can regulate the delivery of numerous compounds that are beneficial to biology. During printing, the hydrogel can be injected with a variety of materials, including biological materials and antimicrobial nanoparticles. The main benefits of using 3D printing to create hydrogel wound patches are their exceptional dependability and low cost. For complex and multimaterial designs, typical manufacturing techniques like casting, molding, forming, and machining are not appropriate, even though they function well for large-scale production.

Applications of 3D printed hydrogels in bone and tissue engineering as well as general medicine have been reviewed in a number of publications. This section compiles reviews that are pertinent and were published between 2014 and the present. M. L. Pita-López et al. [9] demonstrated advancements in the synthesis of physically cross-linked chitosan hydrogels for use in tissue engineering applications. The primary focus of the review was the 3D printing methods used to create the chitosan hydrogel. R. Rodríguez-Rodríguez et al. conducted a study on the latest advancements in composite hydrogel production for biomedical applications [10]. Among the applications were tissue

DOI: 10.1201/9781003504948-5

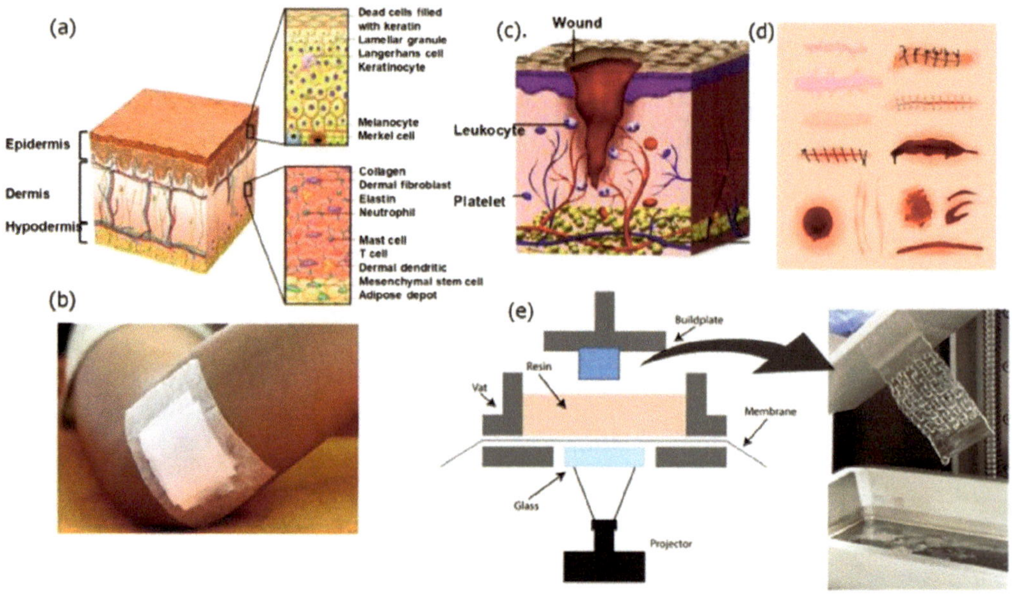

FIGURE 5.1 (a) Skin structure, (b) conventional wound dressing, (c) wound structure, (d) kinds of wound and curing systems, (e) VAT polymerization 3D printing system, and the resulting hydrogel wound dressing [8].

engineering, drug delivery, and wound care; 3D printing technologies were left out of the review. R. Pugliese et al. [11] highlighted recent advancements in the field of 3D printing hydrogels and their biomedical applications. They evaluated the different hydrogels and bioinks based on their biocompatibility, degradation rate, cost-effectiveness, and printability. J. Li et al. conducted a review on hydrogel designs and the creation of hydrogel–bioink materials for 3D printing [12]. There was discussion of inkjet and stereolithography printing techniques. Applications for 3D-printed materials were investigated in the areas of wearable electronics, drug screening, tissue engineering, and regenerative medicine. M. Rajabi et al. examined recent advancements in 3D-printed chitosan-based hydrogels for biological applications [13]. The review included an overview of the physical and chemical procedures utilized to create hydrogels based on chitosan as 3D printable inks. Bakhtiari et al.'s research focused on the characteristics and possible uses of 3D-printed nanocomposite hydrogels in medicine [14]. Recent developments in 3D polysaccharide printing were evaluated by A. Zamboulis et al. [15], who emphasized the technology's potential for drug delivery. The review concentrated on printing cellulose and its derivatives, chitosan, and sodium alginate using extrusion-based printing and fused deposition modeling. Electrically conductive hydrogels were studied by S. Agarwala et al., with a particular emphasis on composite hydrogels and their synthesis techniques [16]. In addition to discussing 3D printing as an improved method of creating, utilizing, and printing conductive polymers, the development, design, and constituent parts of conductive hydrogels were emphasized. The purpose of the review was to help readers choose the best process for making the right kind of composite conductive hydrogels. Researchers S. S. Athukorala et al. examined electrically conductive hydrogels created via 3D printing [17]. The paper included an overview of novel conductive polymers that might be 3D printed, such as polythiophene, polyaniline, and polypyrrole. The review made the fundamentals of electric conductivity and tunable physicochemical properties more understandable. Advanced 3D printing technology and its applications in medicine were showcased by A. I. Pérez-Sanpablo and colleagues [18]. The benefits and drawbacks of contemporary 3D printing technology were examined, as well as difficulties pertaining to its application in the medical field. G. M. Pau l[19] discussed recent advancements in the use of 3D printing in medicine. The limitations of 3D printing in medical applications were examined in the review. A review of 3D bioprinted alternatives was carried out by F. Fayyazbakhsh et al. [20]. The review concentrated on using three different 3D printing mechanisms—inkjet, microextrusion, and laser-assisted bioprinters—as well as bioprinting materials. Advances in 3D printing for the production of biological and pharmaceutical devices for skin applications, drug distribution, and simple dressings were demonstrated by R. S. de Oliveira et al. [21]. The hydrogels under investigation were synthetic, natural, and composite. Hydrogel actuators were made using 3D printing technology, which was studied by W. Sun et al. [22].

5.1.1 Background, Significance, and Importance of Wound Healing

5.1.1.1 Skin Structure

The skin protects itself from environmental, chemical, and biological threats [23]. It also keeps the body temperature constant and prevents water loss, or dehydration. Healthy skin has a slightly acidic pH, which aids in warding off dangerous microorganisms and free radicals that may speed up the aging process. Sweat, prolonged time in the sun, antibacterial soaps and gels, and exposure to environmental conditions (such as humidity and air pollution) can all affect the pH of the skin. When the skin is injured, the tissues release exudate, which causes the pH to shift to an alkaline state.

5.1.1.1.1 Epidermis

The skin that covers the body's exterior and provides protection is called the epidermis. With keratinocytes composing the epithelium and keratinized tissue and basal keratinocytes in the lowest layer, the skin acts as a barrier [24]. To ensure that the epidermis renews itself, the basal keratinocytes multiply regularly to reconstruct the entire layer. The epidermis and dermis are separated by a foundational membrane that lies beneath it.

5.1.1.1.2 Dermis

Underneath the epidermis is the dermis layer, which has a thickness of 1–4 mm. The fibroblasts, which are basic cells, are responsible for the dermis's mechanical strength and elasticity [24]. The dermis layer contains neutrophils, fibroblasts, mast cells, and dermal dendritic cells. Sweat glands, hair follicles, sebaceous glands, and nerve endings are a few of the dermis's features [25]. Large networks of blood, nerve, and lymphatic channels are also present [24]. The two layers that comprise the dermis are the reticular and papillary dermis [24]. The uppermost layer, known as the papillary dermis, is made up of blood vessels and connective tissues and gives the epidermis nourishment [24]. The reticular layer, found under the papillary dermis, is composed of thick collagen and elastic fibers, which provide the skin with strength and elasticity [24].

5.1.1.1.3 Hypodermis

Underneath the dermis is the hypodermis, a subcutaneous layer primarily composed of fat. The hypodermis aids in the body's ability to retain heat by acting as insulation between the skin and the skeletal bones [25]. Every individual and every portion of the body has a unique hypodermis thickness. The hypodermis today serves a variety of purposes, although it was once thought to be a tissue that accumulated fat. These include regulating body temperature, generating hormones, and supplying defense.

5.1.1.2 Skin Wounds

Four types of chronic wounds have been recognized by the Wound Healing Society: pressure ulcers, diabetic ulcers, venous ulcers, and arterial insufficiency ulcers. Among the traits these populations have in common are chronic reactive oxygen species (ROS), senescent fibroblasts, ongoing infection, and insufficient or defective stem cells [26].

5.1.1.2.1 Common Skin Wounds and Clinical Treatment

Disruption to the skin's surface is referred to as a skin wound. The standard clinical procedure for an open split or gaping wound is suturing (stitches). The process of wound healing is complex and dynamic, involving the rebuilding of tissue layers and cellular structures. Depending on whether an injury is acute or chronic, different healing processes and rates apply. The most prevalent forms of wounds, their causes, clinical interventions, and distinguishing features are listed in Tables 5.1 and 5.2.

5.1.1.3 Acute and Chronic Skin Wounds

Wounds can be caused by physical, chemical, or thermal trauma. Skin wounds are classified into two groups based on their healing length and ability: acute and chronic.

5.1.1.3.1 Acute Wounds

Eight to twelve weeks may pass between acute wounds, during which there may be a lot of exudate, severe infections, pain, and tissue necrosis [30, 31]. The majority of mechanical injuries that result in acute wounds include gunshot penetration, surgical incisions, and contact from friction between the skin and hard objects like knives. They can be brought on by radiation, caustic materials, heat trauma, chemicals, and burn injuries [32]. Acute wounds are distinguished by common microorganisms, low inflammatory cytokines, and high mitogenic activity [33].

5.1.1.3.2 Chronic Wounds

Chronic wounds require a minimum of 12 weeks to heal [34]. Recurrent skin damage, exposure to physiological factors like diabetes, insufficient angiogenesis or innervation, or cellular migration are the main causes of these wounds [35]. Infections, cancers, inadequate primary care, and other patient-related variables are correlated [35]. The three most prevalent forms of chronic wounds are diabetic foot ulcers, pressure ulcers, and venous leg ulcers [36]. Chronic wounds are more likely to

TABLE 5.1
Common Types of Wounds

Type of Skin Wound	Cause
Puncture	Caused by a sharp/pointed object.
Laceration	The skin is sliced, ripped, or totally removed (avulsion).
Pressure injury	Created by prolonged pressure on a bony component of the body.
Abrasion	The skin is removed by rubbing or scraping. Minor cuts simply cause damage to the epidermis.
Thermal	Caused by being exposed to severe heat or cold.
Chemical	Caused by exposure to powerful acids or bases.
Incision	A surgical wound or an intentional cut in the skin.

Source: [16].

TABLE 5.2
Conventional Clinical Skin Closures

Type of Skin Closure Strips	Characteristics	Reference
Skin glue	Helps to keep the wound together and promotes healing.	[27]
Sutures	Stitches are placed under the skin in deep wounds to help the injury heal.	[28]
Skin grafts	Utilized when the skin surrounding the wound is too damaged to mend together.	[29]

occur in patients with weaker immune systems, bacterial infections, and other medical conditions. Moreover, diabetes and cancer may make patients more susceptible to persistent wounds. The application of wound dressings helps accelerate the healing of both acute and chronic wounds. Chronic wounds are characterized by high microbial contamination, elevated inflammatory cytokines, and a deteriorating, nonfunctional matrix [37]. Table 5.3 compiles the approaches for detecting various wound types.

5.1.1.4 Skin Wound Healing Process

The quick healing of wounds involves certain biological processes. As an injury starts to heal, blood platelets are stimulated to assist leukocyte recruitment and form a blood clot. Next, neutrophils and macrophages remove foreign infections, wounded or dead cells, and other debris from the wound site. After that, fibroblasts move, proliferate, and initiate the angiogenesis process. In order to reconstruct the dermal tissue, the epidermis then regenerates, granulation tissue is created, and extracellular matrix proteins are deposited [40]. Last, fibroblasts and newly formed capillaries are removed. The following sections describe the four stages of wound healing (Figure 5.2c).

5.1.1.4.1 Hemostasis (Blood Clotting)

Hemostasis is a physiological response wherein inflammatory cells and platelets congregate at the site of injury. The release of clotting factors occurs after it [41]. A few minutes after the damage, blood platelets gather, turn amorphous, and concentrate at the wound site (Figure 5.2c(i)). Leukocyte recruitment and the initiation and development of inflammation both depend on platelets [41].

5.1.1.4.2 Inflammation

Inflammation occurs during the hemostasis stage, which follows the development of the fiber clotting system. Blood monocytes and lymphocytes generate tissue macrophages then release growth factors at the site of the wound [42]. During this stage, immune cells—neutrophils and macrophages in particular—are discharged into the wound site (Figure 5.2c(ii)) [43]. At the location of the wound, a variety of peptides and growth factors secreted by platelets and inflammatory cells promote the development of fibroblasts as well as angiogenesis [44].

TABLE 5.3
Some Methodologies for Identifying a Chronic or Acute Wound

Gauge Used	Measurement	Indication	Reference
iDr or mobile app	3D imaging of the wound (Figure 5.2a)	With the help of a smartphone video, surface feet per minute (SFM), and the optical imaging concept, iDr is able to accurately and noninvasively recreate a 3D wound model and estimate its volume and size in 3D digital space. Using volume and area data from recorded histories, iDr can help physicians assess how well wounds heal during therapy.	[38]
Matrix metalloproteinase (MMP)	Collecting wound fluids (22 samples) and chronic wounds (25 samples) from various etiologies, such as mixed vascular disease, decubitus, and diabetic foot ulcers (see Figure 5.2b).	Chronic wounds had a higher median MMP Eq/mL (22.8 μg) than acute wounds (0.76 μg) ($p < 0.001$).	[39]

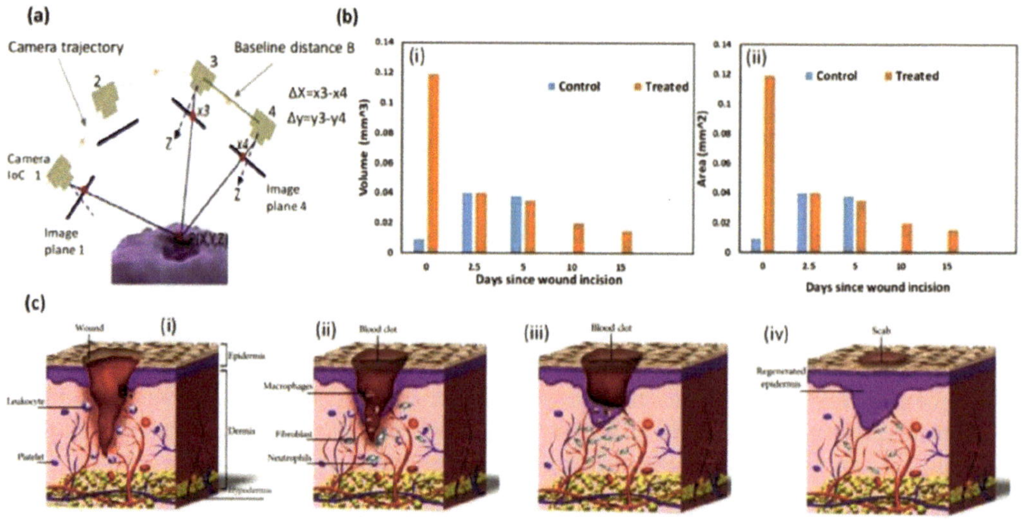

FIGURE 5.2 (a) iDr or mobile app; (b) matrix metalloproteinase (MMP); (c) skin wound healing process: (i) hemostasis (blood clotting), (ii) inflammation, (iii) tissue growth (proliferation), and (iv) tissue remodeling (maturation) [8,38, 39].

5.1.1.4.3 Tissue Growth (Proliferation)

The tissue growth phase endures for two to three days after the damage, or until the wound closes. At this point, fibroblasts mature into myofibroblasts, which close the wound by binding its edges together [45]. During the releasing phase, fibroblasts at the wound site are further stimulated to multiply. This stage reconstructs the components of cutaneous tissue by granulation tissue formation and extracellular matrix protein deposition, primarily collagen [45]. A new blood vessel network is encouraged to grow by enhanced angiogenesis (Figure 5.2c(iii)). Epithelialization is the process wherein epithelial cells migrate from the edges of the wound to cover the defect. During the proliferative phase, the granulated tissues are built from epithelial cells and fibroblasts, and the keratinocytes are transformed into the extracellular matrix (ECM) [46].

5.1.1.4.4 Tissue Remodeling (Maturation)

In the maturation phase, wound contraction reaches its peak and unnecessary collagen fibers are eliminated in the dermis. Eighty percent of the wound's initial maximum mechanical strength is retained after healing [47]. Dermal and epidermal structures in humans are formed by the combination of fibroblasts, keratinocyte cells, and collagen matrix (Figure 5.2c(iv)). In the past, scientists created dermal and epidermal-like structures by introducing fibroblasts and keratinocytes into the collagen matrix to attain compatibility. For example, researchers used 3D printing to introduce a composite containing embedded fibroblasts and keratinocytes into the collagen matrix. McCarty and Percival [48] provide more information about the wound healing mechanisms.

5.1.2 Challenges of Chronic Wounds

Chronically treated wounds do not heal properly or quickly. Chronic wounds frequently stop healing at the inflammatory stage of the healing process. Chronic wounds are characterized by high levels of proinflammatory cytokines, proteases, and ROS, along with senescent cells, persistent infection, and a lack of functional stem cells, despite molecular variations in their genesis. Recurrent tissue injury is caused by microorganisms and platelet-derived factors like TGF-β or ECM fragment molecules. This leads to an ongoing inflow of immune cells. As a result, there is an increase in the proinflammatory cytokine cascade and protease levels. In acute wounds, proteases are strictly controlled by their inhibitors. Prolonged wounds result in elevated protease levels that surpass inhibitory thresholds, resulting in the breakdown of growth factors and receptors, as well as the dissolution of the extracellular matrix. In addition to keeping the lesion from entering the proliferative phase, the proteolytic breakdown of the ECM attracts additional inflammatory cells, speeding up the cycle of inflammation [48].

ROS, which immune cells create in small quantities, acts as a pathogen defense. Nonetheless, the prevailing hypoxic and inflammatory environment in chronic wounds encourages the production of ROS, which results in the breakdown of ECM proteins and cell death. Proteases and inflammatory cytokines are stimulated more by this series of events [49]. In an animal model, it has been suggested that the use of powerful antioxidants decreases ROS to normal levels, hence reversing wound chronicity and improving healing [50].

Moreover, senescent cell populations having diminished secretory and proliferative capacities are present in chronic wounds, rendering them immune to the normal signals that promote wound healing [15]. It has been demonstrated that fibroblasts from pus and veins are less proliferative and senescent. The inability of a wound to heal is directly linked to this decreased proliferative capability [51]. According to currently known data, senescent keratinocytes, endothelial cells, fibroblasts, and macrophages are also present in chronic wounds [52]. Senescent cells from chronic wounds might arise from oxidative stress, cell cycle arrest related to DNA damage, or abnormal metabolic changes in diabetic individuals. These factors can create anomalies in intracellular biochemical pathways, including the GSK-3β/Fyn/Nrf2 pathway [53], which can delay the healing process.

5.1.3 Function of 3D Printing in Tissue Engineering

For humans, tissue engineering (TE) is still a relatively new field of study. It incorporates aspects of biology, medicine, materials engineering, and mechanics. Developing strategies to facilitate the repair of injured organs and tissues—especially those that were previously thought to be nonregenerative—is the primary goal of tissue engineering. This type of tissue and organ damage occurs frequently in clinical practice. These are typically severe abnormalities of the bones, nerves, or skin. These anomalies are most frequently due to trauma, then directly related to tumor activity or tumor removal [54]. Typically, after being created in vitro (in a bioreactor), partially regenerated tissue (cell culture) is transplanted at the site of the defect. To achieve a uniform distribution of cells in the defect space, porous materials known as TE scaffolds are utilized. These scaffolds provide an adequate substrate for grown cells while also allowing free access to nutrients and drainage of cell metabolic products. Tissue scaffolds also play a significant role in restoring the mechanical function of injured tissue (organ). As a result, they should have the proper rigidity. It is also envisaged that the implanted scaffold will be completely resorbed by the time the tissue defect has entirely healed. Scaffolds are often made of biodegradable polymers, either natural (chitosan or cellulose) or manufactured. Ceramic materials, such as β-TCP and hydroxyapatite, are commonly utilized in polymeric matrices to enhance biocompatibility. TE scaffolds must meet a variety of standards. It also turns out that the scaffold's function is influenced by a variety of elements, ranging from pure biological to mechanical [54].

There are several methods for creating TE scaffolds. These include certain traditional methods such as solvent casting, phase separation, and electrospinning that provide little control over the shape of the scaffold. Furthermore, they are not highly replicable. These constraints do not apply to additive manufacturing (AM) processes [54], often known as 3D printing methods. Furthermore, the technology of 3D printing has made the design of computer-assisted TE scaffolds easier to implement. For TE applications, a wide variety of 3D printing techniques are already accessible. They enable the creation of TE scaffolds from a variety of materials, including metals, ceramics, polyesters, and hydrogels.

Generally speaking, the ability of 3D printing to generate complex structures that would be too expensive to produce by injection molding processes is one of its main advantages. The extraordinarily high resolution that 3D printers have been refined for can be advantageous to tissue engineering. There have been recorded attempts to modify industrial printers in order to print scaffolds for tissue engineering. These days, TE structures may be created using 3D printing technology to regenerate a variety of tissues, such as skin [55], cartilage [56], vascular networks [57], and entire organs [58].

Because 3D printing creates scaffolds that are specifically tailored to the requirements of the wound site, it speeds up the regeneration process, which is essential for wound healing. Carefully constructed from biomaterials, these scaffolds mimic the extracellular matrix seen in nature, offering mechanical support and directing cell activity. 3D-printed scaffolds, which incorporate bioactive chemicals and growth factors, can stimulate cell adhesion, proliferation, and differentiation, thereby promoting tissue regeneration. Furthermore, the ability to create complex structures with interconnected porosity improves nutrient and oxygen flow, which promotes vascularization and tissue integration. The customized aspect of 3D printing allows for the creation of patient-specific wound dressings, promoting faster healing and reducing the risk of infection. As a result, 3D printing has great promise for changing wound care by providing novel methods for personalized, efficient, and effective tissue regeneration [59].

5.2 HYDROGELS IN HEALING WOUNDS

Hydrogels are complex three-dimensional structures composed of chains of hydrophilic polymers that swiftly inflate in the presence of water, becoming a partly solid substance. Since water makes up more than 90% of the hydrogel framework, it is possible to keep the wound's surface moist, which promotes tissue repair [60]. Hydrogels have several qualities that make them appropriate for use as wound dressings. Sufficient wound covering and safeguarding are facilitated by mechanical protection, form flexibility, and firm adhesion [61]. The advantage of hydrogel-based dressings is their high degree of customization, which enables the incorporation of cells, biomolecules, growth hormones, and antibacterial and antimicrobial agents [62]. The goal of this augmentation is to hasten the healing and contraction of wounds. Any hydrophilic polymer combined with a certain cross-linking procedure can be used to make a hydrogel. These water-soluble polymers can be produced artificially or naturally. Because of their highly adjustable physical qualities and adhesive properties, synthetic materials have particular advantages. Compared to synthetic polymers, natural polymers are more biocompatible and biodegradable [63]. There has been significant development in blending natural and synthetic polymers to create blended hydrogels. Furthermore, the incorporation of nanomaterials in situ has resulted in the production of "smart" nanogels with tailored functionality, making hydrogels more suitable for treating deep or irregular wounds due to in situ induction [64]. Hydrogels are a type of therapeutic intervention that has the potential to significantly improve the quality of life for many individuals suffering from wounds and their accompanying complications.

Chemicals known as hydrogels have a wide range of uses in the field of skin regeneration. When hydrophilic polymer chains are cross-linked chemically or physically, three-dimensional polymeric structures known as hydrogels are produced [65]. Numerous techniques, including radiation, freeze-thawing, and chemical procedures, can be used to manufacture them. Because ionic and hydrogen interactions or molecular entanglements preserve the structural integrity of hydrogels, these materials are referred to as "reversible" or "physical" gels. When covalent bonds are present, they are called "permanent" or "chemical" gels. These networks have the capacity to expand with water until they find equilibrium while keeping their original structure. This results in a significant ability to absorb wound exudates, promote oxygen flow, and maintain high moisture content at the injury site. This speeds up the healing process [66].

Due to their diverse physical characteristics, hydrogels can be created in an extensive range of sizes and forms, enabling the full covering of wounds with irregular geometries [67]. Because hydrogels are biocompatible and biodegradable, they can serve as a temporary template while chronic wounds reconstruct and re-epithelialize. Hydrogels also show strong bioadhesiveness, which is essential for ensuring stability over the long run. This property maintains the wound at the proper moisture content and enhances hemostasis [68]. Additionally, the flexible structure that hydrogels offer allows for the addition of a wide range of substances, such as medications, auxiliary biomolecules, and antibacterial and antimicrobial agents, enhancing their overall effectiveness in the healing of wounds. Therefore, the best materials to utilize as dressings to cover skin wounds are those based on hydrogel.

5.2.1 Hydrogel Properties and Characteristics

Long chains of polymers that have been cross-linked to create a network structure make up hydrogels. These polymers might be artificial, like polyethylene glycol (PEG) or polyacrylamide, or natural, like alginate or collagen. The cross-linking process, which determines the hydrogel's mechanical characteristics and stability, can be accomplished by covalent bonding or physical interactions. Hydrogels are a unique family of materials with a wide range of biomedical uses due to their tunable mechanical characteristics (viscoelasticity, brittleness, and fluffiness) and biodegradability, biocompatibility, and hydrophilicity. Their inherent ability to respond to various stimuli (temperature, electric field, magnetic field, biological molecules, and ionic strength) offers additional functionalities for controlled drug delivery. Notably, the mucoadhesive and bioadhesive characteristics of hydrogels prolong drug residence time, making them well-suited as drug carriers.

Recently, hydrogels have garnered significant interest in the field of energy storage and conversion due to their inherent flexibility and semisolid nature [69]. Traditional energy storage devices are plagued by limitations such as significant weight, mechanical rigidity, and environmental concerns associated with leakage risks from liquid electrolytes. These electrolytes are not only expensive but also pose potential health hazards due to their organic composition. Hydrogel electrolytes, on the other hand, offer a biocompatible, biodegradable, cost-effective, and environmentally friendly alternative. Their semisolid nature combined with inherent flexibility makes them ideal candidates for practical applications in energy storage devices, particularly supercapacitors.

The unique properties of hydrogels enable the development of flexible, stretchable, and elastic electrochemical devices. This allows for functionality under various mechanical deformations (stretching, bending, folding, and twisting), a significant advantage over traditional, rigid devices. Additionally, the self-healing nature of hydrogels makes them particularly attractive for wearable and portable electronics [69]. This self-healing ability, also known as damage management, contributes to the development of smart and lightweight electronic devices. Hydrogels represent a versatile class of materials with remarkable properties that hold immense promise for advancements in both biomedical and energy storage applications. Their unique characteristics address key limitations in these fields, paving the way for the development of innovative and user-friendly devices.

Their properties include [69]:

High water content: One of the unique qualities of hydrogels is their ability to absorb large amounts of water, frequently hundreds or even thousands of times their dry weight. This exceptional water absorption is thought to be caused by the hydrophilic functional groups present in the hydrogel network.

Biocompatibility: Some hydrogels are able to live alongside living tissues without creating any negative side effects. They are perfect for biomedical uses, including tissue engineering and medication delivery, because of this characteristic.

Biodegradability: A number of hydrogels are biodegradable, which means that biological processes can naturally break them down. This property is essential for biological applications where the substance must eventually break down and leave the body.

Stimuli responsiveness: Hydrogels can be designed to respond to various external stimuli, such as temperature, pH, light, electric field, and magnetic field. This responsiveness allows for controlled release of drugs or changes in the hydrogel's mechanical properties [69].

Mechanical properties: Hydrogels exhibit a wide range of mechanical properties depending on their composition and cross-linking density. They can be soft and elastic, rigid and brittle, or possess viscoelastic behavior (combining viscous and elastic properties).

Swelling behavior: Hydrogels undergo swelling when exposed to water, increasing in size due to water absorption. The extent of swelling depends on the hydrogel's properties and the surrounding environment.

Self-healing: Some hydrogels possess self-healing properties, meaning they can repair minor damage through the internal reorganization of their network. This characteristic is advantageous for applications where the hydrogel may experience wear and tear.

Injectability: Certain hydrogels can be formulated for injection into specific tissues. This property makes them suitable for localized drug delivery or minimally invasive surgical procedures.

Transparency: Hydrogels can be designed to be transparent, allowing for visualization of internal components or tissues encapsulated within them. This is beneficial for monitoring drug delivery or cell growth within the hydrogel.

Conductivity: The electrical conductivity of hydrogels can be tuned by incorporating conductive polymers or ions into the network. This property enables applications in areas like biosensors and bioelectronics.

Characteristics of hydrogels include the following [69]:

Network structure: Hydrogels consist of a three-dimensional network of polymer chains cross-linked together. This network structure defines the hydrogel's properties, such as water uptake and mechanical strength.

Pore size: In applications like drug administration and cell encapsulation, the size and distribution of pores within the hydrogel network are critical factors. Smaller pores can regulate the rate at which medications are released, whereas larger pores enable molecules to diffuse more quickly.

Surface charge: The surface charge of a hydrogel, determined by the presence of charged functional groups, can influence protein adsorption, cell adhesion, and drug-loading capacity [1].

Biodegradation rate: The rate at which a hydrogel biodegrades is an essential characteristic for biomedical applications. It can be controlled by the type of polymer used and the cross-linking density.

Stimulus specificity: The specificity of a hydrogel's response to a particular stimulus is crucial for controlled drug delivery or other responsive applications. Ideally, the hydrogel should only respond to the desired stimulus.

Swelling kinetics: The rate and extent of swelling determine the release profile of drugs or other molecules encapsulated within the hydrogel. Controlled swelling kinetics are essential for sustained or targeted delivery.

Cytocompatibility: For biomedical applications involving interaction with living cells, cytocompatibility is crucial. This characteristic refers to the hydrogel's ability to support cell growth and function without causing toxicity.

Drug loading capacity: The amount of drug a hydrogel can encapsulate is a key factor for drug delivery applications. This capacity depends on the hydrogel's properties and the drug's characteristics.

Mechanical stability: Long-term functioning in biomedical applications depends on a hydrogel's capacity to preserve its structural integrity under physiological conditions (body temperature, pressure, etc.).

Processability: A variety of methods, such as chemical and physical ones, can be used to produce hydrogels. The ultimate qualities and properties of the hydrogel product are determined by the procedure selected.

Scientists and engineers can create hydrogels with certain functions for a variety of uses in biomedicine, energy storage, and other domains by comprehending these qualities and traits. Hydrogels have enormous potential that keeps expanding as long as research and development are conducted.

5.2.2 Biocompatibility and Mechanical Properties

Wound healing is only one of the many uses for hydrogels, an intriguing family of materials. They are very well suited for application in biomedicine because of their special qualities, especially in tissue engineering and wound treatments. The biocompatibility and mechanical characteristics of hydrogels for wound healing will be examined in this section, along with how these characteristics affect the materials' ability to aid in healing and tissue regeneration [70].

A material's biocompatibility is an essential component when it comes to biomedical applications. It describes a material's capacity to carry out its intended role in a particular application without endangering living things. The main reason hydrogels have such a good reputation for biocompatibility is that their networks of water-rich polymers resemble the ECM that is naturally present in living tissues. Because hydrogels have a high water content, they mimic living tissues in their ability to exchange nutrients and waste products while also keeping the wound site hydrated [70]. Hydrogels are biocompatible because of their potential to reduce inflammation and the immune system's reaction at the wound site. Unlike some other materials, such as metals or synthetic polymers, hydrogels are generally nontoxic and nonimmunogenic, meaning they do not trigger an immune response when implanted in the body. This is crucial for wound healing, as inflammation and immune reactions can interfere with the healing process and prolong recovery time.

Moreover, hydrogels can be engineered to break down gradually into biocompatible byproducts that the body can absorb or eliminate, releasing any growth factors or medications contained at the wound site. By preventing the hydrogel from building up in the body and removing the necessity for removal operations, this controlled degradation lowers the risk of problems and pain for patients [70]. Hydrogels' mechanical characteristics, in addition to their biocompatibility, are crucial to their effectiveness in wound healing. The mechanical properties of hydrogels refer to their ability to withstand mechanical forces, such as compression, tension, and shear, which are encountered in the wound environment. These properties can be tailored to match those of the surrounding tissue, providing mechanical support and protection to the wound site.

Moreover, the strength of hydrogels is crucial for providing structural support to the wound site and protecting it from external forces. While some wounds may require a softer, more flexible hydrogel, others may benefit from a stronger, more resilient hydrogel that can withstand higher mechanical loads. Hydrogels' mechanical strength can be modified to match the unique needs of various wound types by varying the cross-linking density and polymer concentration. Porosity, which influences a hydrogel's capacity to take in and hold onto fluids as well as promote cell infiltration and tissue regeneration, is another crucial mechanical characteristic. Hydrogels with high porosity can take up extra fluid from the wound and keep it wet, which promotes healing. Additionally, the interconnected pores in hydrogels provide pathways for the migration of cells, nutrients, and oxygen to the wound site, promoting tissue regeneration and vascularization.

The viscoelastic behavior of hydrogels is also critical for wound healing, as it influences their ability to dissipate mechanical forces and prevent tissue damage. Viscoelastic hydrogels exhibit both elastic (reversible deformation) and viscous (irreversible deformation) behavior, allowing them to absorb and dissipate energy from mechanical stresses. This is particularly important in dynamic wound environments where the tissue undergoes repeated deformation during movement. Hydrogels' adhesion qualities are also crucial for guaranteeing that they stick to the wound surface and stay there during the healing process. By adding bioadhesive molecules like polysaccharides or cell adhesion peptides, hydrogels can be altered to improve their adhesive qualities. By strengthening adhesion and preventing the hydrogel from being moved by movement or exudate, these molecules facilitate contact between the hydrogel and the wound surface.

5.2.3 Mechanisms of Action in Wound Healing

5.2.3.1 Hemostasis

Shortly following an injury, hemostasis occurs [71]. Vasoconstriction occurs to prevent exsanguination, and platelets adhere, activate, and aggregate at the site of injury. When extravascular collagen, such as type I collagen, is present, platelets become activated. They can detect this stimulation thanks to certain integrin receptors, which are cell surface receptors that facilitate a cell's interaction with the extracellular matrix. Platelets cling to collagen by aggregating and releasing sticky glycoproteins and soluble mediators, including growth factors and cyclic AMP. Important glycoproteins such as fibronectin, thrombospondin, fibrinogen, and von Willebrand factor are released by platelet alpha granules. The release of clotting components during platelet aggregation results in the formation of a fibrin clot at the site of injury [72]. The aggregated platelets become caught in the fibrin web, forming the majority of the clot. Their membranes provide a surface for inactive clotting enzyme proteases to bind, activate, and speed the clotting cascade. Growth factors including PDGF, TGF-β, TGF-α, bFGF, IGF-1, and VEGF are released by platelet alpha granules. Neutrophils and monocytes are drawn from the vasculature by PDGF and TGF-β, which start the inflammatory response. Complement breakdown product C5a and bacterial waste product f-Met-Leu-Phe both provide extra chemotactic cues for neutrophil recruitment to the site of damage. Angiogenesis is triggered by endothelial cell activation by VEGF, TGF-α, and bFGF. Following PDGF's activation and recruitment of fibroblasts to the wound site, these cells start to create extracellular matrix proteins called glycosaminoglycans and collagen, which facilitate cellular motility and interactions with the matrix supporting framework. Thus, the healing process begins with hemostasis, platelet deposition at the site of damage, and interactions of soluble mediators and growth factors with the extracellular matrix, which provide the groundwork for subsequent healing activities [71, 73].

5.2.3.2 Inflammation

After injury, the following stage of wound healing is called inflammation, which lasts for 24 hours in the case of normal wounds and far longer in the event of chronic wounds that don't heal. Histamine, enzymes, and other active amines are released in granules by mast cells. The classic indicators of inflammation around the wound site, such as tumor (swelling), dolor (pain), calor (heat), and rubor (redness), are produced by these granules. The most important cells during the inflammatory phase are neutrophils, monocytes, and macrophages. In addition to clearing the wound of debris and infection, they also discharge soluble mediators that aid in the recruitment and activation of fibroblasts and epithelial cells. They clear the wound of debris and infection and release soluble mediators, such as growth factors (PDGF, TGF-β, TGF-α, IGF-1, and FGF) and proinflammatory cytokines (such as IL-1, IL-6, IL-8, and TNF-α), that aid in the recruitment and activation of fibroblasts and epithelial cells to prepare the wound for the next stage of healing. A third significant class of small regulatory proteins, known as chemokines, has been discovered in addition to growth factors and cytokines. These proteins are collectively called chemoattractant cytokine(s) [74, 75] for their biological characteristics (interferon-inducible protein of 10 kDa [IP-10]), isolation (platelet factor-4 [PF-4]), or regulation upon activation (normal T-cell produced and released, [RANTES]). Based on the arrangement of cysteine residues close to the N-terminus, it became apparent that the roughly

40 chemokines could be divided into four main groups when their biochemical characteristics were determined. In actuality, a more organized system of naming based on these four primary divisions has been revived lately. Chemokines primarily regulate leukocyte transport in both developing and healthy organisms. In inflammatory conditions, they also influence the recruitment and activation of neutrophils, lymphocytes, macrophages, eosinophils, and basophils.

5.2.3.3 Neutrophils

Neutrophils are the first inflammatory cells to respond when platelets release soluble mediators and start the coagulation cascade. They phagocytose and destroy bacteria, but they also remove foreign objects and devitalized tissue as the first line of defense against infection. The extravasation of inflammatory cells into a wound is mostly dependent on adhesion molecules (integrins, cadherins, and cell adhesion molecules [CAMS]) and receptors (integrins) that are connected to the plasma membranes of circulating leukocytes and vascular endothelial cells [76, 77]. Because of their poor adherence to the endothelial cell walls through their selectin molecules, leukocytes first slow down and begin rolling across the surface of these cells. Chemoattractants, such as growth factors, cytokines, or bacterial metabolites, may cause leukocytes to roll. After activation, leukocytes connect firmly to endothelial cells by the binding of their integrin receptors to ligands, such as VCAM and ICAM, generated on active endothelial cells. Chemotactic signals outside the venule then trigger leukocytes to squeeze between the endothelial cells and migrate into the damaged tissue. There, they recognize and attach to extracellular matrix components using their integrin receptors. The inflammatory cells release collagenase and elastase in order to penetrate the endothelial cell basement membrane and reach the ECM at the wound site. Furthermore, neutrophils generate and release inflammatory mediators including TNF-α and IL-1, which draw and activate fibroblast and epithelial cells. Neutrophils create oxygen-free radicals that destroy phagocytized bacteria after migrating into the wound site. Neutrophil collagenase, along with neutrophil elastase, two proteases that degrade damaged extracellular matrix components, are also released in significant quantities by them. A wound's chronicity may be exacerbated by the persistent recruitment of neutrophils and the subsequent release of proteases, cytokines, and ROS by these cells. Usually, after two to three days, the process of apoptosis depletes the neutrophils in the wound, and tissue monocytes take their place.

5.2.3.4 Macrophages

The control of healing is largely dependent on activated macrophages, and healing cannot occur normally in the absence of macrophages. About 24 hours after an injury, circulating monocytes are drawn to the wound site and develop into macrophages. They extravasate through the methods outlined for neutrophils, and in response to growth factors, chemokines, cytokines, and soluble fragments of extracellular matrix components created by the proteolytic breakdown of collagen and fibronectin, they are stimulated to differentiate into activated tissue macrophages. Tissue macrophages play two roles in the healing process, much like neutrophils do. They move throughout the area of the wound, consuming and eliminating germs and using elastase and released MMPs to remove devitalized tissue. By secreting inhibitors for proteases, macrophages are able to more tightly control the proteolytic degradation of injured tissue than neutrophils. In addition to their phagocytic role, macrophages are important mediators of the transition from the inflammatory to the proliferative phases of healing. They secrete a variety of growth factors and cytokines, including PDGF, TGF-β, TGF-α, FGF, IGF-1, TNFα, IL-1, and IL-6 [71]. Certain soluble mediators promote angiogenesis, whereas other soluble mediators attract and excite fibroblasts, which subsequently produce, deposit, and organize the new tissue matrix. The inflammatory phase of the wound is coming to a conclusion, and the proliferative phase is starting. The decline in macrophages and the absence of neutrophils are signs of this.

5.2.3.5 Phase of Proliferation

During the proliferative phase, a new fibrin matrix consisting of collagen fibers, proteoglycans, and fibronectin replaces the temporary one, restoring the tissue's structure and function [72]. An important step in the healing process is angiogenesis, or the formation of new capillaries that substitute previously damaged vessels and restore circulation. Two more critical processes in this healing stage are the development of granulation tissue and epithelialization. During the proliferative stage of the healing process, fibroblasts are crucial.

5.2.3.5.1 Fibroblast Migration

Numerous soluble mediators released by platelets and later by macrophages cause fibroblasts to penetrate the wound. Fibroblasts need to recognize and interact with specific matrix components in order to move across the extracellular matrix. Although fibroblasts are often inactive and sporadically distributed in the normal dermis, they are extremely active and numerous in the granulation tissue and the temporary matrix around the wound. To facilitate their migration and accumulation from the ECM into the wound site, they must transform and release proteases [71].

Fibroblasts first bind to matrix components such as fibrin, vitronectin, and fibronectin using their integrin receptors before beginning to migrate [71]. Integrin receptors bind to certain amino acid sequences (such as arginine–glycine–aspartic acid [RGD]) or binding sites in these matrix elements. The fibroblast cell, with one end of the cell still linked to the matrix component, extends a cytoplasmic projection in search of another binding site. When the cell gets to the second site, the first one gets free (perhaps because there is protease activity around), and it uses its cytoskeleton's actin fiber network to move forward.

Which direction fibroblasts migrate in is determined by the gradient in the concentration of chemotactic growth factors, cytokines, and chemokines, as well as by the orientation of the fibrils inside the extracellular matrix and provisional matrix. Fibroblasts usually migrate along these fibrils rather than across them. Fibroblasts produce proteolytic enzymes locally to facilitate movement through the matrix. The fibroblasts generate three different varieties of MMP: stromelysin (MMP-3), which has several protein substrates in the extracellular matrix; collagenase (MMP-1); and gelatinases (MMP-2 and MMP-9) that break down gelatin substrates [71].

5.2.3.5.2 Collagen and Extracellular Matrix Production

Collagen, proteoglycans, and other components of granulation tissue are mostly synthesized and deposited by fibroblasts. PDGF and TGF-β are two of the most important growth factors that regulate fibroblast activity. The primary sources of PDGF, a growth factor that stimulates fibroblast proliferation, chemotaxis, and collagenase production, are platelets and macrophages. Both macrophages and platelets emit TGF-β, which is known to be the main regulatory signal that controls the extracellular matrix's deposition. Through its ability to stimulate the transcription of genes linked to fibronectin, collagen, and proteoglycans, TGF-β increases the total amount of matrix protein generated. Simultaneously, TGF-β inhibits the release of proteases that break down the matrix and promotes the creation of tissue inhibitors of metalloproteinases (TIMP) to further prevent matrix breakdown. Recent research suggests that many of the effects of TGF-β on the formation of the extracellular matrix are mediated by a novel growth factor called connective tissue growth factor (CTGF) [78].

After migrating into the matrix, the fibroblasts undergo another morphological transformation, settle down, start to multiply, and start producing collagen, elastin, and proteoglycans that make up granulation tissue. Collagen is produced when fibroblasts adhere to the cables of the temporary fibrin matrix. Thus far, at least 20 distinct varieties of collagen have been found. High amounts of type III collagen, as well as other extracellular matrix proteins and proteoglycans, are first generated. The collagen mRNA is connected to polyribosomes on the endoplasmic reticulum, where new collagen chains are made, following transcription and processing. One crucial stage in this process is the hydroxylation of proline and lysine residues. The process of glycosylation involves additional modifications to the developing protein chains that come together to form the unique triple helical structure of the fibrillar collagen molecule. Hydroxyproline is a vital component of collagen because it plays a major role in preserving the triple helix form of collagen molecules. Collagen that has reached complete hydroxylation has a higher melting point. Scurvy, which is a vitamin C deficiency that results in low hydroxyproline levels, causes the collagen triple helix to shift shape and denature (unwind) much more quickly at lower temperatures. In order to promote the best possible wound healing, wound care specialists should ensure that patients are receiving enough nutritional support through a diet rich in protein and vitamin C [71].

Ultimately, the short, nonhelical regions at the N and C termini of procollagen molecules are cleaved by proteases as they are released into the extracellular environment for additional processing. Collagen fibrils are created when the collagen molecules spontaneously assemble into bigger bundles known as collagen fibers. These bundles are formed by the collagen molecules arranging themselves side by side and head to tail. Lysyl oxidase, a significant enzyme, interacts with collagen molecules in the extracellular spaces to create persistent, covalent cross-links. An increasing number of these intra- and intermolecular cross-links are incorporated into the collagen molecules as the molecules age and mature. Collagen gains its strength and stability from this crucial cross-linking process; the older the collagen, the more cross-links have been formed [71]. Dermal collagen possesses tensile strength per weight that is on par with steel's. In healthy tissue, it is a potent and well-organized molecule. Conversely, collagen fibers growing in scar tissue have a random form and are much smaller. Scar tissue will always burst first because it is typically weaker than the regular tissue around it.

5.2.3.6 Granulation

During the remodeling stage of healing, granulation tissue serves as a temporary substitute for the regular dermis, eventually forming a scar. It differs from unwounded dermis due to its extremely dense network of blood vessels and capillaries, high fibroblast and macrophage cellular density, and amorphously structured collagen fibers [71]. Moreover, its metabolic rate is higher than that of the normal dermis, indicating that it is more active than the dermis during cellular migration, division, and protein synthesis.

5.2.3.7 Epithelialization

Three fundamental processes underlie the healing of all cutaneous wounds: contraction, the deposition of connective tissue matrix, and epithelialization. Open wounds heal through contraction; tissue moves toward the center of the wound as a result of cell–matrix contact. As was previously mentioned, the process of forming a new extracellular matrix involves the deposition of collagen, proteoglycans, and attachment proteins. The process known as "epithelialization" occurs when epithelial cells in the skin's remaining appendages, such as hair follicles and sebaceous glands, lose their ability to prohibit contact and start to move into the wound region through the process of epiboly. The basal layer's cells start to multiply as migration moves forward, supplying more epithelial cells.

The detachment of epithelial cells and alterations to their internal structure, as well as migration, proliferation, and differentiation, are among the stages involved in the process of epithelialization [79]. The cuboidal basal keratinocytes, which are closest to the dermis, and the flattened, hexagonal, tough keratinocytes in the top layer are two of the five layers of differentiated epithelial cells that make up the entire adult epidermis. Basal epithelial cells are the only cells that have the ability to multiply. These basal cells are connected to the cells around them by desmosomes, which are intercellular connectors, and to the basement membrane via hemidesmosomes, which are attached to the basement membrane. Growth factors that attach to these epithelial cells' receptors throughout the healing process, such as TGF-α, keratinocyte growth factor, and epidermal growth factor (EGF), encourage migration and proliferation. Desmosomes and hemidesmosomes disintegrate in response to growth factor binding, allowing cells to split apart in order to migrate. Subsequently, the expression of integrin receptors causes the typically cuboidal basal epithelial cells to flatten and migrate as a monolayer across the recently deposited granulation tissue, adhering to the collagen fibers. The moving monolayer apron of cells receives new cells from the proliferation of basal epithelial cells close to the wound margin (actively migrating cells cannot proliferate). Proteolytic enzymes (MMPs) are released and generated by epithelial cells in close proximity to the monolayer's leading edge. These enzymes enable the cells to penetrate eschar, surface necrosis, and scabs. Migration will continue until the epithelial cells meet up with other migrating cells to form a confluent sheet. The entire epithelial monolayer enters a proliferative state after this point of contact, and the stratified layers of the epidermis are reestablished and begin to develop to restore barrier function. TGF-β is one growth factor that may speed up the maturation of the epidermal layers, causing them to differentiate and keratinize more quickly. Reestablished are the hemidesmosome attachments to the newly created basement membrane as well as the intercellular desmosomes. Although epithelialization is the clinical hallmark of healing, the granulation tissue still needs to rebuild after this process is complete [71].

5.2.3.8 Remodeling

The granulation tissue becomes a scar and gains tensile strength during the final stage of the healing process, called remodeling. The amount of glycosaminoglycans (GAGs) and the water that binds to them decreases as granulation tissue ages, and it also undergoes a decrease in capillaries due to aggregation into larger vessels. The granulation tissue's metabolic activity and cell density decrease with maturation. In order to increase tensile strength, collagen also changes in type, quantity, and organization. Type I collagen, which is the main fibrillar collagen in skin, supersedes high levels of type III collagen. The tensile strength of normal tissue is only around 25% present in a freshly epithelialized incision [71]. Normal, uninjured tissue is always stronger than tissue that has been repaired or healed. The reorganization of randomly deposited collagen fibers during granulation and the improvement of covalent cross-linking of collagen molecules by the enzyme lysyl oxidase, which is secreted into the ECM by fibroblasts, are the main mechanisms that increase tissue tensile strength. The tensile strength of the repaired tissue will gradually improve over several months or longer due to changes in the structure of collagen, reaching a maximum of approximately 80% of normal tissue.

Vital growth factors, chemokines, and cytokines regulate cellular activity at different phases of wound healing. The way that cells interact with the ECM through their integrin receptors and adhesion molecules influences how active the cells are. Fibroblasts, vascular endothelial cells, and epidermal cells produce MMPs that help in cell migration, whereas neutrophils and macrophages release proteolytic enzymes that remove denatured ECM components and aid in the remodeling of early scar tissue [71].

5.2.4 Uses for Dressings for Wounds

A pad or other piece of material is put on a wound to promote healing and provide protection from further harm. This is referred to as a compress or dressing [80]. In contrast to a bandage, which is usually used to keep a dressing in place, a dressing is meant to make direct contact with the tissue. A dressing can have a number of purposes, all aimed at promoting healing and providing protection from further harm, depending on the type, size, and location of the wound. Important goals for a dressing are to stop bleeding to hasten the clotting process by helping to seal the wound; prevent infection and to shield the wound from bacteria and external trauma; absorb exudate, blood, plasma, and other fluids that have leaked from the wound in order to keep them contained and avoid maceration; reduce pain through compression, medication-assisted analgesia, or just by shielding the area from additional impact; and decrease psychological stress by hiding a healing wound from the patient's and other people's eyes. The ultimate goal of a dressing is to aid in wound healing by offering a moist, sterile, and breathable environment that encourages granulation and epithelialization. Consequently, there will be a lower chance of infection, faster wound healing, and less scarring [81].

Contemporary dressings include hydrocolloids, hydrogels, plastic films, gels, foams, dry or impregnated gauze, and alginates. They provide a variety of physical settings that are suitable for different kinds of wounds: gas exchange and permeability, especially with regard to oxygen and water vapor; maintaining the proper temperature to encourage healing; and eliminating slough from a wound by mechanical debridement. For example, film dressings are unable to absorb exudate, hydrocolloids maintain the wound moisture, while dry gauzes absorb exudate, drying the wound.

Pressure dressings are frequently applied following skin grafts and to treat burns. They exert pressure and keep the tissue from being saturated with fluid [81]. By impregnating topical antiseptic compounds, dressings can also control the chemical environment of a lesion, typically with the goal of preventing infection. Antiseptics that are frequently used include castor oil, povidone-iodine, and dressings made of boracic lint [82]. To stop bacterial infections, bandages are frequently used in conjunction with antibiotics. Another antiseptic is medical-grade honey; there is some evidence that honey dressings work better for treating infected postoperative wounds than gauze and regular antiseptic [83]. Bioelectric dressings have the potential to combat some bacteria that are resistant to antibiotics [84] and hasten the healing process [85].

Analgesics are frequently infused into dressings to lessen pain. The effectiveness of such topical drugs can be affected by the physical characteristics of a dressing. To accelerate their rate of skin absorption, occlusive dressings can be applied. These dressings are composed of materials that are resistant to moisture, such as plastic or latex. Typically, dressings are fastened with bandages or adhesive tape. Nowadays, many dressings are made in the form of an "island" with an adhesive backing that is ready to be applied right away; they are referred to as island dressings [85].

5.2.4.1 Passive Products

Often, these remedies are limited to treating clean, dry, superficial wounds with little or no exudate. They can also be used as secondary dressings, which are supplementary dressings used to absorb additional wound drainage or to keep the primary dressing in place. Examples include cotton wool, gauze, lint, and plasters or adhesive bandages. The main objective is to prevent the wound from becoming contaminated by bacteria. They serve as secondary dressings as well. The woven and nonwoven fibers of polyester, rayon, and cotton make up gauze dressings. Despite their ability to absorb wound drainage, gauze dressings require regular replacement. Excessive wound discharge might cause the gauze to adhere to the site and cause pain while trying to remove it. Bandages are made from materials like cotton wool, cellulose, or polyamide. Cotton bandages can be used as a backup dressing in case of venous ulcers, even though compression bandages provide efficient compression. Nonetheless, tulle gras dressing soaked with paraffin oil is advised for a clean, superficial wound [86].

5.2.4.2 Interactive Products

The term "interactive products" refers to a variety of products, including hydrogel, hydrocolloid, hydrofiber, alginate, and semipermeable film and foam dressings. They not only keep the wound from becoming contaminated by bacteria, but they also maintain moisture in the wound environment to aid in healing [86].

Semipermeable dressing for films is a polyurethane film that is transparent. It permits the passage of carbon dioxide, oxygen, and water vapor into and out of the dressing. It also has a role in autolytic debridement, which is a less painful method of removing dead tissue from wounds in the operating room than manual debridement. Because of its extreme flexibility and elasticity, it clings tightly to the skin. Because the dressing is transparent, it may be inspected without needing to be removed. Owing to its restricted absorption ability, this type of dressing is limited to superficial wounds with minimal drainage [86].

Dressing using semipermeable foam is composed of an exterior layer with adhesive boundaries that is hydrophobic (repelled by water) and foam with hydrophilic (attracted to water) qualities. The wound is shielded from external fluid contamination by the hydrophobic layer. The inner hydrophilic layer, meanwhile, has the capacity to take up a moderate quantity of wound discharge. As a result, this kind of dressing works well on wounds with granulation tissue and excessive discharge. It's not necessary to use secondary dressings. But it needs to be changed often, and it doesn't work well on dry wounds. One prevalent ingredient in foam is silicone. The foam has the ability to conform to the contours of the wound [86].

Polyvinyl pyrrolidine and methacrylate are examples of synthetic polymers used to make hydrogel dressing. Due to its high water content, the wound receives moisture and a cooling effect. It is simple to remove the dressing from the wound without causing any harm. Moreover, the dressing doesn't irritate. As a result, pressure ulcers, burn wounds, and dry necrotic wounds are all treated with it. Infected wounds and wounds with a lot of discharge should not be treated with it [86].

Hydrocolloid dressing is made up of an outer waterproof layer and an interior colloidal layer. Pectin, gelatin, and carboxymethylcellulose are among the gel-forming substances it includes. The dressing retains the wound discharge to create a gel that creates a moist environment for the healing of the wound. It digests necrotic tissues, absorbs wound discharge, and shields the wound from bacterial invasion. It is mostly used as a backup dressing. Neuropathic ulcers and wounds with excessive drainage, however, are not treated with it [86].

Alginate dressing is composed of alginic acid salts, either calcium or sodium. This dressing has a strong capacity to absorb wound drainage. Blood and dressing ions can combine to create a film that shields the wound from bacterial infection. Deep wounds with exposed bone, third-degree burn wounds, and dry wounds should not be treated with this treatment. Alginate dressing can cause wounds to dry up quickly, which is another reason why it needs to be replaced [86].

Consisting of sodium carboxymethyl cellulose, hydrofiber dressings have a high absorption capacity, which enables them to create a gel and shield the skin from maceration [87].

5.2.4.3 Bioactive Products

Biomedical improvements in the treatment of acute, chronic, and various types of wounds have been driven by advances in our understanding of wounds. Numerous biologics, skin replacements, biomembranes, and scaffolds have been created to promote wound healing via a variety of methods [88].

5.3 SMART HYDROGELS FOR WOUND HEALING

When used in the right conditions, hydrogels' dynamic equilibrium with water or other biological fluids permits the absorption and release of drugs. The hydrogel matrix's water content influences the polymeric structures' physical and chemical properties, as well as their biocompatibility, in the drug delivery industry. Since hydrogels reduce the possibility of inflammatory reactions from mechanical frictions with surrounding tissues during implant procedures, they are an intriguing tool in the field of regenerative medicine. These matrices can also be made into injectable solutions, which is a big advantage over conventional surgical techniques [89].

Hydrogels have attracted increasing attention in the medical field over the past few decades because they can be made to undergo discrete or continuous volume transformation in response to minuscule changes in environmental stimuli like pH, temperature, electric field, solvent substance, salt concentrations/ionic resilience, light/photon, pressure, and coupled magnetic or electric fields. These extraordinarily flexible and adaptable polymers, also known as "smart" or "stimuli-responsive" hydrogels, are able to detect changes in their environment and respond by changing their structure (a rise or fall in the intensity of swelling) without the need for an external driving force. This "volume-changing" phenomenon is especially helpful for applications involving medication delivery since these changes in the environment can cause the release of drugs.

For the creation of self-regulating drug delivery systems with improved therapeutic efficacy, smart hydrogels are excellent options. Since temperature and pH have biological and physiological significance, they are the most often used triggers to initiate the therapeutic activity of the hydrogel [90]. Three forces play a major role in hydrogels' capacity to expand in response to external environmental factors: (1) the interaction between the polymer and solvent; (2) the elasticity of the polymer; and (3) the ion osmotic force, also known as electrostatic forces [91]. As the solvent's molecules infiltrate the polymer network in the first instance, translational entropy increases as a result of the contact, causing swelling. The elastic chains that entwine the cross-links within the polymer network determine the second force, known as polymer elasticity. This is specifically influenced by the cross-link density increasing with temperature and the mean number of chains per cross-link. Although the initial models created by Flory and Rehner could forecast polymer swelling by considering these two forces, a third component necessitates extra adjustment [92]. The third force, also known as the ion osmotic force or electrostatic force, needs to be taken into account when there are ionized groups present.

Analogous to the Gibbs–Donnan effect, counterions have to be pulled into the hydrogel while coions are concurrently released for electroneutrality to be enforced everywhere except at the polymer interface [89].

5.3.1 Definition and Characteristics

Smart hydrogel wound dressings adapt to skin injuries and react to changes in the wound environment, such as biomaterials responding to stimuli and having the capacity to heal naturally [93]. By adding active microparticles, smart hydrogels can be incorporated with nanoparticles to enhance personalized functionality in drug delivery, which are adaptable to different types of healing conditions of individuals, helping to regulate new skin proliferation, diminished fibrosis, and modification of inflammation associated with tissue regeneration [94, 95]. Smart hydrogels are interlinked by using various physical or chemical techniques based on their characteristics, including benefits of flexible usage without assistance, good patient compliance, and reasonable manufacturing expenses. Additionally, smart spraying hydrogels improve the administration and absorption of therapeutic doses by permeating into the injured area of the wound [96]. Smart hydrogel has certain characteristics such as biodegradability, pH variation, biocompatibility, swelling, deswelling, hydrophilicity, super absorbency, viscoelasticity, etc. [97]. In contrast, these smart hydrogel polymers can respond to variations in pH, ionic strength, or temperature by exhibiting significant modifications in their properties during inflammation, the transformation of sol to gel, chemical structure, permeability, or mechanical strength unlike conventional hydrogel polymers [95, 98, 99].

5.3.2 Drug Delivery and Tissue Regeneration

Smart hydrogels have great potential to be used in drug delivery systems due to their rigidity, physicochemical properties, simplistic preparation methods, and biocompatibility. By preventing drugs from building up in nontarget tissues, smart hydrogels can be used in drug delivery systems to decrease side effects, maintain desired therapeutic benefits within the concentration of a single dose, and decrease the frequency of dosing [100]. Hydrogels are generated with adequate pores for effective tissue engineering applications, containing the cells in pores by liberation of the extracellular matrix and other growth factors to regenerate the tissue. Hydrogels are effective at the removal of the cellular byproducts as well as supplementing these byproducts with nutrients and growth factors. Conversely, hydrated smart hydrogels can imitate the chemical and physical conditions of the extracellular matrix of the body. The combination of nanostructure with injectable hydrogels associated with the extracellular matrix results in a well-defined physical integration [98, 101].

5.3.3 Stimuli-Responsive Hydrogels

Mechanism of the stimuli-response hydrogel includes volume change, phase transition, and structural change. There are certain procedures of hydrogel generation such as formation of primary, secondary bound water, and dissolution of interconnected mesh in which the design and control of stimuli-responsive hydrogels greatly depend on the interactions between gelato and gelato solvent. Moreover, the physicochemical characteristics of stimuli-responsive hydrogels are dependent upon the interconnected links and chemical behavior of the gelator, such as greater cross-linking degrees resulting in a smaller pore size, lower swelling ratio, and high mechanical strength in a single hydrogel mesh [102]. Multistimuli-responsive hydrogels improve the liberation of in situ drugs by releasing their constituents under a variety of conditions and have proven to be useful in chemotherapy [103]. Stimuli-responsive hydrogel can be adaptable (changes in solubility and matrix shape and the capacity to absorb water or physiological liquid) in response to external stimuli such as temperature, pH, or magnetic ions [104].

5.3.3.1 Temperature-Responsive Hydrogels

The presence of hydrophobic groups and hydrophilic chemical functional groups in the monomers defines thermoresponsive polymer hydrogels. Characteristics of temperature-mediated hydrogels include inverse temperature dependence and a nonlinear connection between temperature and size, which comprises retreating polymer solutions as the lower critical solution temperature gradually rises. These temperature-sensitive polymer hydrogels can be mixed with a wide range of polysaccharides such as polyethylene glycol, hyaluronic acid, chitosan, alginate, chondroitin sulfate, cellulose, and dextran [98, 105].

5.3.3.2 pH-Responsive Hydrogels

pH-sensitive drugs improve their drug liberation by the ionization of the functional groups in the polymeric mesh. pH-sensitive hydrogels with weak acidic or basic groups within the polymer mesh can swell or contract at a pH above or below the pKa, causing the mesh to absorb or lose water. Bacterial infections reacting to a particular endogenous stimulus can be treated with the release of bioactive materials from this hydrogel. It works because when bacteria have colonized a wound, the pH of healthy skin is higher than normal, which is generally between pH 4 and pH 6. Smart hydrogels incorporated with the bioactive material enhance the release of drugs in response to pH variations in the wound [106]. Borax hydrogel can be degraded and rebuilt in response to pH variations by utilizing the dynamic borate ester in the hydrogel mesh, accomplishing controlled drug release [107].

5.3.3.3 Other Stimuli-Responsive Hydrogels

5.3.3.3.1 Light-Responsive Hydrogels

The creation of light-responsive hydrogels primarily employs photochemical cleavage of the polymer backbones at interconnected link sites, photoisomerization by modification of the density of the interconnected mesh, charged condition of the gel, and addition of photothermal agents causing phase transition of temperature-responsive hydrogels and heat generation due to light exposure. Whereas photocleavage results in irreversible hydrogels, photoisomerization produces reversible hydrogels. Conversely, near-infrared-responsive hydrogels have deep tissue penetration and are safe [108].

5.3.3.3.2 Magnetic-Responsive Hydrogels

Magnetic-sensitive hydrogels are mixed with paramagnetic iron oxide nanoparticles. They enhance the effectiveness of therapy or chemotherapy via temperature regulating mechanisms such as having the capacity to significantly raise the localized temperature due to exposure to the magnetic field, providing deep tissue penetration, and decreasing the incidence of invasive methods [108].

5.3.3.3.3 Ionic-Strength-Sensitive Hydrogels

Ionic-strength-sensitive hydrogels are hydrogels that exhibit alterations in their functioning when exposed to cations of calcium, sodium, and potassium, as well as zwitterionic polymers. They can be used to treat cancer by, for example, adding doxorubicin poly hydrogels, which provide ionic strength sensitivity by increasing the hydrogel's swelling ratio in proportion to an increase in the surrounding medium's ionic strength. Hydrogel swelling can be promoted and electrostatic interactions can be inhibited by amide, sodium, and chloride ions [108].

5.3.3.3.4 Electro-Responsive Hydrogels

Activation of electro-responsive polymers occurs due to contact with external electric fields. Electro-responsive hydrogel contains multifunctional properties which are safe, reversible, induce rapid recovery, and flexible to manipulate for incorporation of other biomaterials. The two main components of electro-responsive hydrogel are electro-responsive particles and a non-electric hydrogel polymer such as polylactic acid, silicones, and polyurethanes. Soft elastomeric polymers are commonly used because of their greater denaturation ability, biocompatibility, and mechanical and thermal stabilities [109].

5.3.4 Role in Enhancing Wound Healing

Smart hydrogels have the ability to interact with the surrounding tissue, reflect changes that occur during the healing process, manage infection, prevent the progression of chronic wounds, and promote self-healing [110]. Because hydrogels are biodegradable and biocompatible, they can be used as a temporary template while chronic wounds remodel and re-epithelialize. Hydrogels also exhibit an adequate bioadhesive quality, which is essential for maintaining stability over time. This characteristic promotes hemostasis and keeps the wound's moisture content at its ideal level. To further improve their overall effectiveness in promoting wound healing, hydrogels provide a flexible framework for incorporating different components like medications, antibacterial and antimicrobial agents, and other supplementary biomaterials demonstrating greater suitability for wound dressings [111].

The process of creating biodegradable hydrogels usually involves using heat to cause the chemical structure of protein-based polymers, such as collagen and gelatin, to join together. These biomaterials' intrinsic cell recognition molecules are useful for cell adhesion and regulated enzymatic biodegradation during tissue remodeling and cell growth. Natural hydrogels based on polysaccharides, like hyaluronic acid, alginate, dextran, and chitosan, are widely used as biodegradable materials in the treatment of chronic wounds, promoting granulation, migration, and the development of new blood vessels, as well as proper degradation behavior during [112].

Bioadhesive hydrogels can provide stability and help dressings adhere to a wet wounded area. Increased stretchability and flexibility of these hydrogels provide good patient compliance by providing comfort to the wounds. For shallow and small chronic wounds, highly adhesive hydrogel dressings such as gallic acid and chitosan were found to be helpful, whereas less adhesion was preferred for large and deep chronic wounds, improving antibacterial, antioxidant activities, and bioadhesion capacity [112].

Antibacterial agents such as penicillin, gentamicin, cephalosporin, and vancomycin, are incorporated with hydrogels to create antibacterial hydrogels for treating chronic wounds by preventing the synthesis of bacterial protein. Antifungal hydrogels are generated by polymeric immobilization of antifungal medications. These antimicrobial substances work by preventing the synthesis of bacterial protein. To avoid drug resistance, the antibacterial agents must be strictly released from the hydrogel gradually by modifying the swelling, degradation, or rigidity characteristics of the hydrogel and by adding reactive chemical functional groups to the hydrogels that change due to variations in temperature and pH [112].

5.4 UTILIZING 3D PRINTING TECHNOLOGY TO FABRICATE HYDROGELS

The process of creating hydrogel formulations for 3D printing requires fine-tuning a number of factors to guarantee appropriate printability, structural integrity, and printing method compatibility. Hydrogel formulation design for 3D printing requires a careful balance of rheological characteristics, gelation kinetics, and print fidelity [113]. Gelation kinetics affect the printing process's temporal features, whereas rheological properties control the hydrogel's flow behavior during printing. Achieving the best print fidelity, especially in biomedical applications, necessitates carefully balancing these features to guarantee precise layer deposition and overall effectiveness of 3D-printed hydrogel constructions. Additionally, additives are essential in modifying the properties of hydrogels to suit particular application needs, allowing structures with optimal mechanical, biological, and functional qualities to be created. It is important to remember that the study of material flow and deformation is known as rheology. Viscosity and shear-thinning behavior are important rheological characteristics in the context of hydrogels for 3D printing. While shear-thinning action enables the hydrogel to reduce viscosity under shear stress during printing, promoting smooth flow and layer deposition, the right viscosity guarantees proper extrusion via the printer nozzle. The relationship between printing behavior and rheological qualities has been clarified by a number of researchers. This correlation was investigated by Bom et al. [114] who emphasized the significance of comprehending rheological parameters in order to get good 3D printing results. Amorim et al.'s research [113] contributed to our basic knowledge of the behavior of printing materials under shear stress by shedding light on the shear rheology of inks used in extrusion-based 3D bioprinting. In a similar vein, Kokol et al. [115] examined the effects of flow- and horizontally-induced chilling rates on gelatine hydrogel rheological properties during 3D cryo-printing. In order to improve the rheological characteristics of alginate hydrogels for extrusion-based bioprinting applications, Kim et al. [116] concentrated on adding carrageenan to the mixture. Townsend et al. [117] highlighted the flow behavior prior to cross-linking and the importance of precursor rheology for hydrogel insertion in medical applications and 3D bioprinting. Zhou et al.'s [118] introduction of gelatin methacryloyl's microbial transglutaminase-induced regulated cross-linking demonstrated how to modify rheological characteristics for optimal 3D printing.

On the other hand, the process through which a liquid hydrogel transforms into a gel and solidifies the material is known as gelation. Gelation kinetics is the study of the transition time of a hydrogel from a liquid to a solid state in 3D printing. The gelation speed is a significant factor that influences both the printing speed and the overall printing time. Ideally, the printing speed and the gelation kinetics should match for proper layer-by-layer deposition. Furthermore, print fidelity refers to how well the printed structure matches the digital layout in terms of precision and correctness. The rheological characteristics of the hydrogel, the gelation kinetics, and the overall printing parameters are some of the elements that affect print fidelity. Sheikhi et al.'s [119] study concentrated on the 3D printing of self-supporting jammed microgels and offered a different method that takes conductivity, cross-linking, and shape fidelity into account.

5.4.1 Synopsis of 3D Printing

Additive manufacturing, or 3D printing, has become a cutting-edge technology with applications in a wide range of fields, including manufacturing, aerospace, healthcare, and the arts. This summary explores the principles, capabilities, and uses of the wide range of 3D printing technology.

5.4.1.1 Stereolithography (SLA)

Hydrogel-based stereolithography (SLA) 3D printing is a state-of-the-art additive manufacturing method that uses light to produce complex three-dimensional structures out of hydrogel materials. The foundations of conventional stereolithography are expanded upon by this technology, which modifies them to take into account the special characteristics of hydrogels, which are generally made up of networks of water-based polymers. A hydrogel resin that might include cross-linkers, photo-initiators, and other additives is the first step in the process. Using a carefully calibrated light source, usually a laser or projector, this technique selectively solidifies certain areas of the hydrogel resin. The build platform descends gradually as each layer solidifies, enabling the 3D item to be constructed bit by bit. Because stereolithography is so precise and finely detailed, it is a good fit for applications in microfluidic systems, biomedical devices, and tissue engineering. To improve structural stability, extra processes like rinsing and curing might be needed after printing. Numerous studies have looked into the use of stereolithography in the field of 3D printing to produce hydrogels with special qualities and abilities [120]. Kalossaka et al. [121] investigated the synthesis of three-dimensional nanocomposite hydrogels with lattice vascular networks, indicating the possibility of fabricating complex structures with this method. Karakurt et al.'s study [122] concentrated on the controlled release features of ascorbic acid-loaded hydrogels printed using SLA 3D printing. The paper clarifies how SLA is used in medication delivery systems.

It has also been described how SLA 3D printing affects the features of PEGDMA hydrogels, offering insights into how the printing process and hydrogel properties interact [123]. Magalhães et al. [124] investigated the creation of 3D hydrogel structures using inexpensive stereolithography technology, which may have uses in biomaterials and biomedicine. Alketbi et al. [125] reported on the effects of PEGDA photopolymerization in micro-stereolithography on 3D-printed hydrogel structures, addressing elements of structural integrity and swelling behavior. More adaptable and useful hydrogel applications are made possible by Sun et al.'s [126] novel stereolithographic 3D printing technique for hydrogels, which focuses on obtaining large mechanical tunability and self-weldability.

5.4.1.2 Digital Light Processing (DLP)

Digital light processing (DLP)-based 3D printing for hydrogels is a novel and accurate additive manufacturing process that uses hydrogel materials to create complex 3D objects. This process expands on the ideas of photopolymerization, which involves using a digital light source—typically a projector or UV LED—to selectively cure a liquid resin that contains hydrogel components. This procedure creates the desired object layer by layer, with each layer being precisely projected with light patterns to solidify it. Hydrogel DLP 3D printing is praised for its exceptional speed, excellent resolution, and capacity to produce intricate geometries with exquisite detail. It is especially useful in biological applications, such as tissue engineering, implant customization, and drug delivery systems.

New opportunities and uses have arisen as a result of recent developments in hydrogel DLP 3D printing. Numerous research teams have significantly advanced this developing topic. Hosseinabadi et al. [127] examined the crucial elements of optical design and ink material selection in DLP 3D printing, emphasizing the significance of accuracy and efficiency in hydrogel printing. Ding et al. [128] have provided a succinct assessment of the current status of DLP 3D printing for hydrogels, including the major advancements and future potential of this technology. On the other hand, Dong et al. [129] concentrated on employing DLP 3D printing to produce robust supramolecular hydrogels, highlighting their possible utility as impact-absorption components. Cafiso et al.'s [130] investigation into the 3D printing of hydrogels made entirely of cellulose using DLP demonstrated the adaptability and sustainability of this method. In the biomedical field, Wang et al. [131] effectively created antibacterial hydrogels by DLP 3D printing, utilizing hybrid nanospheres and sustainable resin for possible uses in medical applications.

This system's clever design created a sequential semi-interpenetrated network by adding chemical covalent network precursors to a solution containing a linear polymer. This linear polymer is essentially locked inside the cross-linked matrix following polymerization. Poly(ethylene glycol) diacrylate (PEGDA), acrylic acid (AAc), aqueous solutions of unaltered, non-cross-linked polyvinyl alcohol (PVA), and a water-compatible photoinitiator constructed from diphenyl (2,4,6-trimethylbenzoyl)phosphine oxide (TPO) were combined to create the photocurable ink. It is noteworthy that in the case of DLP-based 3D printing technology, post-processing steps like further curing and rinsing may be required to enhance structural stability following printing. It is important to keep in mind that material selection and compatibility with the photopolymerization process are important considerations [132]. Despite these challenges, DLP-based 3D printing remains a powerful method for creating complex hydrogel structures that push the boundaries of precision and complexity in a variety of fields, including science and medicine.

5.4.2 Hydrogel Printing Methodologies

Hydrogel-based 3D printing has revolutionized biomedical engineering by offering precise control over the fabrication of complex structures with tailored properties. The methodologies covered include stereolithography, digital light processing, extrusion-based printing, inkjet printing, and bioprinting.

5.4.2.1 Stereolithography (SLA)

SLA is a 3D printing technique that utilizes photopolymerization to create objects layer by layer [132]. The process involves the following steps:

Preparation of photopolymer resin: A photosensitive resin is prepared by mixing a photoinitiator with a liquid monomer or pre-polymer. The formulation can be tailored to achieve the desired mechanical properties and biocompatibility.

Layer-by-layer printing: The printing platform is lowered into the resin bath, and a UV laser selectively cures the resin according to the computer-aided design (CAD) model. This process solidifies the resin, forming one layer of the desired object.

Object retrieval and post-processing: Once the printing is complete, the object is removed from the resin bath. It may undergo post-processing steps such as rinsing in a solvent to remove excess resin and curing under UV light to improve mechanical strength.

Recent advancements in SLA have focused on improving print speed, resolution, and material options. For example, researchers have developed biocompatible resins such as GelMA (gelatin methacryloyl) and PEGDA, which offer excellent cell viability and support tissue growth [133].

5.4.2.2 Digital Light Processing (DLP)

DLP is a similar technique to SLA but uses a digital light projector to cure entire layers of resin simultaneously [132]. The process involves the following steps:

Preparation of resin: Similar to SLA, a photosensitive resin is prepared with a photoinitiator and monomer or pre-polymer.

Layer-by-layer printing: The resin is spread in a thin layer onto the build platform, and the entire layer is exposed to UV light from the digital light projector. The UV light selectively cures the resin, solidifying the desired structures.

Object retrieval and post-processing: After printing, the object is removed from the resin and may undergo post-processing steps such as rinsing and curing.

Recent advancements in DLP have focused on improving material properties and process reliability. Novel photoinitiators and resin formulations have been developed to enhance print accuracy and mechanical strength. Additionally, DLP systems with multiple light sources enable the printing of multimaterial structures with precise control over material composition [134].

5.4.3 Utilization in the Healing of Wounds

Hydrogels are appropriate for applications involving wound healing because of a number of their characteristics. A significant water content, biocompatibility, adjustable mechanical characteristics, and the capacity to encapsulate bioactive substances like medications and growth hormones are a few of these [123]. The ECM found in tissues is modeled by the high water content of hydrogels, which creates a moist environment that encourages cell migration, proliferation, and differentiation. Hydrogels can also be engineered to mimic the mechanical characteristics of genuine tissues, which is essential for encouraging tissue integration and cell adhesion.

Numerous hydrogel varieties, including those made of natural polymers like collagen, alginate, and gelatin, as well as synthetic polymers like polyethylene glycol and polyvinyl alcohol, have been studied for use in wound healing applications. Although natural polymers may have low mechanical strength and stability, they offer great biocompatibility and bioactivity. Synthetic polymers, on the other hand, can be engineered to have precise mechanical and degradation properties but may lack biological cues for cell adhesion and proliferation [131]. By combining natural and synthetic polymers, researchers can develop hybrid hydrogels that combine the advantages of both types.

5.4.4 Scaffold Design Customization and Precision

Scaffold design is a crucial aspect of construction and maintenance projects, providing temporary structures to support workers and materials during building, renovation, or repair work. The ability to create customized scaffolds is a defining feature of 3D-printed hydrogels for wound healing. Through CAD software, healthcare professionals can generate digital models of a wound based on imaging techniques like digital photography or 3D scanning [133]. This digital representation allows for the design of a hydrogel scaffold that perfectly matches the wound's unique shape, size, and depth. The benefits of such customization include:

Optimal wound coverage: A custom-designed scaffold ensures complete coverage of the wound bed, preventing the formation of dead space or pockets that could harbor bacteria.

Reduced pressure on periwound tissue: By precisely conforming to the wound topography, the scaffold minimizes pressure on the surrounding healthy tissue, promoting blood flow and accelerating healing.

Improved patient comfort: A well-fitting scaffold provides superior patient comfort, minimizing pain and discomfort associated with movement.

Beyond altering the hydrogel's shape and size, 3D printing enables the construction of complex structures inside the material. With this degree of accuracy, scaffolds that closely resemble the ECM's intricate network of proteins and polysaccharides—which gives cells structural and metabolic support—can be engineered [134]. Here's how thoughtful scaffold design might improve the healing of wounds:

Enhanced cell adhesion and migration: By incorporating specific biomolecules or cell adhesion peptides into the hydrogel ink, researchers can design scaffolds that promote cell attachment and migration, crucial steps in wound closure [135].

Controlled drug delivery: 3D printing allows for the precise localization of therapeutic agents within the scaffold. This enables the controlled release of drugs directly at the wound site, maximizing their efficacy and minimizing systemic side effects [136].

Directed angiogenesis: Wound healing necessitates the formation of new blood vessels (angiogenesis) to deliver oxygen and nutrients to the healing tissue. By incorporating pro-angiogenic factors or creating microchannels within the scaffold, researchers can promote directed blood vessel growth, accelerating healing [137].

5.4.5 Benefits of Hydrogel 3D Printing

Hydrogel 3D printing allows for the creation of highly customized wound dressings tailored to the specific needs of each patient [138]. Healthcare professionals can make dressings that perfectly fit the size and shape of the wound, improving wound coverage and contact with the injured tissue. This is accomplished by carefully regulating the hydrogel's composition and design. The capacity to directly insert therapeutic chemicals, such as growth factors, antibacterial agents, and anti-inflammatory medications, into the dressing is one of the main benefits of hydrogel 3D printing. These medicinal substances can be gradually delivered under regulated conditions to treat the wound site over time and hasten its recovery. Because hydrogels are very water-containing, they aid in keeping the wound site wet [139]. This moisture retention is essential for promoting cell migration, proliferation, and tissue regeneration. Hydrogel dressings created through 3D printing can be designed to optimize moisture retention, ensuring an ideal environment for wound healing. Hydrogels are often composed of biocompatible materials, making them well-suited for use in wound healing applications. Hydrogel dressings fabricated using 3D printing techniques can be further optimized for biocompatibility, minimizing the risk of adverse reactions and promoting tissue integration. Hydrogel 3D printing allows for the fabrication of porous structures that serve as scaffolds for cell growth and tissue regeneration. These porous structures provide a supportive environment for cells to adhere, proliferate, and differentiate, leading to faster wound closure and tissue repair. Hydrogel dressings created through 3D printing can be designed with specific surface properties to enhance adherence to the wound bed [140]. This ensures that the dressing stays in place, even in challenging anatomical locations, promoting uninterrupted contact between the dressing and the wound site. Hydrogel dressings are typically soft, flexible, and easy to handle, making them well-suited for wound care applications. 3D printing allows for the fabrication of dressings with tailored shapes and sizes, facilitating easy application and conforming to the contours of the wound site. Additionally, hydrogel dressings can be easily removed without causing trauma to the healing tissue. Some hydrogels exhibit hemostatic properties, meaning they can help control bleeding by promoting blood clotting. By incorporating hemostatic agents into hydrogel dressings [141], 3D printing can create wound dressings that not only facilitate healing but also prevent excessive bleeding, particularly in cases of traumatic injuries or surgical wounds. Hydrogel dressings can be designed to contain antimicrobial agents, such as silver nanoparticles or antimicrobial peptides, which help prevent infection at the wound site. By releasing these agents in a controlled manner, hydrogel dressings created through 3D printing can effectively reduce the risk of infection and promote a sterile wound environment conducive to healing. Advances in 3D printing technology have made it possible to produce hydrogel dressings in a scalable and cost-effective manner. Mass production of customized wound dressings using 3D printing can significantly reduce manufacturing costs and shorten production lead times, making advanced wound care more accessible to patients [142]. Hydrogel 3D printing offers a high degree of versatility in terms of material selection, design complexity, and functionality. This versatility allows for the creation of multifunctional wound dressings capable of addressing various aspects of wound healing, such as inflammation, angiogenesis, and scar reduction [143, 144]. With the advancement of telemedicine and remote monitoring technologies, hydrogel dressings created through 3D printing could enable more effective wound management in remote or underserved areas. Patients can receive customized dressings tailored to their needs without the need for frequent visits to healthcare facilities, improving access to care and reducing healthcare costs [8].

5.5 CURRENT DEVELOPMENTS AND NOVELTIES

Although the usage of hydrogels for 3D printing has advanced recently in the biomedical field, there are still some obstacles that need to be removed to maximize their potential. Enhancing cell viability and function, reducing structural complexity, and enhancing resolution are some of these problems. Other issues include addressing ethical and legal issues related to therapeutic translation.

5.5.1 Incorporating Bioactive Agents

Different bioactive materials, which are biologically stimulated and safely processed, are incorporated to be made into hydrogel bioinks [145]. Hydrogel bioinks are necessary for the successful construction of a skin tissue model. In order to satisfy the requirements of biomimicry, bioinks need to possess specific attributes such as biofunctionality, mechanical stability, adjustable rheological properties, increased shape fidelity after printing, and biocompatibility to promote cell growth [146]. Hydrogel bioinks, including 2-hydroxyethyl methacrylate, sodium alginate, polyethylene glycol dimethacrylate, resins, chitosan, gelatin, and carboxymethyl cellulose, are made by incorporating synthetic polymers. Natural polymers are a combination of various polymers that are incorporated to create hydrogel inks. Polymers and additional bioactive materials can be modified via different mechanisms to promote swift healing and reduce skin irritation [147].

Alginate has been enhanced by combining it with different organic and inorganic polymers and nanoparticles [148]. For instance, alginate hydrogel layer incorporated onto a poly lactic-co-glycolic acid membrane has proved to create a human skin imitating bilayer membrane, promoting cell adhesion and proliferation while resisting bacterial invasion and preserving the inner moisture content. According to in vivo applications, it has increased generation of new blood vessels and the accumulation of collagen I and III, which further promotes the renewal of skin layers [149]. Moreover, hydrogel consisting of alginate embedded with isolated bacteriophage has proved to have antibacterial properties by effectively suppressing the bacterial growth of the targeted *E. coli* gene and gradually releasing lytic phages in response to the emergence of drug-resistant bacteria. A 3D-printed hydrogel scaffold made of gelatin and dopamine-modified alginate has been cross-linked with calcium or copper ions in order to treat infectious bacterial wounds. In this case, copper ions have provided a photothermal effect in addition to a sustained release from the hydrogel, which enhanced its antibacterial activity against *S. aureus* and *E. coli* and promoted angiogenesis and wound healing [150].

Since gelatin has poor mechanical qualities, it is combined with other polymers despite being a very beneficial bioactive substance. One interesting option for creating 3D-printed wound dressings is gelatin. A sticky peptide sequence that promotes cell attachment, matrix metalloproteinase cleavages, and the flexibility of gelatin as a result of methacrylation in gelatin methacrylate are the causes of gelatin breakdown [151].

Chitosan–chondroitin sulfate has been studied to be incorporated into a usable hydrogel scaffold form with additional materials such as incorporating heteropolysaccharide pectin with chitosan due to its certain significant properties such as signaling of growth factor, biocompatibility, biodegradability, lack of toxicity, and antibacterial activity. These hydrogels have proved to be a potential wound healing dressing by demonstrating compression resistibility, a high degree of swelling, and ability to create in situ cross-links [152].

With their resemblance to extracellular matrix components, cellulose derivatives have been found to be promising for the creation of hydrogel scaffolds because of their biocompatibility, non-cytotoxicity, porous microstructure, preferred mechanical properties, and tunable architecture. Moreover, it has been demonstrated that the more pliable nanocellulose derived from wood is biocompatible and supports crucial cellular functions in different cultivated cell lines. The cellulose derivative carboxymethyl cellulose has been applied as a hydrogel dressing in wound healing applications due to its improved mechanical properties, biocompatibility, and increased stability.

Carboxymethyl cellulose has also been incorporated with a cationic polypeptide called ε-polylysine to create hydrogel dressings with porous structures for large irregular wounds, exhibiting antibacterial and antioxidant properties as well as poor mechanical properties. Cellulose nanofibrils could enhance 3D printability and the perseverance of shape integrity. Alginate hydrogels have been reported to be reinforced by cellulose nanocrystals and utilized as an ingredient in hydrogel inks [153].

Poly(N-isopropylacrylamide) is a superporous hydrogel particularly appealing for the management of chronic wounds due to its ability to gel at body temperature because of its low critical solution temperature, increased porosity, and cell adhesion. Moreover, incorporating poly(N-isopropylacrylamide) with hybrid interconnected networks has shown effective wound closure due to its administration as injectable materials [154]. These hydrogels are created when ethylene monomers react with hydroxyl and carboxyl groups in polymers and natural polysaccharides [155]. Inhibiting the growth of scar tissue and accelerating the healing of infected wounds are two benefits of using 3D-printed nanocomposite superporous hydrogels made of polyacrylamide and hydroxypropyl methylcellulose cross-linked with Ag NPs. A non-cytotoxic hydrogel that is resistant to gram-positive as well as gram-negative bacteria and fungi has been created by combining sodium alginate, poly(N-isopropylacrylamide), methylcellulose, and 2-phenoxyethanol [155]. Utilized in the treatment of diabetic foot ulcers, platelet-rich plasma gel can accelerate cell proliferation and the migration of wound tissues, which is also enriched with various vascular, endothelial, and epithelial growth factors [154, 155].

5.5.2 Therapies in Combination

Hydrogels are ideal replacements for cartilage defects as they can mimic the activity of the extracellular matrix of articular cartilage and contain mechanical characteristics, swelling capacity, and lubricating behavior. Hydrogels can foster cell differentiation by enclosing stem cells and containing protein-associated growth factors. Hydrogel scaffolds have demonstrated effective results for repairing damaged bone and cartilage [99]. By mixing therapeutic agents with chitosan-based hydrogels, a range of therapeutic approaches can be accomplished, including gene therapy, chemotherapy, and combined therapy. Also, chitosan-based hydrogels can be used as nanocarriers to load drugs, cells, proteins, genes, and other materials. They can also be applied in certain new technologies [156]. By applying a uniform layer of polycaprolactone to the 3D-printed alginate–gelatin hydrogel scaffolds, a core fiber scaffold was created. The free diffusion of medications from the core gels may be lessened by the PCL coatings. Afterward, the Gel/PCL core/shell scaffolds were coated with polydopamine (PDA), which greatly enhanced the photothermal effects of the scaffolds. This system's on-demand drug release, enabled by a near-infrared (NIR) laser, was made possible by the core gels' thermally induced sol-gel transition. Photothermal therapy combined with the released drugs (doxorubicin) could effectively prevent or ablate tumors both in vivo and in vitro. Due to the possibility of chemo-photothermal synergistic cancer therapy between the released drugs and the NIR-induced photothermal effect, NIR-triggered drug release hydrogel is one of the most commonly used systems for cancer therapy [157]. When the local pH of an infarcted area is low, thermoresponsive and pH-sensitive hydrogels are frequently employed for cardiac therapies. The responsive polymeric materials exhibit a local response upon stimulation, which may serve as a catalyst for intracellular drug delivery intended for tissue targeting. Furthermore, 3D chemically or physically reversible networks that can restore their original morphology after damage are known as self-healing hydrogels. Dynamic connections control the processes of dissociation and recombination, giving rise to a self-healing characteristic akin to human tissue regeneration. Gallium maltolate, a novel antimicrobial, is loaded into 3D-printed hydrogel wound dressings to guard against chronic wound infection [158]. Y. Hu et al. created a 3D scaffold dressing (SIS/MBG@Exos) that could release bioactive exosomes continuously by building decellularized small intestinal submucosa (SIS) integrated with a mesoporous bioactive glass (MBG) and exosomes using an extrusion-based cryogenic 3D printing technique. It was demonstrated that the hydrogel scaffolds made of SIS/MBG@Exos had a 3D structure with the right amount of porosity, biocompatibility, and hemostasis ability. The outcomes of the scaffolds' testing on diabetic wounds demonstrated that hydrogel scaffolds can hasten the healing of diabetic wounds by boosting blood flow and encouraging the wound's angiogenesis process [158].

One promising approach to treating osteoarthritis and articular cartilage defects is tissue engineering. Mesenchymal stem cells (MSCs) and three-dimensional scaffolds work together in a mutualistic fashion that is essential for the regeneration of articular cartilage. To illustrate a few points, tissue engineering incorporates process-influencing modifiers like mechanical stimulation and functional biomolecule delivery systems, which are innovative and potentially beneficial for enhancing the regeneration process. The milieu in which tissue repair occurs is composed of a variety of signals that are released by the body in reaction to damage. In order to address the delivery issues and enable a regulated release of bioactive molecules to particular target sites, it appears that controlled delivery systems are necessary for cartilage repair [99]. Intelligent and clever biomaterials offer intriguing, renewable, and environmentally friendly potential answers to today's problems. When treating disorders including birth defects, injuries, and obesity (which is linked to other ailments like heart disease, type 2 diabetes, and chronic illnesses), these can be employed in medical settings in lieu of reconstructive procedures. They can even be utilized to prevent skin damage caused by UV rays [159]. Smart nanogel is one kind of hydrogel that is utilized to treat cancer and a number of other illnesses, such as diabetes, autoimmune diseases, neurological disorders, immunizations, viral infections, along with skin care [160]. Hydrogel 3D printing is critical to the creation of scaffolds and structures for tissue engineering.

With the use of this technology, injured tissues or organs can eventually regenerate by adhering, proliferating, and differentiating on a scaffold. Xu et al. introduced a novel glucose-responsive antioxidant hybrid hydrogel that aims to enhance diabetic wound healing. This invention showed how the hydrogel could be used to solve significant medical problems. In their review, Tsegay and colleagues highlighted the potential for enhanced patient care and recovery with smart 3D-printed hydrogel skin wound bandages [158]. Dressings based on hydrogel and featuring 3D-printed structures offer the perfect setting for wound healing. These dressings have the ability to encourage cell migration, support tissue regeneration, and maintain an environment that is moist and protective. Hydrogel dressings manufactured through 3D printing, for instance, can be used to cover burn injuries to promote wound healing, prevent infection, and lessen discomfort and scarring. Growth factors and antimicrobial agents can be added to hydrogel-based wound dressings to hasten the healing of chronic wounds such as diabetic ulcers.

Xiong et al. developed DLP-based 3D-printed wearable sensors with self-adhesion and self-healing properties using polymerizable rotaxane hydrogels. According to their research, bile acid and acrylated β-cyclodextrin can self-organize into a polymerizable pseudorotaxane when host and guest identities are correctly identified. Next, acrylamide is used to photopolymerize this pseudorotaxane to create conductive polymerizable rotaxane hydrogels. There was evidence of strong skin-to-PR-Gel adhesion [161]. Furthermore, it was demonstrated that the hydrogels promoted the growth of granulation tissue, angiogenesis, and the development of a high-density, ordered network of collagen fibers, demonstrating their potential for use as dressings for diabetic wounds [162].

5.5.3 Use in Clinical Settings

Ren et al. used graphene oxide quantum dots and citric acid to cross-link tannic acid/keratin hydrogel wound dressings in vivo tests on rat models. According to the findings, skin wounds healed completely in 16 days. X. Zhang et al. created macroporous gels in an alkaline environment by utilizing modified alginate microgels and methacrylate hyaluronic acid. Tested on rat models, these gels demonstrated efficacy for cell migration during blood vessel growth. Probiotic treatment of chitosan nanogels by Y. Ashori et al. aided in wound healing in rat models during in vivo experiments. After chitin hydrogel and nano ZnO were homogenized, the mixture was freeze-dried to produce microporous composite bandages. The antibacterial, blood-clotting, and swelling properties of the nanocomposite bandages were improved. PR-Gel's self-healing capabilities were further investigated using structural damage and recovery studies that involved continuous strain sweeps with alternating low (1%) and high (500%) oscillatory excitations. The rate of damage recovery for PR-Gel is demonstrated by the storage modulus (G′) and loss modulus (G″) under continuous strain sweeps. This process is repeatable, which highlights the reproducible restoration of the PR-Gel network [132]. Another example was given by Alonso et al., who developed a hydrogel of polyvinyl alcohol and gelatin reinforced with ceramic particles for use in dental applications. Furthermore, 3D printing can help patients heal from fractures or abnormalities in their bones. The creation of a 3D-printed PCL/β-tricalcium phosphate mandibular prosthesis was described in detail by Park et al. Scaffolds containing tissue-derived mesenchymal stem cells (TMSCs) were implanted into rabbits in order to evaluate the impact on bone regeneration. Osteogenesis was efficiently encouraged by the implanted material, indicating that the 3D prosthetic provides an environment that is conducive to bone regeneration and attachment [163]. Several preclinical investigations using rabbits, dogs, pigs, rats, and mice have demonstrated the efficacy of 3D-printed hydrogels in bone tissue engineering. These results suggest that 3D-printed hydrogel composites with improved printability and mechanical strength can significantly improve bone regeneration by promoting tissue vascularization and decreasing inflammation [164]. Strong cell attachment and biocompatibility were demonstrated by a scaffold composed of spermatogonial stem cells and dECM in both in vitro and in vivo mice models [165]. Recently, the implantation of a scaffold made of alginate, gelatin, and MSCs has been studied in rats. This approach has been shown to increase endometrial receptivity functional indicators and histomorphology, which in turn largely restored the injured endometrium's capacity to implant embryos and sustain pregnancy. Research on a midgestation rabbit model revealed that an in situ printed scaffold made of GelMA and PEGDA had a good sealing effect against the premature rupture of membranes [165]. Electroactive hydrogel scaffolds made of oxidized hyaluronic acid and chitosan have been printed onto partially resected rats' livers to promote in vivo cell migration, proliferation, and differentiation [161]. But after 12 weeks, it was shown that a dual-layer tubular scaffold loaded with MSCs and made of gelatin, PEGDA, PLGA, and methacrylic anhydride enhanced bile duct healing and biliary epithelial regeneration in mice [166]. The dECM hybrid encapsulation approach was utilized to develop both in vitro and in vivo biocompatibility. Moreover, due to beta-cell edge intercellular connections, the pancreatic islet-like aggregates in this system showed structural maturation and functional augmentation when studied in rats. The 15-week improvement in streptozotocin-induced hyperglycemia symptoms in immunocompetent mice administered subcutaneous injections of GelMA-encapsulated islets occurred without immunosuppression. More recently, research has demonstrated that printed organoids made of mouse islet cells and GelMA can maintain islet cell activity while improving glucose sensitivity. In a mouse model of volumetric muscle loss, it has recently been demonstrated that photopolymerized GelMA constructions printed directly from a handheld 3D printer can significantly increase muscle regeneration [165]. In a different study, critical neovascularization in the rat brain region was established by in vivo label-free photoacoustic imaging using a patch made of ECM, methacrylated hyaluronic acid, and angiogenic growth factors. Additionally, rat models were used to demonstrate the usefulness of printed hydrogels in nerve restoration. It has been suggested that GelMA, PEGDA, and platelets work together to support peripheral nerve healing. Likewise, successful axonal regeneration and recovery were induced by a regenerative conduit made of PLCL and gelatin [167]. GelMA and methacrylated hyaluronic acid nerve-like fibers also showed remarkable functional outcomes from the development of immunological regulation, angiogenesis, neurogenesis, and neural relay constructions [168].

5.5.4 Limitations and Barriers to Adoption and Future Directions

Because they are typically fragile, hydrogels present challenges when handling and loading. Spatial inhomogeneity, the term for the irregular cross-link density distribution that hydrogels frequently display, is a feature that lowers the hydrogels' strength and is therefore undesirable. Hydrogels are not recommended for the treatment of wounds with moderate to high exudation because they are not very absorbent and may require a secondary dressing to keep them in place. When hydrogels are used as contact lenses, immunological and inflammatory reactions result in lens deposition, hypoxia, dehydration, and eye reactions. Because of their delicate nature and soft structures, these materials belong to a class that also has low mechanical strength. Hydrogels are both biocompatible and biodegradable. Natural polymer-based hydrogels are typically biocompatible and biodegradable. One drawback of the natural polymer hydrogels used in tissue engineering is that, because they mimic and resemble the natural tissue they replace, they are challenging to isolate from biological tissue. Their versatility is also limited. However, one drawback of synthetic hydrogels is that they might not be biocompatible or biodegradable. The difficulty of sterilizing these systems is their other drawback. Because hydrogels contain water, which can amplify the negative effects of the sterilization processes, sterilizing hydrogels can be more difficult than sterilizing other types of polymers. Another disadvantage of hydrogels is the difficulty in controlling the rate of degradation. Depending on the application, the hydrogels' degradation profile varies. Biodegradable hydrogels are typically used as a temporary ECM in scenarios like cell cultivation until a new tissue is added, in which case the hydrogels' faster degradation profile would be appropriate. In order to deliver nutrients and bioactive molecules necessary for new tissue growth in drug delivery applications, the hydrogels must stay in the target site for a significantly longer period of time. According to the intended polymer structure, biodegradable hydrogels will completely break down after a predetermined amount of time [99]. Nonetheless, certain intelligent 3D biomaterials integrated into hydrogel are not limited to clinical use, as they can also be leveraged to address environmental issues like climate change. For example, new approaches can be made to lessen coral bleaching, which upsets the symbiotic connection between the coral and its zooxanthellae and exposes the coral to illnesses that may even kill it [159]. Despite the many flexible qualities and uses of hydrogel 3D printing, a number of issues remain, including scalability, post-printing processing, resolution constraints, etc. High-resolution printing is still difficult because of hydrogels' intricate rheological characteristics. Certain factors, such as nozzle clogging and the tendency of hydrogels to spread upon deposition, make it difficult to precisely deposit fine features. In order to improve mechanical stability, several hydrogel 3D printing techniques also call for post-processing procedures like cross-linking or curing, which can be time-consuming and result in variability in the finished product. It's also difficult to scale up hydrogel 3D printing to create complex structures in large quantities. One major challenge is to maintain consistent resolution, gelation kinetics, and print quality over a larger volume [132]. Resolution, post-printing processing, and scalability are the three main issues that hydrogel 3D printing is currently facing. Promising developments and trends are ahead of us. Hydrogel 3D printing holds the potential to revolutionize industries and make significant advancements in tissue engineering, personalized medicine, and other fields as materials, techniques, and applications continue to advance. Novel hydrogel formulations with enhanced mechanical, biocompatibility, and printability are anticipated to be introduced in the future. It is possible that hybrid hydrogels—which blend synthetic and natural components—will proliferate and enable the development of materials that are specifically suited for given uses. It's likely that researchers will create faster and more precise printing methods. This could include developments in light-based curing techniques for SLA and DLP, enhanced droplet formation in inkjet-based systems, and improved nozzle design. Technological developments in multimaterial 3D printing may make it possible to combine hydrogels with other materials, like ceramics and polymers, to create intricate and useful hybrid structures. It is probable that hydrogel 3D printing will investigate the direct integration of bioactive agents into printed structures. This may result in the development of constructs that have integrated therapeutic qualities, thereby obviating the requirement for functionalization after printing. The investigation of complex scaffold architectures will persist, with the aim of improving cellular responses, nutrient transport, and tissue integration. Investigating gradient structures and hierarchical designs may be part of this. It is possible that the field will grow into personalized medicine, where patients' specific hydrogel formulations and structures can be customized. Point-of-care 3D printing may make it possible to create medication delivery systems and implants tailored to each patient. More work will be put into the clinical translation and regulatory approval of hydrogel-based medical implants and devices as hydrogel 3D printing advances [169]. The non-physiological pH environment present before or during the gelation process is the primary drawback of pH responsive hydrogels. During the bioprinting process, prolonged exposure to an acidic or basic environment is fatal to cells [170]. Even so, platelet-rich plasma (PRP) gel experiences fast growth factor release, necessitating repeated administration that prolongs wound healing, increases treatment costs, and worsens patient suffering.

5.6 CONCLUSION

In conclusion, the creation of 3D-printed intelligent hydrogels is a major development in the field of wound healing. With careful engineering and design, these advanced materials offer a multimodal solution to the problems associated with tissue regeneration and repair. Hydrogel matrices in conjunction with 3D printing technology provide a flexible platform that can be tailored to specific wound types, patient needs, and healing stages. One of the most appealing aspects of smart 3D-printed hydrogels is their ability to mimic the extracellular matrix found in nature and provide an environment that is conducive to cellular proliferation, differentiation, and tissue remodeling. Through the integration of bioactive compounds, including growth factors, cytokines, and antimicrobial agents, into the hydrogel network, scientists can accurately regulate the wound microenvironment to expedite the healing process while concurrently reducing the likelihood of infection. Additionally, these hydrogels' adjustable mechanical qualities allow for customization based on the anatomical location and degree of tissue injury. This flexibility guarantees the best possible coverage and support for the wound, improving patient comfort and encouraging positive results. Furthermore, hydrogels' natural biocompatibility and biodegradability reduce the possibility of negative reactions and promote a smooth integration into host tissues. The incorporation of intelligent features, like stimulus responsiveness and medication administration capabilities, expands the therapeutic potential of hydrogels manufactured through 3D printing. By responding to environmental cues like pH, temperature, or enzymatic activity, future developments in material science, biology, and additive manufacturing processes have a great deal of potential to further the evolution of intelligent 3D-printed hydrogels. We may expect even more complex hydrogel formulations that provide previously unheard-of levels of accuracy, effectiveness, and patient-centered care as researchers continue to delve further into understanding the nuances of wound healing and tissue regeneration. Smart 3D-printed hydrogels, which provide a synergistic combination of biocompatibility, adaptability, and therapeutic efficacy, essentially mark a paradigm shift in wound care. These innovative materials have the potential to completely change the way we treat wounds and bring in a new era of faster healing and better patient outcomes by combining technology and creativity.

REFERENCES

1. ReportLinker. (2021, November 11). The global advanced wound care market is projected to reach USD 12.8 billion by 2026 from USD 9.4 billion in 2021, at a CAGR of 6.2%. *Yahoo Finance*. Retrieved from https://uk.finance.yahoo.com
2. Sen, C. K. (2019). Human wounds and its burden: An updated compendium of estimates. *Advances in Wound Care*, 8(2), 39–48. https://doi.org/10.1089/wound.2019.0946
3. Guest, J. F., Ayoub, N., McIlwraith, T., Uchegbu, I., Gerrish, A., Weidlich, D., . . Vowden, P. (2016). Health economic burden that different wound types impose on the UK's national health service. *International Wound Journal*, 14(2), 322–330. https://doi.org/10.1111/iwj.12603
4. Schiestl, C., Stiefel, D., & Meuli, M. (2010). Giant naevus, giant excision, eleg(i)ant closure? Reconstructive surgery with integra artificial skin® to treat giant congenital melanocytic naevi in children. *Journal of Plastic, Reconstructive & Aesthetic Surgery*, 63(4), 610–615. https://doi.org/10.1016/j.bjps.2009.01.050

5. Schiestl, C., Neuhaus, K., Biedermann, T., Böttcher-Haberzeth, S., Ernst, R., & Meuli, M. (2010). Novel treatment for massive lower extremity avulsion injuries in children: Slow, but effective with good cosmesis. *European Journal of Pediatric Surgery*, 21(02), 106–110 https://doi.org/10.1055/s-0030-1267234
6. Böttcher-Haberzeth, S., Kapoor, S., Meuli, M., Neuhaus, K., Biedermann, T., Ernst, R., & Schiestl, C. (2011). Osmotic expanders in children: No filling - no control - no problem? *European Journal of Pediatric Surgery*, 21(03), 163–167. https://doi.org/10.1055/s-0030-1270460
7. Barbour, J. R., Schweppe, M., & Seung-Jun, O. (2008). Lower-extremity burn reconstruction in the child. *The Journal of Craniofacial Surgery/the Journal of Craniofacial Surgery*, 19(4), 976–988. https://doi.org/10.1097/scs.0b013e318175f35a
8. Hayat, M., Nawaz, A., Chinnam, S., Muzammal, M., Latif, M. S., Yasin, M., Ashique, S., Zengin, G., & Farid, A. (2023, July 1). Formulation development and optimization of herbo synthetic gel: In vitro biological evaluation and in vivo wound healing studies. *Process Biochemistry*, 130, 116–126.
9. Pita-López, M. L., Fletes-Vargas, G., Espinosa-Andrews, H., & Rodríguez-Rodríguez, R. (2021). Physically cross-linked chitosan-based hydrogels for tissue engineering applications: A state-of-the-art review. *European Polymer Journal/European Polymer Journal*, 145, 110176. https://doi.org/10.1016/j.eurpolymj.2020.110176
10. Rodríguez-Rodríguez, R., Espinosa-Andrews, H., Velasquillo-Martínez, C., & García-Carvajal, Z. Y. (2019). Composite hydrogels based on gelatin, chitosan and polyvinyl alcohol to biomedical applications: A review. *International Journal of Polymeric Materials*, 69(1), 1–20. https://doi.org/10.1080/00914037.2019.1581780
11. Agarwal, R., Malhotra, S., Gupta, V., & Jain, V. (2022). The application of three-dimensional printing on foot fractures and deformities: A mini-review. *Annals of 3D Printed Medicine*, 5, 100046. https://doi.org/10.1016/j.stlm.2022.100046
12. Li, J., Wu, C., Chu, P. K., & Gelinsky, M. (2020). 3D printing of hydrogels: Rational design strategies and emerging biomedical applications. *Materials Science & Engineering. R, Reports*, 140, 100543. https://doi.org/10.1016/j.mser.2020.100543
13. Rajabi, M., McConnell, M., Cabral, J. D., & Ali, M. A. (2021). Chitosan hydrogels in 3D printing for biomedical applications. *Carbohydrate Polymers*, 260, 117768. https://doi.org/10.1016/j.carbpol.2021.117768
14. Bakhtiari, S. S. E., Bakhsheshi-Rad, H. R., Karbasi, S., Razzaghi, M., Tavakoli, M. R., Ismail, A. F., . . . Berto, F. (2021). 3-dimensional printing of hydrogel-based nanocomposites: A comprehensive review on the technology description, properties, and applications. *Advanced Engineering Materials*, 23(10). https://doi.org/10.1002/adem.202100477
15. Zamboulis, A., Michailidou, G., Koumentakou, I., & Bikiaris, D. N. (2022). Polysaccharide 3D printing for drug delivery applications. *Pharmaceutics*, 14(1), 145. https://doi.org/10.3390/pharmaceutics14010145
16. Agarwala, S. (2020). Electrically conducting hydrogels for health care: Concept, fabrication methods, and applications. *International Journal of Bioprinting*, 6(2), 273. https://doi.org/10.18063/ijb.v6i2.273
17. Athukorala, S. S., Tran, T. S., Balu, R., Truong, V. K., Chapman, J., Dutta, N. K., & Choudhury, N. R. (2021). 3D printable electrically conductive hydrogel scaffolds for biomedical applications: A review. *Polymers*, 13(3), 474. https://doi.org/10.3390/polym13030474
18. Siree, K. G., Amulya, T. M., Kumar, T. M. P., Sowmya, S. S., Divith, K., Ramu, B. G., & Gowrav, M. P. (2021). Transfiguring healthcare: Three-dimensional printing in pharmaceutical sciences; Trends during COVID-19: A review. *Journal of Pharmaceutical Research International*, 207–218. https://doi.org/10.9734/jpri/2021/v33i56b33946
19. Medical applications for 3D printing: Recent developments. (2018, February 1). Retrieved from https://pubmed.ncbi.nlm.nih.gov/30228688/
20. Fayyazbakhsh, F., & Leu, M. (2020). A brief review on 3D bioprinted skin substitutes. *Procedia Manufacturing*, 48, 790–796. https://doi.org/10.1016/j.promfg.2020.05.115
21. De Oliveira, R. T., Fantaus, S. S., Guillot, A. J., Melero, A., & Beck, R. C. R. (2021). 3D-Printed products for topical skin applications: From personalized dressings to drug delivery. *Pharmaceutics*, 13(11), 1946. https://doi.org/10.3390/pharmaceutics13111946
22. Sun, W., Schaffer, S., Dai, K., Yao, L., Feinberg, A. W., & Webster-Wood, V. A. (2021). 3D Printing hydrogel-based soft and biohybrid actuators: A mini-review on fabrication techniques, applications, and challenges. *Frontiers in Robotics and AI*, 8. https://doi.org/10.3389/frobt.2021.673533
23. Vig, K., Chaudhari, A. A., Tripathi, S., Dixit, S., Sahu, R., Pillai, S. R., . . . Singh, S. R. (2017). Advances in skin regeneration using tissue engineering. *International Journal of Molecular Sciences*, 18(4), 789. https://doi.org/10.3390/ijms18040789

24. Metcalfe, A. D., & Ferguson, M. W. J. (2006). Tissue engineering of replacement skin: the crossroads of biomaterials, wound healing, embryonic development, stem cells and regeneration. *Journal of the Royal Society Interface*, 4(14), 413–437. https://doi.org/10.1098/rsif.2006.0179
25. Pereira, R. F., Sousa, A., Barrias, C. C., Bayat, A., Granja, P. L., & Bártolo, P. (2017). Advances in bioprinted cell-laden hydrogels for skin tissue engineering. *Biomanufacturing Reviews*, 2(1). https://doi.org/10.1007/s40898-017-0003-8
26. Uccioli, L., Izzo, V., Meloni, M., Vainieri, E., Ruotolo, V., & Giurato, L. (2015). Non-healing foot ulcers in diabetic patients: general and local interfering conditions and management options with advanced wound dressings. *Journal of Wound Care*, 24(Sup4b), 35–42. https://doi.org/10.12968/jowc.2015.24.sup4b.35
27. Streitz, M. J. (2023, October 5). How to cleanse, irrigate, debride, and dress wounds. Retrieved from https://www.msdmanuals.com/professional/injuries-poisoning/how-to-care-for-wounds-and-lacerations/how-to-cleanse,-irrigate,-debride,-and-dress-wounds
28. Gottrup, F. (1999). Wound closure techniques. *Journal of Wound Care*, 8(8), 397–400. https://doi.org/10.12968/jowc.1999.8.8.25906
29. Andreassi, A., Bilenchi, R., Biagioli, M., & D'Aniello, C. (2005). Classification and pathophysiology of skin grafts. *Clinics in Dermatology*, 23(4), 332–337. https://doi.org/10.1016/j.clindermatol.2004.07.024
30. Bielefeld, K. A., Amini-Nik, S., & Alman, B. A. (2012). Cutaneous wound healing: recruiting developmental pathways for regeneration. *Cellular and Molecular Life Sciences*, 70(12), 2059–2081. https://doi.org/10.1007/s00018-012-1152-9
31. Li, W., Pappas, A., Zhang, L., Ruvolo, E., & Cavender, D. (2013). IL-11, IL-1α, IL-6, and TNF-α are induced by solar radiation in vitro and may be involved in facial subcutaneous fat loss in vivo. *Journal of Dermatological Science*, 71(1), 58–66. https://doi.org/10.1016/j.jdermsci.2013.03.009
32. Percival, N. (2002). Classification of wounds and their management. *Surgery*, 20(5), 114–117. https://doi.org/10.1383/surg.20.5.114.14626
33. Hadian, Y., Bagood, M. D., Dahle, S. E., Sood, A., & Isseroff, R. R. (2019). Interleukin-17: Potential target for chronic wounds. *Mediators of Inflammation*, 2019, 1–10. https://doi.org/10.1155/2019/1297675
34. Wound emergencies: The importance of assessment, documentation, and early treatment using a wound electronic medical record. (2009, May 1). Retrieved from https://pubmed.ncbi.nlm.nih.gov/19471049/
35. Moore, K., McCallion, R. L., Searle, R., Stacey, M., & Harding, K. (2006). Prediction and monitoring the therapeutic response of chronic dermal wounds. *International Wound Journal*, 3(2), 89–98. https://doi.org/10.1111/j.1742-4801.2006.00212.x
36. Chandika, P., Ko, S., & Jung, W. (2015). Marine-derived biological macromolecule-based biomaterials for wound healing and skin tissue regeneration. *International Journal of Biological Macromolecules*, 77, 24–35. https://doi.org/10.1016/j.ijbiomac.2015.02.050
37. Morton, L. M., & Phillips, T. J. (2016). Wound healing and treating wounds. *Journal of the American Academy of Dermatology*, 74(4), 589–605. https://doi.org/10.1016/j.jaad.2015.08.068
38. Yee, A., Patel, M., Wu, E., Yi, S., Marti, G., & Harmon, J. (2016, June). IDr: An intelligent digital ruler app for remote wound assessment. 2016 IEEE First International Conference on Connected Health: Applications, Systems and Engineering Technologies (CHASE) (pp. 380–381). IEEE .
39. Trengove, N., Stacey, M., Macauley, S. P., Bennett, N. T., Gibson, J., Burslem, F., . . . Schultz, G. S. (1999). Analysis of the acute and chronic wound environments: The role of proteases and their inhibitors. *Wound Repair and Regeneration*, 7(6), 442–452. https://doi.org/10.1046/j.1524-475x.1999.00442.x
40. Rossaint, J., Margraf, A., & Zarbock, A. (2018). Role of platelets in leukocyte recruitment and resolution of inflammation. *Frontiers in Immunology*, 9. https://doi.org/10.3389/fimmu.2018.02712
41. Greaves, N. S., Ashcroft, K. J., Baguneid, M., & Bayat, A. (2013). Current understanding of molecular and cellular mechanisms in fibroplasia and angiogenesis during acute wound healing. *Journal of Dermatological Science*, 72(3), 206–217. https://doi.org/10.1016/j.jdermsci.2013.07.008
42. Versteeg, H. H., Heemskerk, J. W. M., Levi, M., & Reitsma, P. H. (2013). New fundamentals in hemostasis. *Physiological Reviews*, 93(1), 327–358. https://doi.org/10.1152/physrev.00016.2011
43. Pereira, R. F., Sousa, A., Barrias, C. C., Bayat, A., Granja, P. L., & Bártolo, P. (2017b). Advances in bioprinted cell-laden hydrogels for skin tissue engineering. *Biomanufacturing Reviews*, 2(1). https://doi.org/10.1007/s40898-017-0003-8
44. Midwood, K. S., Williams, L. V., & Schwarzbauer, J. E. (2004). Tissue repair and the dynamics of the extracellular matrix. *International Journal of Biochemistry & Cell Biology*, 36(6), 1031–1037. https://doi.org/10.1016/j.biocel.2003.12.003
45. desJardins-Park, H. E., Foster, D. S., & Longaker, M. T. (2018). Fibroblasts and wound healing: An update. *Regenerative Medicine*, 13(5), 491–495. https://doi.org/10.2217/rme-2018-0073

46. Bowden, L. G., Byrne, H. M., Maini, P. K., & Moulton, D. E. (2016). A morphoelastic model for dermal wound closure. *Biomech Model Mechanobiol*, *15*(3), 663–681. https://doi.org/10.1007/s10237-015-0716-7
47. Yee, A., Harmon, J. W., & Yi, S. (2016). Quantitative monitoring wound healing status through three-dimensional imaging on mobile platforms. *The Journal of the American College of Clinical Wound Specialists*, 8(1–3), 21–27. https://doi.org/10.1016/j.jccw.2017.11.00
48. McCarty, S., & Percival, S. L. (2013). Proteases and delayed wound healing. *Advances in Wound Care*, 2(8), 438–447. https://doi.org/10.1089/wound.2012.0370
49. Schreml, S., Szeimies, R., Prantl, L., Karrer, S., Landthaler, M., & Babilas, P. (2010). Oxygen in acute and chronic wound healing. *British Journal of Dermatology/British Journal of Dermatology, Supplement*, 163(2), 257–268. https://doi.org/10.1111/j.1365-2133.2010.09804.x
50. Dhall, S., Danh, C., Garcia, M., Jane, K., Mirebrahim, S. H., Lyubovitsky, J. G., . . . Martins-Green, M. (2014). Generating and reversing chronic wounds in diabetic mice by manipulating wound redox parameters. *Journal of Diabetes Research*, 2014, 1–18. https://doi.org/10.1155/2014/562625
51. Stanley, A., & Osler, T. M. (2001). Senescence and the healing rates of venous ulcers. *Journal of Vascular Surgery*, 33(6), 1206–1211. https://doi.org/10.1067/mva.2001.115379
52. Bourguignon, L. Y. (2014). Matrix Hyaluronan-Activated CD44 signaling promotes keratinocyte activities and improves abnormal epidermal functions. *The American Journal of Pathology*, 184(7), 1912–1919. https://doi.org/10.1016/j.ajpath.2014.03.010
53. Telgenhoff, D., & Shroot, B. (2005). Cellular senescence mechanisms in chronic wound healing. *Cell Death and Differentiation*, 12(7), 695–698. https://doi.org/10.1038/sj.cdd.4401632
54. Zaszczyńska, A., Moczulska-Heljak, M., Gradys, A., & Sajkiewicz, P. (2021). Advances in 3D printing for tissue engineering. *Materials (Basel, Switzerland)*, 14(12), 3149. https://doi.org/10.3390/ma14123149
55. Lee, V., Singh, G., Trasatti, J. P., Bjornsson, C. S., Xu, X., Tran, T. N., . . . Karande, P. (2014). Design and fabrication of human skin by Three-Dimensional Bioprinting. *Tissue Engineering. Part C Methods*, 20(6), 473–484. https://doi.org/10.1089/ten.tec.2013.0335
56. Markstedt, K., Mantas, A., Tournier, I., Ávila, H. M., Hägg, D., & Gatenholm, P. (2015). 3D Bioprinting human chondrocytes with nanocellulose–alginate bioink for cartilage tissue engineering applications. *Biomacromolecules*, 16(5), 1489–1496. https://doi.org/10.1021/acs.biomac.5b00188
57. Cui, H., Zhu, W., Nowicki, M., Zhou, X., Khademhosseini, A., & Zhang, L. G. (2016). Hierarchical fabrication of engineered vascularized bone biphasic constructs via dual 3d bioprinting: Integrating regional bioactive factors into architectural design. *Advanced Healthcare Materials/Advanced Healthcare Materials*, 5(17), 2174–2181. https://doi.org/10.1002/adhm.201600505
58. Laronda, M. M., Rutz, A. L., Xiao, S., Whelan, K. A., Duncan, F. E., Roth, E. W., . . . Shah, R. N. (2017). A bioprosthetic ovary created using 3D printed microporous scaffolds restores ovarian function in sterilized mice. *Nature Communications*, 8(1). https://doi.org/10.1038/ncomms15261
59. Umur, E., Bayrak, E., Arslan, F. D., Bulut, S., Baysoy, E., Kaleli-Can, G., & Ayan, B. (2023). Advances in three dimensional bioprinting for wound healing: A Comprehensive review. *Applied Sciences*, 13(18), 10269. https://doi.org/10.3390/app131810269
60. Xiang, J., Shen, L., & Hong, Y. (2020). Status and future scope of hydrogels in wound healing: Synthesis, materials and evaluation. *European Polymer Journal*, 130, 109609.
61. Ahmed, E. M. (2015). Hydrogel: Preparation, characterization, and applications: A review. *Journal of Advanced Research*, 6(2), 105–121. https://doi.org/10.1016/j.jare.2013.07.006
62. Aswathy, S. H., Narendrakumar, U., & Manjubala, I. (2020). Commercial hydrogels for biomedical applications. *Heliyon*, 6(4), e03719. https://doi.org/10.1016/j.heliyon.2020.e03719
63. Sheokand, B., Vats, M., Kumar, A., Srivastava, C. M., Bahadur, I., & Pathak, S. R. (2023). Natural polymers used in the dressing materials for wound healing: Past, present and future. *Journal of Polymer Science*, 61(14), 1389–1414. https://doi.org/10.1002/pol.20220734
64. Chandel, A. K. S., Kannan, D., Nutan, B., Singh, S., & Jewrajka, S. K. (2017). Dually crosslinked injectable hydrogels of poly(ethylene glycol) and poly[(2-dimethylamino)ethyl methacrylate]-b-poly(N-isopropyl acrylamide) as a wound healing promoter. *Journal of Materials Chemistry. B*, 5(25), 4955–4965. https://doi.org/10.1039/c7tb00848a
65. Koehler, J., Brandl, F., & Goepferich, A. (2018). Hydrogel wound dressings for bioactive treatment of acute and chronic wounds. *European Polymer Journal/European Polymer Journal*, 100, 1–11. https://doi.org/10.1016/j.eurpolymj.2017.12.046
66. Van Vlierberghe, S., Dubruel, P., & Schacht, E. (2011). Biopolymer-based hydrogels as scaffolds for tissue engineering applications: A review. *Biomacromolecules*, 12(5), 1387–1408. https://doi.org/10.1021/bm200083n

67. Bilici, Ç., Can, V., Nöchel, U., Behl, M., Lendlein, A., & Okay, O. (2016). Melt-processable shape-memory hydrogels with self-healing ability of high mechanical strength. *Macromolecules*, 49(19), 7442–7449. https://doi.org/10.1021/acs.macromol.6b01539
68. Arif, M. M., Khan, S. M., Gull, N., Tabish, T. A., Zia, S., Khan, R. U., . . . Butt, M. A. (2021). Polymer-based biomaterials for chronic wound management: Promises and challenges. *International Journal of Pharmaceutics*, 598, 120270. https://doi.org/10.1016/j.ijpharm.2021.120270
69. Bashir, S., Hina, M., Iqbal, J., Rajpar, A. H., Mujtaba, M., Alghamdi, N. A., . . . Ramesh, K. (2020). Fundamental concepts of hydrogels: Synthesis, properties, and their applications. *Polymers*, 12(11), 2702. https://doi.org/10.3390/polym12112702
70. Ikada, Y. (2001). Biocompatibility of hydrogels. In *Elsevier eBooks* (pp. 388–407). https://doi.org/10.1016/b978-012394690-4/50094-3
71. Schultz, G. S., Chin, G. A., Moldawer, L., et al. (2011). Principles of wound healing. In R. Fitridge & M. Thompson (Eds.), *Mechanisms of vascular disease: A reference book for vascular specialists* (p. 23). University of Adelaide Press. Retrieved from: https://www.ncbi.nlm.nih.gov/books/NBK534261/
72. Gailit, J., & Clark, R. A. (1994). Wound repair in the context of extracellular matrix. *Current Opinion in Cell Biology*, 6(5), 717–725. https://doi.org/10.1016/0955-0674(94)90099-x
73. Bennett, N. T., & Schultz, G. S. (1993). Growth factors and wound healing: Biochemical properties of growth factors and their receptors. *The American Journal of Surgery*, 165(6), 728–737. https://doi.org/10.1016/s0002-9610(05)80797-4
74. Rumalla, V. K., & Borah, G. (2001). Cytokines, growth factors, and plastic surgery. *Plastic and Reconstructive Surgery/PSEF CD Journals*, 108(3), 719–733. https://doi.org/10.1097/00006534-200109010-00019
75. Gillitzer, R., & Goebeler, M. (2001). Chemokines in cutaneous wound healing. *Journal of Leukocyte Biology*, 69(4), 513–521.
76. Frenette, P. S., & Wagner, D. D. (1996). Adhesion molecules — blood vessels and blood cells. *New England Journal of Medicine/the New England Journal of Medicine*, 335(1), 43–45. https://doi.org/10.1056/nejm199607043350108
77. Ps, F., & Dd, W. (1996). Adhesion molecules. *New England Journal of Medicine/the New England Journal of Medicine*, 334(23), 1526–1529. https://doi.org/10.1056/nejm199606063342308
78. Connective tissue growth factor mediates transforming growth factor beta-induced collagen synthesis: Down-regulation by cAMP. (1999, October 1). Retrieved from https://pubmed.ncbi.nlm.nih.gov/10506580/
79. O'Toole, E. A. (2001). Extracellular matrix and keratinocyte migration. *Clinical and Experimental Dermatology*, 26(6), 525–530. https://doi.org/10.1046/j.1365-2230.2001.00891.x
80. Chapter 3 First aid equipment, supplies, rescue, and transportation. (n.d.). Retrieved from https://medical.tpub.com/14295/css/Chapter-3-First-Aid-Equipment-Supplies-Rescue-And-Transportation-115.htm
81. Eaglstein, W. H. (2001). Moist wound healing with occlusive dressings: A clinical focus. *Dermatologic Surgery*, 27(2), 175–182. https://doi.org/10.1046/j.1524-4725.2001.00299.x
82. Gallant A. E. (1897). VI. Report upon the use of a mixture of castor oil and balsam of Peru as a surgical dressing. *Annals of Surgery*, 26(3), 329–339.
83. Jull, A., Cullum, N., Dumville, J. C., Westby, M. J., Deshpande, S., & Walker, N. (2015). Honey as a topical treatment for wounds. *Cochrane Library*, 2015(6). https://doi.org/10.1002/14651858.cd005083.pub4
84. Kim, H., Makin, I. R. S., Skiba, J. B., Ho, A., Housler, G., Stojadinovic, A., & Izadjoo, M. (2014). Antibacterial efficacy testing of a bioelectric wound dressing against clinical wound pathogens. *The Open Microbiology Journal*, 8(1), 15–21. https://doi.org/10.2174/1874285801408010015
85. Banerjee, J., Ghatak, P. D., Roy, S., Khanna, S., Sequin, E. K., Bellman, K., . . . Sen, C. K. (2014). Improvement of human keratinocyte migration by a Redox active bioelectric dressing. *PLoS One*, 9(3), e89239. https://doi.org/10.1371/journal.pone.0089239
86. Thangaraj, P., Padma, V. V., & Santhini, E. (2015). Wound dressings – a review. *Biomedicine*, 5(4). https://doi.org/10.7603/s40681-015-0022-9
87. Wiegand, C., Tittelbach, J., Hipler, U., & Elsner, P. (2015). Clinical efficacy of dressings for treatment of heavily exuding chronic wounds. *Chronic Wound Care Management and Research*, 101. https://doi.org/10.2147/cwcmr.s60315
88. Vyas, K. S., & Vásconez, H. C. (2014). Wound healing: Biologics, skin substitutes, biomembranes and scaffolds. *Healthcare*, 2(3), 356–400. https://doi.org/10.3390/healthcare2030356
89. Parodi, A., Khaled, S. M., Yazdi, I. K., Evangelopoulos, M., Furman, N. E. T., Wang, X., . . . Tasciotti, E. (2015). Smart hydrogels. In *Springer eBooks* (pp. 1–13). https://doi.org/10.1007/978-94-007-6178-0_104-2
90. Chilkoti, A., Dreher, M. R., Meyer, D. E., & Raucher, D. (2002). Targeted drug delivery by thermally responsive polymers. *Advanced Drug Delivery Reviews*, 54(5), 613–630. https://doi.org/10.1016/s0169-409x(02)00041-8

91. Erman, B., & Flory, P. J. (1986). Critical phenomena and transitions in swollen polymer networks and in linear macromolecules. *Macromolecules*, 19(9), 2342–2353. https://doi.org/10.1021/ma00163a003
92. Hoti, G., Caldera, F., Cecone, C., Pedrazzo, A. R., Anceschi, A., Appleton, S. L., . . . Trotta, F. (2021). Effect of the cross-linking density on the swelling and rheological behavior of ester-bridged B-cyclodextrin nanosponges. *Materials*, 14(3), 478. https://doi.org/10.3390/ma14030478
93. Dong, R., & Guo, B. (2021). Smart wound dressings for wound healing. *Nano Today*, 41, 101290. https://doi.org/https://doi.org/10.1016/j.nantod.2021.101290
94. Stan, D., Tanase, C., Avram, M., Apetrei, R., Mincu, N.-B., Mateescu, A. L., & Stan, D. (2021). Wound healing applications of creams and "smart" hydrogels. *Experimental Dermatology*, 30(9), 1218–1232. https://doi.org/https://doi.org/10.1111/exd.14396
95. Asadi, N., Pazoki-Toroudi, H., del Bakhshayesh, A. R., Akbarzadeh, A., Davaran, S., & Annabi, N. (2021). Multifunctional hydrogels for wound healing: Special focus on biomacromolecular based hydrogels. *International Journal of Biological Macromolecules*, 170, 728–750. https://doi.org/https://doi.org/10.1016/j.ijbiomac.2020.12.202
96. Tavakoli, S., & Klar, A. S. (2020). Advanced hydrogels as wound dressings. *Biomolecules*, 10(8). https://doi.org/10.3390/biom10081169
97. Bashir, S., Hina, M., Iqbal, J., Rajpar, A. H., Mujtaba, M. A., Alghamdi, N. A., Wageh, S., Ramesh, K., & Ramesh, S. (2020). Fundamental concepts of hydrogels: Synthesis, properties, and their applications. *Polymers*, 12(11), 2702. https://doi.org/10.3390/polym12112702
98. Mantha, S., Pillai, S., Khayambashi, P., Upadhyay, A., Zhang, Y., Tao, O., Pham, H. M., & Tran, S. D. (2019). Smart hydrogels in tissue engineering and regenerative medicine. *Materials*, 12(20). https://doi.org/10.3390/ma12203323
99. Amukarimi, S., Ramakrishna, S., & Mozafari, M. (2021). Smart biomaterials—A proposed definition and overview of the field. *Current Opinion in Biomedical Engineering*, 19, 100311. https://doi.org/https://doi.org/10.1016/j.cobme.2021.100311
100. Bordbar-Khiabani, A., & Gasik, M. (2022). Smart hydrogels for advanced drug delivery systems. *International Journal of Molecular Sciences*, 23(7). https://doi.org/10.3390/ijms23073665
101. Salehi, S., Naghib, S. M., Garshasbi, H. R., Ghorbanzadeh, S., & Zhang, W. (2023). Smart stimuli-responsive injectable gels and hydrogels for drug delivery and tissue engineering applications: A review. *Frontiers in Bioengineering and Biotechnology*, 11. https://doi.org/10.3389/fbioe.2023.1104126
102. Li, Z., Zhou, Y., Li, T., Zhang, J., & Tian, H. (2022). Stimuli-responsive hydrogels: Fabrication and biomedical applications. *VIEW*, 3(2), 20200112. https://doi.org/https://doi.org/10.1002/VIW.20200112
103. Mateti, T., K, L., Laha, A., & Thakur, G. (2023). A critical analysis of the recent developments in multi-stimuli responsive smart hydrogels for cancer treatment. *Current Opinion in Biomedical Engineering*, 25, 100424. https://doi.org/https://doi.org/10.1016/j.cobme.2022.100424
104. Kasiński, A., Zielińska-Pisklak, M., Oledzka, E., & Sobczak, M. (2020). Smart hydrogels – synthetic stimuli-responsive antitumor drug release systems. *International Journal of Nanomedicine*, 15, 4541–4572. https://doi.org/10.2147/IJN.S248987
105. Brighenti, R., & Cosma, M. P. (2022). Mechanics of multi-stimuli temperature-responsive hydrogels. *Journal of the Mechanics and Physics of Solids*, 169, 105045. https://doi.org/https://doi.org/10.1016/j.jmps.2022.105045
106. Haidari, H., Kopecki, Z., Sutton, A. T., Garg, S., Cowin, A. J., & Vasilev, K. (2021). pH-Responsive "smart" hydrogel for controlled delivery of silver nanoparticles to infected wounds. *Antibiotics*, 10(1). https://doi.org/10.3390/antibiotics10010049
107. Yang, G., Zhang, Z., Liu, K., et al. (2022). A cellulose nanofibril-reinforced hydrogel with robust mechanical, self-healing, pH-responsive and antibacterial characteristics for wound dressing applications. *Journal of Nanobiotechnology*, 20, 312. https://doi.org/10.1186/s12951-022-01523-5
108. Li, L., Scheiger, J. M., & Levkin, P. A. (2019). Design and applications of photoresponsive hydrogels. *Advanced Materials*, 31(26), 1807333. https://doi.org/https://doi.org/10.1002/adma.201807333
109. Yarali, E., Baniasadi, M., Zolfagharian, A., Chavoshi, M., Arefi, F., Hossain, M., Bastola, A., Ansari, M., Foyouzat, A., Dabbagh, A., Ebrahimi, M., Mirzaali, M. J., & Bodaghi, M. (2022). Magneto-/ electro-responsive polymers toward manufacturing, characterization, and biomedical/ soft robotic applications. *Applied Materials Today*, 26, 101306. https://doi.org/https://doi.org/10.1016/j.apmt.2021.101306
110. Dong, R., & Guo, B. (2021). Smart wound dressings for wound healing. *Nano Today*, 41, 101290. https://doi.org/https://doi.org/10.1016/j.nantod.2021.101290
111. Gounden, V., & Singh, M. (2024). Hydrogels and wound healing: Current and future prospects. *Gels (Basel, Switzerland)*, 10(1), 43. https://doi.org/10.3390/gels10010043

112. Firlar, I., Altunbek, M., McCarthy, C., Ramalingam, M., & Camci-Unal, G. (2022). Functional hydrogels for treatment of chronic wounds. *Gels (Basel, Switzerland)*, 8(2), 127. https://doi.org/10.3390/gels8020127
113. Amorim, P. A., D'ávila, M. A., Anand, R., Moldenaers, P., Van Puyvelde, P., & Bloemen, V. (2021). Insights on shear rheology of inks for extrusion-based 3D bioprinting. *Bioprinting*, 22, e00129. https://doi.org/10.1016/j.bprint.2021.e00129
114. Bom, S., Ribeiro, R., Ribeiro, H. M., Santos, C., & Marto, J. (2022). On the progress of hydrogel-based 3D printing: Correlating rheological properties with printing behaviour. *International Journal of Pharmaceutics,* 615, 121506. https://doi.org/10.1016/j.ijpharm.2022.121506
115. Kokol, V., Pottathara, Y. B., Mihelčič, M., & Perše, L. S. (2021). Rheological properties of gelatine hydrogels affected by flow- and horizontally-induced cooling rates during 3D cryo-printing. *Colloids and Surfaces. A, Physicochemical and Engineering Aspects*, 616, 126356. https://doi.org/10.1016/j.colsurfa.2021.126356
116. Kim, M. H., Lee, Y. W., Jung, W., Oh, J., & Nam, S. Y. (2019). Enhanced rheological behaviors of alginate hydrogels with carrageenan for extrusion-based bioprinting. *Journal of the Mechanical Behavior of Biomedical Materials/Journal of Mechanical Behavior of Biomedical Materials*, 98, 187–194. https://doi.org/10.1016/j.jmbbm.2019.06.014
117. Townsend, J. M., Beck, E. C., Gehrke, S. H., Berkland, C., & Detamore, M. S. (2019). Flow behavior prior to crosslinking: The need for precursor rheology for placement of hydrogels in medical applications and for 3D bioprinting. *Progress in Polymer Science*, 91, 126–140. https://doi.org/10.1016/j.progpolymsci.2019.01.003
118. Zhou, M., Lee, B. H., Tan, Y., & Tan, L. P. (2019). Microbial transglutaminase induced controlled crosslinking of gelatin methacryloyl to tailor rheological properties for 3D printing. *Biofabrication*, 11(2), 025011. https://doi.org/10.1088/1758-5090/ab063f
119. Sheikhi, M., Rafiemanzelat, F., Ghodsi, S., Moroni, L., & Setayeshmehr, M. (2022). 3D printing of jammed self-supporting microgels with alternative mechanism for shape fidelity, crosslinking and conductivity. *Additive Manufacturing*, 58, 102997. https://doi.org/10.1016/j.addma.2022.102997
120. Kantaros, A. (2022). 3D printing in regenerative Medicine: technologies and resources utilized. *International Journal of Molecular Sciences*, 23(23), 14621. https://doi.org/10.3390/ijms232314621
121. Kalossaka, L. M., Mohammed, A. A., Sena, G., Barter, L., & Myant, C. (2021). 3D printing nanocomposite hydrogels with lattice vascular networks using stereolithography. *Journal of Materials Research/Pratt's Guide to Venture Capital Sources*, 36(21), 4249–4261. https://doi.org/10.1557/s43578-021-00411-2
122. Karakurt, I., Aydoğdu, A., Çıkrıkcı, S., Orozco, J. S., & Lin, L. (2020). Stereolithography (SLA) 3D printing of ascorbic acid loaded hydrogels: A controlled release study. *International Journal of Pharmaceutics*, 584, 119428. https://doi.org/10.1016/j.ijpharm.2020.119428
123. Burke, G., Devine, D. M., & Major, I. (2020). Effect of stereolithography 3D printing on the properties of PEGDMA hydrogels. *Polymers*, 12(9), 2015. https://doi.org/10.3390/polym12092015
124. Magalhães, L. S. S. M., Santos, F. E. P. D., De Maria Vaz Elias, C., Afewerki, S., De Sousa, G. F., Furtado, A. S. A., . . . Lobo, A. O. (2020). Printing 3D hydrogel structures employing Low-Cost stereolithography technology. *Journal of Functional Biomaterials*, 11(1), 12. https://doi.org/10.3390/jfb11010012
125. Alketbi, A. S., Shi, Y., Li, H., Raza, A., & Zhang, T. (2021). Impact of PEGDA photopolymerization in micro-stereolithography on 3D printed hydrogel structure and swelling. *Soft Matter*, 17(30), 7188–7195. https://doi.org/10.1039/d1sm00483b
126. Sun, Z., Lu, Y., Zhao, Q., & Wu, J. (2022). A new stereolithographic 3D printing strategy for hydrogels with a large mechanical tunability and self-weldability. *Additive Manufacturing*, 50, 102563. https://doi.org/10.1016/j.addma.2021.102563
127. Hosseinabadi, H. G., Nieto, D., Yousefinejad, A., Fattel, H., Ionov, L., & Miri, A. K. (2023). Ink material selection and optical design considerations in DLP 3D printing. *Applied Materials Today*, 30, 101721. https://doi.org/10.1016/j.apmt.2022.101721
128. Ding, H., Dong, M., Zheng, Q., & Wu, Z. L. (2022). Digital light processing 3D printing of hydrogels: A minireview. *Molecular Systems Design & Engineering*, 7(9), 1017–1029. https://doi.org/10.1039/d2me00066k
129. Dong, M., Han, Y., Hao, X. P., Yu, H., Yin, J., Du, M., . . . Wu, Z. L. (2022). Digital light processing 3D printing of tough supramolecular hydrogels with sophisticated architectures as impact-absorption elements. *Advanced Materials*, 34(34). https://doi.org/10.1002/adma.202204333
130. Cafiso, D., Septevani, A. A., Noè, C., Schiller, T. L., Pirri, C., Roppolo, I., & Chiappone, A. (2022). 3D printing of fully cellulose-based hydrogels by digital light processing. *Sustainable Materials and Technologies*, 32, e00444. https://doi.org/10.1016/j.susmat.2022.e00444

131. Wang, L., Wang, Q., Slita, A., Backman, O., Gounani, Z., Rosqvist, E., . . . Wang, X. (2022). Digital light processing (DLP) 3D-fabricated antimicrobial hydrogel with a sustainable resin of methacrylated woody polysaccharides and hybrid silver-lignin nanospheres. *Green Chemistry*, 24(5), 2129–2145. https://doi.org/10.1039/d1gc03841a
132. Agrawal, A., & Hussain, C. M. (2023). 3D-printed hydrogel for diverse applications: A review. *Gels*, 9(12), 960. https://doi.org/10.3390/gels9120960
133. Lee, K. Y., & Mooney, D. J. (2001). Hydrogels for tissue engineering. *Chemical Reviews*, 101(7), 1869–1880. https://doi.org/10.1021/cr000108x
134. Gungor-Ozkerim, P. S., İnci, İ., Zhang, Y. S., Khademhosseini, A., & Dokmeci, M. R. (2018). Bioinks for 3D bioprinting: An overview. *Biomaterials Science*, 6(5), 915–946. https://doi.org/10.1039/c7bm00765e
135. Chimene, D., Kaunas, R., & Gaharwar, A. K. (2019). Hydrogel BioInk reinforcement for additive manufacturing: A focused review of emerging strategies. *Advanced Materials*, 32(1). https://doi.org/10.1002/adma.201902026
136. Ouyang, L., Yao, R., Zhao, Y., & Sun, W. (2016). Effect of bioink properties on printability and cell viability for 3D bioplotting of embryonic stem cells. *Biofabrication*, 8(3), 035020. https://doi.org/10.1088/1758-5090/8/3/035020
137. Murphy, S. V., & Atala, A. (2014). 3D bioprinting of tissues and organs. *Nature Biotechnology*, 32(8), 773–785. https://doi.org/10.1038/nbt.2958
138. Skardal, A., & Atala, A. (2014). Biomaterials for Integration with 3-D Bioprinting. *Annals of Biomedical Engineering*, 43(3), 730–746. https://doi.org/10.1007/s10439-014-1207-
139. Gröll, J., Boland, T., Blunk, T., Burdick, J. A., Cho, D., Dalton, P. D., Derby, B., Forgács, G., Li, Q., Mironov, V., Moroni, L., Nakamura, M., Shu, W., Takeuchi, S., Vozzi, G., Woodfield, T., Xu, T., Yoo, J. J., & Malda, J. (2016). Biofabrication: reappraising the definition of an evolving field. *Biofabrication*, 8(1), 013001. https://doi.org/10.1088/1758-5090/8/1/013001
140. Hoare, T., & Kohane, D. S. (2008). Hydrogels in drug delivery: Progress and challenges. *Polymer*, 49(8), 1993–2007. https://doi.org/10.1016/j.polymer.2008.01.027
141. Highley, C. B., Rodell, C. B., & Burdick, J. A. (2015). Direct 3D printing of shear-thinning hydrogels into self-healing hydrogels. *Advanced Materials*, 27(34), 5075–5079. https://doi.org/10.1002/adma.201501234
142. Shin, S. R., Aghaei-Ghareh-Bolagh, B., Gao, X., Nikkhah, M., Jung, S. M., Dolatshahi-Pirouz, A., Kim, S. B., Kim, S. M., Dokmeci, M. R., Tang, X. S., & Khademhosseini, A. (2014). Layer-by-layer assembly of 3D tissue constructs with functionalized graphene. *Advanced Functional Materials*, 24(39), 6136–6144. https://doi.org/10.1002/adfm.201401300
143. Bahram, M., Mohseni, N., & Moghtader, M. (2016). An introduction to hydrogels and some recent applications. *InTech eBooks*. https://doi.org/10.5772/64301
144. Gounden, V., & Singh, M. (2024). Hydrogels and wound healing: Current and future prospects. *Gels*, 10(1), 43. https://doi.org/10.3390/gels10010043
145. Kammona, O., Tsanaktsidou, E., & Kiparissides, C. (2024). Recent developments in 3D-(Bio)printed hydrogels as wound dressings. *Gels*, 10(2). https://doi.org/10.3390/gels10020147
146. Deptuła, M., Zawrzykraj, M., Sawicka, J., Banach-Kopeć, A., Tylingo, R., & Pikuła, M. (2023). Application of 3D- printed hydrogels in wound healing and regenerative medicine. *Biomedicine & Pharmacotherapy*, 167, 115416. https://doi.org/https://doi.org/10.1016/j.biopha.2023.115416
147. Kim, N., Lee, H., Han, G., Kang, M., Park, S., Kim, D. E., Lee, M., Kim, M.-J., Na, Y., Oh, S., Bang, S.-J., Jang, T.-S., Kim, H.-E., Park, J., Shin, S. R., & Jung, H.-D. (2023). 3D-Printed functional hydrogel by DNA-induced biomineralization for accelerated diabetic wound healing. *Advanced Science*, 10(17), 2300816. https://doi.org/https://doi.org/10.1002/advs.202300816
148. Zhang, K., Wang, Y., Wei, Q., Li, X., Guo, Y., & Zhang, S. (2021). Design and fabrication of sodium alginate/Carboxymethyl cellulose sodium blend hydrogel for artificial skin. *Gels*, 7(3). https://doi.org/10.3390/gels7030115
149. Wang, S., Xiong, Y., Chen, J., Ghanem, A., Wang, Y., Yang, J., & Sun, B. (2019). Three dimensional printing bilayer membrane scaffold promotes wound healing. *Frontiers in Bioengineering and Biotechnology*, 7. https://doi.org/10.3389/fbioe.2019.00348
150. Shen, H.-Y., Liu, Z.-H., Hong, J.-S., Wu, M.-S., Shiue, S.-J., & Lin, H.-Y. (2021). Controlled-release of free bacteriophage nanoparticles from 3D-plotted hydrogel fibrous structure as potential antibacterial wound dressing. *Journal of Controlled Release*, 331, 154–163. https://doi.org/https://doi.org/10.1016/j.jconrel.2021.01.024
151. Piola, B., Sabbatini, M., Gino, S., Invernizzi, M., & Renò, F. (2022). 3D Bioprinting of gelatin–xanthan gum composite hydrogels for growth of human skin cells. *International Journal of Molecular Sciences*, 23(1). https://doi.org/10.3390/ijms23010539

152. Zarandona, I., Bengoechea, C., Álvarez-Castillo, E., de la Caba, K., Guerrero, A., & Guerrero, P. (2021). 3D Printed chitosan-pectin hydrogels: From rheological characterization to scaffold development and assessment. *Gels*, 7(4). https://doi.org/10.3390/gels7040175
153. Leppiniemi, J., Lahtinen, P., Paajanen, A., Mahlberg, R., Metsä-Kortelainen, S., Pinomaa, T., Pajari, H., Vikholm-Lundin, I., Pursula, P., & Hytönen, V. P. (2017). 3D-Printable bioactivated nanocellulose–alginate hydrogels. *ACS Applied Materials & Interfaces*, 9(26), 21959–21970. https://doi.org/10.1021/acsami.7b02756
154. Nizioł, M., Paleczny, J., Junka, A., Shavandi, A., Dawiec-Liśniewska, A., & Podstawczyk, D. (2021). 3D Printing of thermoresponsive hydrogel laden with an antimicrobial agent towards wound healing applications. *Bioengineering*, 8(6). https://doi.org/10.3390/bioengineering8060079
155. Malekmohammadi, S., Sedghi Aminabad, N., Sabzi, A., Zarebkohan, A., Razavi, M., Vosough, M., Bodaghi, M., & Maleki, H. (2021). Smart and biomimetic 3D and 4D printed composite hydrogels: Opportunities for different biomedical applications. *Biomedicines*, 9(11). https://doi.org/10.3390/biomedicines9111537
156. Min Mu Xiaoling Li, A. T., & Guo, G. (2019). Multi-functional chitosan-based smart hydrogels mediated biomedical application. *Expert Opinion on Drug Delivery*, 16(3), 239–250. https://doi.org/10.1080/17425247.2019.1580691
157. Liu, C., Wang, Z., Wei, X., Chen, B., & Luo, Y. (2021). 3D printed hydrogel/PCL core/shell fiber scaffolds with NIR-triggered drug release for cancer therapy and wound healing. *Acta Biomaterialia*, 131, 314–325. https://doi.org/https://doi.org/10.1016/j.actbio.2021.07.011
158. Tsegay, F., Elsherif, M., & Butt, H. (2022). Smart 3D printed hydrogel skin wound bandages: A review. *Polymers*, 14(5). https://doi.org/10.3390/polym14051012
159. de León, E. H.-P., Valle-Pérez, A. U., Khan, Z. N., & Hauser, C. A. E. (2023). Intelligent and smart biomaterials for sustainable 3D printing applications. *Current Opinion in Biomedical Engineering*, 26, 100450. https://doi.org/https://doi.org/10.1016/j.cobme.2023.100450
160. Abedi, F., Ghandforoushan, P., Adeli, F., Yousefnezhad, M., Mohammadi, A., Moghaddam, S. v, & Davaran, S. (2023). Development of stimuli-responsive nanogels as drug carriers and their biomedical application in 3D printing. *Materials Today Chemistry*, 29, 101372. https://doi.org/https://doi.org/10.1016/j.mtchem.2022.101372
161. Yang, Y., Yu, Z., Lu, X., Dai, J., Zhou, C., Yan, J., Wang, L., Wang, Z., & Zang, J. (2023). Minimally invasive bioprinting for in situ liver regeneration. *Bioactive Materials*, 26, 465–477. https://doi.org/https://doi.org/10.1016/j.bioactmat.2023.03.011
162. Huang, Q., Wu, T., Guo, Y., Wang, L., Yu, X., Zhu, B., Fan, L., Xin, J. H., & Yu, H. (2023). Platelet-rich plasma-loaded bioactive chitosan@sodium alginate@gelatin shell-core fibrous hydrogels with enhanced sustained release of growth factors for diabetic foot ulcer healing. *International Journal of Biological Macromolecules*, 234, 123722. https://doi.org/https://doi.org/10.1016/j.ijbiomac.2023.123722
163. Ghilan, A., Chiriac, A. P., Nita, L. E., Rusu, A. G., Neamtu, I., & Chiriac, V. M. (2020). Trends in 3D printing processes for biomedical field: Opportunities and challenges. *Journal of Polymers and the Environment*, 28(5), 1345–1367. https://doi.org/10.1007/s10924-020-01722-x
164. Barcena, A. J. R., Dhal, K., Patel, P., Ravi, P., Kundu, S., & Tappa, K. (2024). Current biomedical applications of 3D-printed hydrogels. *Gels*, 10(1). https://doi.org/10.3390/gels10010008
165. de León, E. H.-P., Valle-Pérez, A. U., Khan, Z. N., & Hauser, C. A. E. (2023). Intelligent and smart biomaterials for sustainable 3D printing applications. *Current Opinion in Biomedical Engineering*, 26, 100450. https://doi.org/https://doi.org/10.1016/j.cobme.2023.100450
166. Yan, J., Liu, X., Liu, J., Zhang, X., Zheng, Q., Nan, J., Lin, M., Pan, H., Wang, Y., Cai, X., & Yin, J. (2021). A dual-layer cell-laden tubular scaffold for bile duct regeneration. *Materials & Design*, 212, 110229. https://doi.org/https://doi.org/10.1016/j.matdes.2021.110229
167. Lee, H. S., Jeon, E. Y., Nam, J. J., Park, J. H., Choi, I. C., Kim, S. H., Chung, J. J., Lee, K., Park, J. W., & Jung, Y. (2022). Development of a regenerative porous PLCL nerve guidance conduit with swellable hydrogel-based microgrooved surface pattern via 3D printing. *Acta Biomaterialia*, 141, 219–232. https://doi.org/https://doi.org/10.1016/j.actbio.2022.01.042
168. Yang, J., Yang, K., Man, W., Zheng, J., Cao, Z., Yang, C.-Y., Kim, K., Yang, S., Hou, Z., Wang, G., & Wang, X. (2023). 3D bio-printed living nerve-like fibers refine the ecological niche for long-distance spinal cord injury regeneration. *Bioactive Materials*, 25, 160–175. https://doi.org/https://doi.org/10.1016/j.bioactmat.2023.01.023
169. Agrawal, A., & Hussain, C. M. (2023). 3D-Printed hydrogel for diverse applications: A review. Gels, 9(12). https://doi.org/10.3390/gels9120960
170. Wang, S., Lee, J. M., & Yeong, W. Y. (2015). Smart hydrogels for 3D bioprinting. *International Journal of Bioprinting*, 1. https://doi.org/10.18063/IJB.2015.01.005

6 3D-Printed Drug-Loaded Nanomaterials for Wound Healing

Alejandro Huerta-Saquero,[1] Yaaziel Melgarejo-Ramírez,[2] Alberto Caballero-Ruiz[3,4], and Zaira García Carvajal[5]*

[1] Universidad Nacional Autónoma de México, Baja California, Mexico
[2] Instituto Nacional de Rehabilitación Luis Guillermo Ibarra Ibarra, Mexico City, Mexico
[3] Universidad Nacional Autónoma de México, Mexico City, Mexico
[4] National Laboratory for Additive and Digital Manufacturing MADiT, Mexico City, Mexico
[5] Centro de Investigación y Asistencia en Tecnología y Diseño del Estado de Jalisco, A.C. (CIATEJ), Guadalajara, Mexico
*Corresponding Author

6.1 WOUND HEALING: AN OVERVIEW

The skin is the largest organ in the human body and has several functions [1]. The skin provides a strong barrier against various physical and chemical insults and harmful pathogens. At the same time, it has to facilitate subtle communication with the environment [2, 3]. However, the skin is susceptible to various external stimuli, leading to skin injuries and damage [4]. According to the Wound Healing Society, a wound results from "alteration of the normal anatomical structure and function of the skin" [5].

Wound healing is a complex and dynamic biological process that requires a suitable environment to promote the healing process. One can make the analogy of an orchestra to understand the wound-healing process. To perform a beautiful symphony, everyone has to play their part precisely so the patient can heal successfully [6, 7]. After a skin injury or trauma, many physiologic processes start, both locally and systemically. They aim to restore the skin's integrity, avoid fluid loss, and reduce infection risks. These processes are not linear; most of the time, they overlap in space and time. The intensity of the injury will determine the duration of the wound-healing process. This process comprises four typical phases: hemostasis, inflammation, proliferation, and matrix remodeling [8].

Hemostasis starts within the first 24 hours after an injury or trauma as a response to vascular endothelial damage. Its main objective is to stop hemorrhage from the blood vessels. It starts with vasoconstriction and the subsequent formation of the blood clot. Primary hemostasis consists of platelet aggregation, activation, and blood clot formation. Secondary hemostasis involves activating the coagulation cascade's extrinsic pathway as a response to external traumatic events. This leads to the formation of a fibrin clot, which is physiologically more stable than the blood clot that proceeded [9]. Platelets are the first cells arriving at the wound site; they promote infiltration of immune cells, such as monocytes and neutrophils, from the blood vessels. In response to endothelial damage, they secrete platelet-derived growth factor (PDGF), stimulating endothelial cells and smooth muscle cells to repair blood vessels. Bacterial inhibition is achieved through the expression of Toll-like receptors (TLRs) on its surface.

DOI: 10.1201/9781003504948-6

These receptors constitute the first line of defense against damage and infection, maintain the inflammatory response, and activate innate immunity. Platelets also stimulate keratinocytes from the epidermis, fibroblasts from the dermis, epithelial cells, and macrophages to secrete growth factors [10].

Inflammatory phase. Opposite to hemostasis, vasodilatation and an increase in vascular permeability occur at the wound site between the first 24 hours and days 4–6. Innate inflammation is activated as a response to the release of damage-associated molecular patterns (DAMPs) and pathogen-associated molecular patterns (PAMPs). Vascular permeability increases as endothelial cells elevate the expression of surface selectins. These proteins favor the diapedesis of neutrophils and monocytes from the blood to the wounded tissue, reaching their maximum peak 24 hours after trauma. Neutrophils are recruited by interleukin 1 (IL-1), tumor necrosis factor-alpha (TNF-α), and transforming growth factor beta (TGF-β). They can phagocytose necrotic tissue and pathogens to generate reactive oxygen species (ROS), which sustain an inflammatory state. As monocytes infiltrate the wound site, they differentiate into macrophages due to IL-2, TNF-α, PDGF, and interferon-gamma (IFN-γ) stimulation, secreted by T lymphocytes. Macrophages are key cells for wound recovery and reach their peak 48–92 hours post-injury; they secrete angiogenic mediators such as vascular endothelial growth factor (VEGF), fibroblast growth factor (FGF), endothelial growth factor (EGF), and PDGF. Additionally, IL-1 and TNF-α stimulate inducible nitric oxide synthase (iNOS) expression by macrophages, subsequently elevating nitric oxide levels, EGF, and TGF-β, stimulating epithelial cell proliferation, and chemotaxis. Pro-inflammatory macrophages, activated via the classical pathway, secrete pro-resolving cytokines such as IL-4, IL-10, and IL-13, which promote re-epithelialization. The TNF-α stimulates keratinocytes, fibroblasts, monocytes, and macrophages to secrete matrix metalloproteinases (MMPs) in subsequent matrix remodeling [11].

Proliferation phase. During the proliferation phase, relevant processes such as wound closure (keratinocytes), granulation tissue formation, re-epithelialization, extracellular matrix (ECM) deposition by fibroblasts, and angiogenesis by endothelial cells occur. This phase takes place between days 4 to 6 and 14 post-injury. Platelets and macrophages secrete EGF and TGF-β to activate the proliferation of epithelial cells from the edges to the center of the wound. Fibroblasts promote the formation of granulation tissue, which is rich in fibronectin, immature collagen, and proteoglycans. IL-1 and TNF-α stimulate fibroblasts to secrete keratinocyte growth factor 2 (KGF-2) that induces keratinocyte polarity, migration, proliferation, and differentiation changes to fulfill epithelialization. Angiogenesis emerges as a response of endothelial cells to VEGF, secreted by keratinocytes, macrophages, fibroblasts, and platelets. Hypoxia at the wound site induces endothelial nitric oxide synthase (eNOS) and iNOS expression. Hence, nitric oxide and VEGF levels increase as vasodilation takes place. Finally, fibroblasts migrate to the wound site and differentiate into pro-fibrotic fibroblasts, responsible for synthesizing type III collagen, glycosaminoglycans, and fibronectin. On the other hand, they can also differentiate into myofibroblasts, which will reduce proliferation under TGF-β stimuli and secrete type I collagen to promote contraction and wound closure [12].

Remodeling phase. The remodeling phase is the most relevant phase from the clinical point of view, and it is characterized by organized collagen deposition between days 14 and 1 and more years post-injury. After skin damage, platelets and macrophages synthesize a fibrin and fibronectin-rich matrix to stop bleeding and fluid loss. Then, fibroblasts replace this matrix with glycosaminoglycans, proteoglycans, and other ECM molecules; then, they synthesize type III and type I collagen. At the same time, molecules like TGF-β, IL-1, and IL-6 stimulate fibroblasts to secrete MMPs to improve matrix remodeling. As previously explained, TGF-β stimulates fibroblast differentiation into myofibroblasts that secrete smooth muscle actin alpha (α-SMA) to remodel the extracellular matrix and facilitate wound closure [12].

6.1.1 Three-Dimensional (3D) Structural Microenvironment of Wounds

Most research has focused on identifying key bioactive molecules and chemokines that can influence the stages of the wound healing process in two-dimensional (2D) models. However, the traumatic wound's three-dimensional (3D) structural microenvironment impacts healing [13].

The skin's 3D microenvironment is vital for wound healing. It has been described that cell differentiation, biomechanical interconnections, and the accumulation of growth factors rely on the 3D conformation of the ECM in order to fulfill wound recovery and closure. The 3D structure connects cells to their surroundings, enabling cell-to-cell communication to adapt and respond to external stimuli. In vitro, static 2D environments limit cell interactions to proteins like integrins, fibronectin, or collagens. In contrast, 3D environments result in highly dynamic interactions, focal adhesion formation, and cell migration activation in different modes: mesenchymal, amoeboid, and lobopodial cell migration [14]

The 3D microstructure is evident at the biological level during platelet aggregation, activated to prevent hemorrhage after injury. Endothelial damage activates blood platelets to initiate plug formation and stop hemorrhage. As a result of the coagulation cascade, soluble fibrinogen turns into a network of insoluble fibrin fibers, the most relevant transitory 3D ECM that reestablishes cell–cell interactions and provides a provisional matrix in the wounded tissue, regulates inflammation, and regulates physiological cell activities. This transitory fibrin matrix is also constituted by fibronectin and vitronectin, proteins that conjugate endothelial cells, smooth muscle cells, and other specialized cells. Degradation products of this temporary ECM also stimulate fibroblast proliferation and angiogenesis, giving rise to granulation tissue formation, a stage characterized by the gross granular appearance of the tissue constituted by a dense quantity of macrophages, fibroblasts, and neovasculature embedded in a relaxed matrix of collagen and fibronectin [15].

6.1.2 Wound Care Management

Wound care and soft tissue repair, especially recalcitrant wounds, remain challenging for the healthcare system worldwide [16]. The aging of the population complicates the outlook, adding to the increase in chronic diseases such as vascular diseases and diabetes, and the higher incidence of bacterial resistance to antibiotics [17].

Great efforts have been made to develop alternative therapies that restore the regenerative properties of native skin by restoring skin barrier function in complex situations [6].

Despite significant advances in chronic wound care, wound management remains a significant challenge and financial burden for patients and the healthcare sector. Many products exist to treat different types of wounds, focusing on various aspects of the healing process, but most strategies focus on using dressings. Wound dressings are the gold standard for treating severe skin injuries, as they provide necessary moisture, protect the wound bed from the external environment, and accelerate the healing process [13, 16, 18, 19]. Depending on the type of wound, the appropriate dressing should be used [7, 20]. The type of wound, its size, depth, and location affect the healing process and closure of the wound [21]. When wounds do not heal adequately, they require enormous resources from the medical system, causing suffering to patients since a poorly cared-for wound can cause complications that are exacerbated in cases of diabetes or infections [21–23].

However, innovative and multifunctional dressings that facilitate and promote the healing process in a personalized way according to the wound's location, type, and size have been booming [24]. Furthermore, incorporating therapeutic agents into these systems would improve the treatment of these lesions in an easy, painless way and would favor the integral restoration of damaged skin [5]. In addition, numerous drug delivery systems have been developed for wound healing and skin regeneration [25].

6.2 NANOMATERIALS FOR WOUND HEALING

The development of nanomaterials offers extraordinary alternatives for wound healing. Currently, nanomaterials are being developed as delivery vehicles for bioactive components for wound healing, antimicrobial agents, and growth factors [26].

Nanomaterials for drug delivery have shown promising applications in wound healing by directly releasing therapeutic agents at the wound site [27]. These materials offer several advantages over conventional drug delivery systems, including improved stability, better bioavailability, and targeted delivery [28].

Nanotechnology presents unique and alternative approaches to accelerate the healing of chronic wounds through the interaction of nanomaterials during different stages of the wound healing process [27, 29].

Nanoparticles. Nanoparticles, such as liposomes, polymeric nanoparticles, and dendrimers, can encapsulate drugs and growth factors, protecting them from degradation and allowing sustained release kinetics at the wound site [29]. Liposomes, for example, have been used to deliver antibiotics, anti-inflammatory agents, and growth factors to promote wound healing [30].

Hydrogels. Hydrogels are three-dimensional networks of hydrophilic polymers capable of retaining large amounts of water. They provide a moist environment conducive to wound healing and can be loaded with therapeutic agents for controlled release. Additionally, hydrogels can be designed to respond to specific stimuli, such as pH or temperature, for on-demand drug release [19].

Nanofibers. Electrospun nanofibers have attracted attention for wound healing applications due to their high surface area, porosity, and similarity to the extracellular matrix. Nanofibers can be functionalized with drugs or growth factors and applied as dressings or scaffolds to facilitate tissue regeneration and wound closure [31].

Microneedling. Microneedles are minimally invasive drug delivery devices that penetrate the skin barrier to deliver therapeutic agents directly to the wound bed. They offer precise control over drug release kinetics and can be manufactured from biocompatible materials, such as polymers or metals. Microneedle patches have been explored to deliver antibiotics, analgesics, and growth factors in wound healing applications [31].

Nanostructured surfaces. Nanostructured surfaces can modulate cellular responses at the wound site and promote tissue regeneration [32]. Nanotopographies, nanopatterned substrates, and nanocoatings can improve cell adhesion, migration, and proliferation and ultimately accelerate wound healing [33].

By taking advantage of the unique properties of nanomaterials for drug delivery, researchers aim to develop advanced wound care therapies with greater efficacy and fewer side effects [17, 29, 34, 35].

Among them, nanocomposite-based printable inks have been developed that include carbon nanomaterials, transition metal chalcogenides, and polymeric nanoparticles that form networks composed of nanofibers dispersed into a hydrogel matrix [36].

6.3 ADDITIVE MANUFACTURING

According to the ISO/ASTM 52900:2021 standard, additive manufacturing is defined as the process of joining materials to make parts from 3D model data, usually layer upon layer, as opposed to subtractive or formative manufacturing methodologies.

In addition, according to the same standard, 3D printing is related to the fabrication of objects employing printer heads, nozzles, or other printer technologies; this term is often used in nontechnical contexts as synonymous with additive manufacturing. Although there are many additive manufacturing technologies in the world, which the manufacturers themselves commonly name, the ISO/ASTM 52900:2021 standard brings together all these technologies into seven processes (International Organization for Standardization, 2021), in which the particularity of each process regards the raw material, the principle of deposition, and the method of material curing [37–39].

6.3.1 Additive Manufacturing Processes: A Short Explanation

The binder jetting process uses powder as raw material, and a liquid binder is applied to bind each layer. Due to supply chain disruptions, the need for innovation has emerged both as a priority and a necessity, and 3D printing proved to be a primary, smart, effective, and innovative additive manufacturing method [40]. In the directed energy deposition process, powder particles are deposited onto a surface; in contrast, an energy source fuses them. The material extrusion process can use solid, semisolid, or liquid raw materials, which are extruded and selectively deposited through a nozzle [41]. In the material jetting process, the liquid material is jetted through a nozzle and cured by light or another principle [42]. Some technologies use printer heads [43]. The powder bed fusion process uses an energy source to selectively fuse the powder particles onto a bed [44]; the sheet lamination process is based on material with a sheet shape stacked layer by layer to conform a part [45]. Finally, the vat photopolymerization process consists of liquid photopolymer, which is selectively cured by a light source (laser, image projection, or LED) [46].

6.3.2 3D-Printable Technologies for Wound Healing

Additive manufacturing has revolutionized and transformed traditional approaches in wound management therapies, as they can produce more complex, functional, and personalized three-dimensional architectures with better epithelial tissue mimicking behavior. The first research on organ printing was published in the early 2000s [44], describing a system derived from commercial inkjet devices depositing proteins and cells in 2D arrays. Boland et al. (2003) developed a system able to drop cells layer by layer, providing the opportunity for developing 3D structures. In 2004, Ben Hsieh et al. [47] used bioprinting to describe a system capable of depositing 12 biological fluids to create arrays.

6.3.2.1 Bioprinting and Bioinks

Bioprinting is an emerging technology with applications in making functional tissue constructs to replace injured or diseased tissues [48]. Bioprinting refers to technologies capable of developing three-dimensional living cell arrays using additive manufacturing processes. The three additive manufacturing processes commonly employed in bioprinting are material extrusion, material jetting, and vat photopolymerization [49]. During the bioprinting process, it is generally necessary to encapsulate the desired cell types using a solution of biomaterials or a mixture of several biomaterials to form hydrogels called bioinks [50, 51]. Bioinks are used in 3D bioprinting to create complex tissue structures. Bioinks typically consist of living cells, biomaterials, and other bioactive components and their interactions to create a formulation that supports cell viability, growth, and tissue function [50–52]. Bioinks can be natural or synthetic polymers or mixtures selected for their biocompatible components and their favorable rheological properties to be extruded [53].

The main challenge related to bioprinting is guaranteeing the integration of cells because it depends on multiple factors such as stress [54], temperature [55], and thermal and radiative influence of light [56, 57], among others. The typical bioprinting applications are related to tissue generation, like bone, cartilage, or skin [55, 57–59].

The development and use of bioinks have revolutionized the field of 3D cell culture [51, 60]. However, ensuring terminal sterilization is critical in developing and applying bioinks [53]. The combination of additive manufacturing and the effective administration of treatments at the nanoscale allows the development of 3D models of in vitro study of the traumatic wound [34, 35]. The challenge is to improve the manufacturing of smart dressings with 3D architectures, enabling key functionalities such as accelerated and effective wound healing [18, 61, 62].

Many bioinks based on natural and synthetic polymers have been used as printable materials in wound healing treatment [18, 62]. Table 6.1 summarizes the notable nanomaterials for bioinks used in 3D bioprinting.

In recent years, the interest in bioprinting in wound healing has grown [13, 18, 20, 22, 24, 62–64].

Natural nanomaterials included in bioinks for 3D hydrogels: Natural nanomaterials offer promising characteristics for bioinks used in bioprinting applications due to their biocompatibility, biodegradability, and ability to mimic the extracellular matrix of tissues, promoting cell adhesion,

TABLE 6.1
Notable Nanomaterials for Bioinks Used in 3D Bioprinting

Type of Material for Bioinks	Description	Reference
Natural		
Alginate	Alginate hydrogels have been used extensively as bioinks for 3D bioprinting. Due to its biocompatibility and gelation, alginate is a natural biopolymer that forms hydrogels by ionic cross-linking with calcium ions. Alginate is easily combined with other polysaccharides and additives to enhance performance, including nanomaterials that tune the rheological properties, peptides to encourage cell adhesion, or growth factors to direct stem cell differentiation. Alginate is used as a cell encapsulation material and in tissue engineering scaffolds due to its biocompatibility and degradation.	[76, 77]
Nanocellulose	Derived from plant sources such as wood pulp, nanocellulose possesses excellent mechanical properties, such as high tensile strength and stiffness, making it suitable for bioink formulations. Its nanoscale dimensions mimic the fibrous structure of the extracellular matrix, providing an environment conducive to cell growth. Furthermore, nanocellulose-based bioinks exhibit thinning behavior, facilitating extrusion through printing nozzles. Surface modifications can further improve its bioactivity and compatibility with different cell types.	[78]
Chitosan nanoparticles	Chitosan, a derivative of chitin, is widely investigated for biomedical applications due to its biocompatibility, antimicrobial properties, and biodegradability. When formulated into nanoparticles, chitosan offers improved colloidal stability and controlled release of encapsulated bioactive molecules, such as growth factors or drugs, within printed constructs. Chitosan nanoparticles can be functionalized to promote specific cellular interactions and tissue regeneration.	[79]
Silk nanofibers	Silk fibroin, derived from silkworm cocoons, has attracted attention in tissue engineering due to its remarkable mechanical properties, biocompatibility, and controllable degradation. When processed into nanofibers, silk fibroin provides a structurally biomimetic scaffold for cell attachment, proliferation, and differentiation. Silk-based bioinks can be tailored by adjusting the concentration and processing parameters to achieve the desired rheological properties for bioprinting.	[80]
Hyaluronic acid nanogels	Hyaluronic acid (HA) is a natural polysaccharide found in connective tissues and has excellent biocompatibility and hydration properties. When cross-linked in nanogels, HA forms a 3D network capable of encapsulating bioactive molecules and promoting cell proliferation and migration. HA nanogels can be loaded with growth factors or peptides to modulate cellular responses and tissue regeneration within printed constructs.	[81]
Collagens	Collagens are the most abundant protein found throughout the body. In the healing wound, these collagens are synthesized by fibroblast cells and modified into complex morphologies. Collagen hydrogels have many cell attachment ligands required for cell adhesion and proliferation but have poor mechanical properties.	[82, 83]

(Continued)

TABLE 6.1 (CONTINUED)
Notable Nanomaterials for Bioinks Used in 3D Bioprinting

Type of Material for Bioinks	Description	Reference
Synthetic: Metals and ceramics		
Conductive polymers, such as polyaniline (PANI), polypyrrole (PPy), and poly(3,4-ethylenedioxythiophene) (PEDOT)	There are examples of synthetic metals used in bioinks. These materials offer tunable conductivity, mechanical flexibility, and biocompatibility, making them suitable for various tissue engineering applications. Incorporation of conductive polymers into bioinks can promote cell adhesion, proliferation, and differentiation, especially in electrically excitable tissues such as cardiac and neural tissues.	[84–86]
Ceramic nanoparticles, such as hydroxyapatite (HA), silicon dioxide (SiO2), and titanium dioxide (TiO2)	These nanomaterials can improve hydrogels' mechanical strength, bioactivity, and osteoconductivity, making them promising candidates for bone tissue engineering. Additionally, ceramic nanoparticles can carry bioactive molecules, growth factors, or drugs, allowing controlled release within the 3D scaffold.	[87]
Hybrid nanomaterials		
Hybrid organic–inorganic nanomaterials	Organic components, typically biopolymers such as gelatin, alginate, or hyaluronic acid, provide the biocompatibility and bioactivity of the bioink. These biopolymers mimic the environment of the extracellular matrix, facilitating cell adhesion, proliferation, and differentiation. Inorganic components, such as nanoparticles or nanosheets, offer unique properties such as conductivity, magnetism, or optical activity, which can be exploited for various biomedical applications.	[71]

proliferation, and differentiation [34, 64, 65]. These natural nanomaterials offer versatile platforms to engineer bioinks with tailored properties for various tissue engineering applications [66]. By taking advantage of their unique characteristics and bioactive functionalities, advanced bioprinting strategies can be developed to fabricate complex and functional tissues and organs [57, 58, 63, 66].

Synthetic metals and ceramic nanomaterials included in bioinks for 3D hydrogels: Synthetic ceramic and metallic nanomaterials have attracted attention for their potential applications in bioinks for 3D hydrogels due to their unique properties that can improve the resulting constructs' mechanical, electrical, and structural characteristics [67–70].

The combination of synthetic metals and ceramic nanomaterials in bioinks offers a synergistic approach to tailor the physicochemical properties of 3D hydrogels for specific tissue engineering applications, such as cardiac patches, neuronal scaffolds, bone substitutes, and bioelectronic devices [58, 66, 70, 71].

Hybrid nanomaterials included in bioinks for 3D hydrogels: Hybrid nanomaterials integrated into bioinks for 3D printing hydrogels represent an exciting frontier in tissue engineering and regenerative medicine. These hybrid materials combine the advantageous properties of organic (biological) and inorganic (synthetic) components, allowing for greater mechanical strength, bioactivity, and customized functionalities for wound healing [62].

Hybrid nanomaterials improved the mechanical properties of the materials. For example, they strengthen the mechanical properties of the resulting bioinks and hydrogels. The incorporation of nanoclays [72, 73] or graphene oxide (GO) nanosheets can significantly improve the stiffness and toughness of the hydrogel, which is crucial for promoting cell growth and tissue development [74]. The synergistic combination of organic and inorganic components can overcome the limitations of individual materials, resulting in hydrogels with greater structural integrity and resilience [75].

Hybrid nanomaterials allow precise control over the biofunctionalization of hydrogels. Functional groups on the surface of the nanoparticles can be conjugated with bioactive molecules, growth factors, or drugs, allowing spatiotemporal release within the 3D scaffold. This capability is precious for creating complex tissue constructs with customized biochemical signals to guide cellular behavior and tissue regeneration [68–71].

Khoshnood and Zamanian developed a bioink composed of alginate at 6% and polyethyleneimine (PEI) for the material extrusion process; the concentrations were varied in the range of 3%–6% (w/v) and 5%–9% (w/v), respectively, showing good biocompatibility and cellular functions at concentrations of alginate at 6% and PEI at 7%, with good results in skin regeneration during in vivo study on a rat model [88]. Lin et al. developed a hydrogel based on fish skin decellularized extracellular matrix, which is functional for the material extrusion process and exhibits satisfactory therapeutic effects when used to treat rats with bacterial infection and injury [89]. Ceballos-Santa et al. proposed using a material extrusion process for wound dressing constructed with aloe vera/alginate biomaterial inks, demonstrating its biocompatibility at the proper concentration [90].

Dental implants are the frequently preferred method for oral rehabilitation worldwide [91]. However, it is very common for skin lesions to occur around dental implants that, if not cared for, can cause infections by bacteria and yeast, especially in elderly patients [92].

The InterLatin company, based in Guadalajara, Mexico, launched its first Colibrí 3D printer in 2012, of which it managed to sell 200 units on the same day of its launch [93].

Using the experience acquired in the last 15 years and with an internal research and development department at the forefront, in 2015, they began to develop the first Mexican bioprinter under the Innovation Stimulus Program (PEI) in Mexico (project number 252780). The innovation and scientific research in the project were focused on the oral health of the elderly. The bioprinter prototype by Colibri 3D was used to print bioinks (hydrogels) loaded with nanoemulsions containing lemon essential oil against yeast infection. Figure 6.1 shows the printing process of the hydrogel using the Colibri Bioprinter.

6.3.2.2 Bioinks Loaded with Nanomaterials: The Role of Physicochemical Characteristics

The influence of the physicochemical properties of drug-loaded nanomaterials and bioinks on controlled release during the wound healing process is crucial because they play active roles in hemostasis, antimicrobial inhibition, inflammation, and the promotion of proliferation [35, 63].

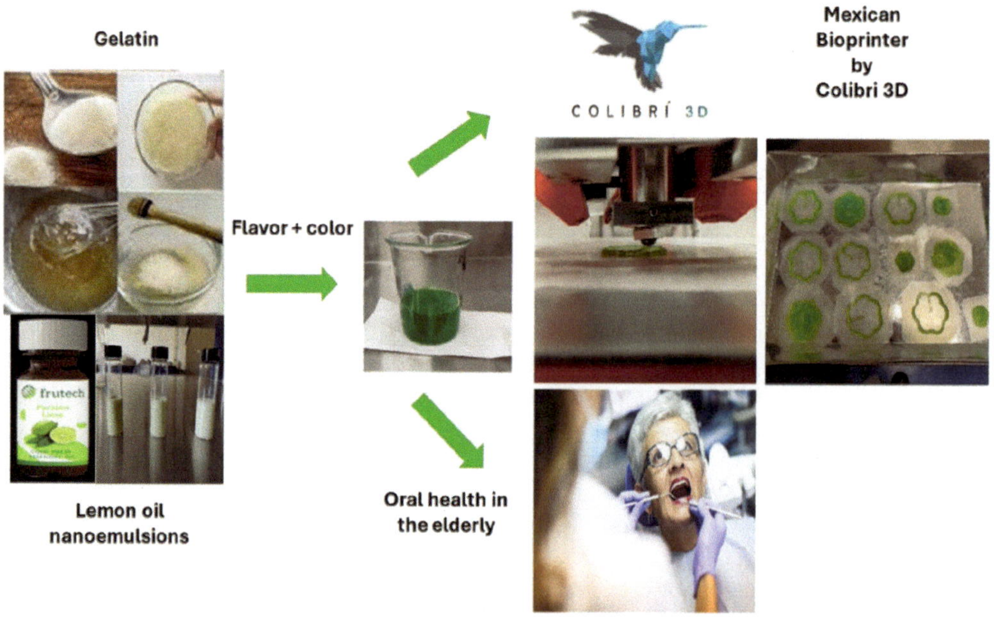

FIGURE 6.1 The printing process uses a bioink (hydrogel) loaded with lemon oil nanoemulsions using the Colibri Bioprinter. (The image was integrated using Biorender and istockphoto free (2024).)

The physicochemical properties of drug-loaded nanomaterials play a crucial role in modulating drug release kinetics and therapeutic efficacy during wound healing [94, 95]. These properties encompass several factors, such as particle size, surface charge, morphology, composition, and drug loading capacity. Understanding how these properties influence controlled drug release is essential to optimize nanomaterial design in wound healing applications [17, 25, 29, 65, 96].

1. Particle size. Nanomaterials with smaller particle sizes exhibit higher surface-to-volume ratios, leading to higher drug loading capacity and faster release kinetics. Additionally, smaller particles can penetrate deeper into the wound bed, ensuring more effective drug delivery to specific wound sites [97, 98].
2. Surface load. The surface charge of nanomaterials, determined by their zeta potential, influences their interaction with biological systems and affects drug release profiles. Positively charged nanomaterials can adhere more strongly to negatively charged cell membranes, which could lead to sustained drug release. In contrast, negatively charged nanoparticles may exhibit faster release kinetics due to repulsion from negatively charged tissues [99–101].
3. Morphology. The shape and structure of nanomaterials affect their interaction with biological components and subsequent drug release. For example, nanofibers or nanotubes offer a large surface area for drug loading and can provide sustained release profiles by controlling drug diffusion through their porous structure. In contrast, spherical nanoparticles can release drugs more rapidly due to their greater mobility and higher surface–volume ratio [99–101].
4. Composition. The choice of materials used in manufacturing nanomaterials can significantly influence drug release kinetics and biocompatibility. Biodegradable polymers such as poly(lactic-co-glycolic acid) (PLGA) or chitosan allow controlled release of drugs as they degrade over time, releasing encapsulated drugs sustainably. Inorganic nanoparticles such as mesoporous silica or gold nanoparticles offer tunable drug release profiles and can be functionalized to improve therapeutic and targeting efficacy [94, 95, 102].
5. Drug loading capacity. The ability of nanomaterials to efficiently encapsulate and retain drugs within their structure is critical for sustained release. The high loading capacity of the drug ensures that therapeutic concentrations are maintained at the wound site for prolonged periods, promoting improved tissue regeneration and accelerated wound healing [102, 103].

In the case of bioinks, the microstructure of the gel, the rheology, and the chemical composition influence cell functions and behavior while influencing printability and post-printing stability [104].

1. The microstructure of gels also impacts oxygen and nutrient diffusion and waste removal [105].
2. The rheological properties of bioink materials are crucial for their handling and use. Higher shear viscosity means smaller, well-formed droplets with high post-printing shape fidelity while inflicting more significant shear stress on cells during dispensing and increasing the risk of nozzle clogging [54].
3. The printability of hydrogels for bioprinters is crucial for fabricating high-resolution tissue models. The printability properties consider droplet formation during dispensing and wetting behavior at the droplet–substrate interface [104].

By systematically optimizing these physicochemical properties, researchers can design bioinks containing drug-loaded nanomaterials with tailored release profiles to address specific requirements of the wound healing process, ranging from promoting the resolution of inflammation and tissue regeneration to preventing infection and minimizing the formation of scars.

6.4 FUTURE DIRECTIONS AND CHALLENGES

Hybrid nanomaterials integrated into bioinks for 3D hydrogel printing offer a versatile platform to engineer complex tissue constructs with improved mechanical, biological, and functional properties, holding great promise for regenerative medicine and beyond.

The design and selection of the exploratory or confirmatory evaluation of nanomaterials and 3D technologies together are still challenging [98, 102, 103].

Furthermore, hybrid nanomaterials expand the scope of 3D bioprinting applications. For example, incorporating magnetic nanoparticles enables remote manipulation and alignment of printed constructs using external magnetic fields. Similarly, conductive nanomaterials such as carbon nanotubes or metal nanoparticles facilitate the development of bioelectronic interfaces to interact with cells or tissues, opening avenues for bioelectronic devices and biosensors integrated into living systems [95, 98, 102].

Despite promising advances, challenges remain in optimizing hybrid nanomaterials' long-term stability and biocompatibility within bioinks. Further research is needed to elucidate their interactions with cells and tissues, their degradation kinetics, and potential immunogenicity. Furthermore, the standardization of manufacturing techniques and regulatory considerations are essential to translate these innovative approaches from the laboratory to clinical applications [5, 62, 63].

Moreover, nanomaterials and nanocomposite-based hybrid bioinks for 3D bioprinting technology are still challenging because of the limitations in the clinical approaches and ethical and regulatory considerations for real potential solutions for the patients [71]. Figure 6.2 shows the main goals of combining nanomaterials and 3D technologies for wound healing.

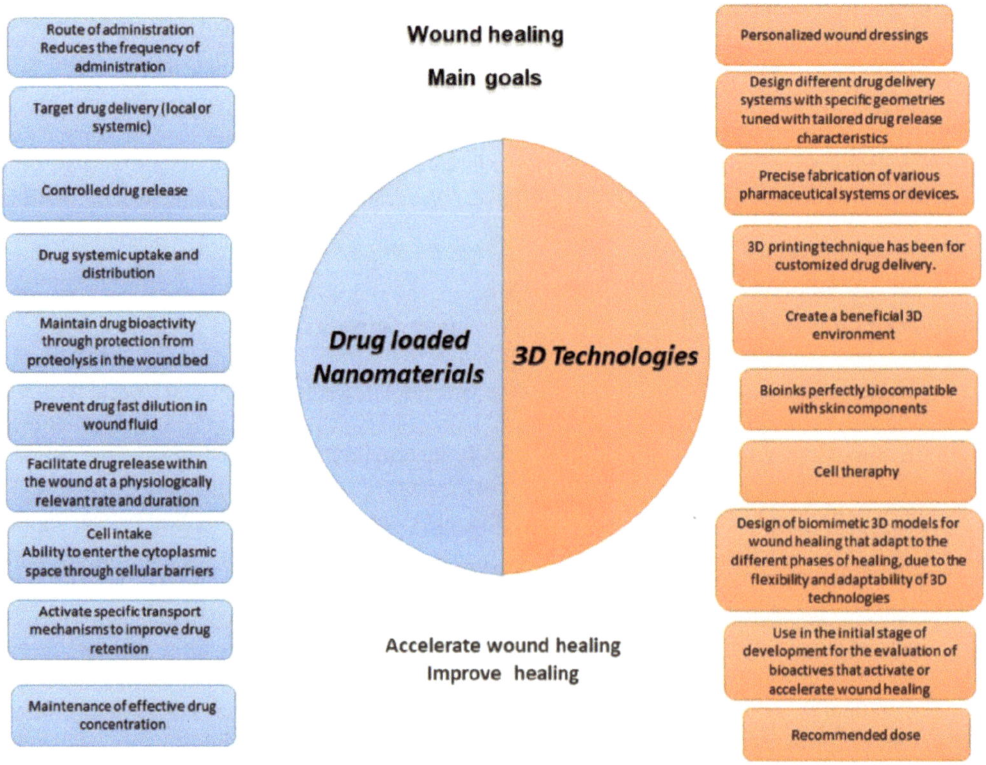

FIGURE 6.2 The main goals for using nanomaterials and 3D technologies for wound healing.

6.5 CONCLUSION

Wound healing is a highly complex process, and it is essential to understand the etiology of each wound type and develop similar delivery systems for individualized treatment using 3D technologies and nanomaterials together to accelerate and improve healing. Toward this goal, combining different biomaterials to create composite systems capable of extruding therapeutics agents on a nanoscale for controlled delivery with precise release kinetics will be necessary for wound healing. Moreover, the design of biomimetic 3D models for wound healing that adapt to the different phases of healing due to the flexibility and adaptability of 3D technologies could be optimal for a particular application or circumstance.

REFERENCES

1. H. Yousef, M. Alhajj, and S. Sharma, "Anatomy, skin (Integument), epidermis," *StatPearls*, Nov. 2022, Accessed: May 07, 2024. Available from: https://www.ncbi.nlm.nih.gov/books/NBK470464/.
2. A. K. Dąbrowska, F. Spano, S. Derler, C. Adlhart, N. D. Spencer, and R. M. Rossi, "The relationship between skin function, barrier properties, and body-dependent factors," *Skin Res Technol*, vol. 24, no. 2, pp. 165–174, May 2018, doi: 10.1111/SRT.12424.
3. "The skin's barrier - preview & related info | Mendeley." Accessed: May 28, 2024. Available from: https://www.mendeley.com/catalogue/05d249fb-6788-3b42-acdd-2d48fab19885/.
4. M. Takeo, W. Lee, and M. Ito, "Wound healing and skin regeneration," *Cold Spring Harb Perspect Med*, vol. 5, no. 1, 2015, doi: 10.1101/CSHPERSPECT.A023267.
5. R. Tiwari and K. Pathak, "Local drug delivery strategies towards wound healing," *Pharmaceutics*, vol. 15, no. 2, Feb. 2023, doi: 10.3390/PHARMACEUTICS15020634.
6. E. M. Tottoli, R. Dorati, I. Genta, E. Chiesa, S. Pisani, and B. Conti, "Skin wound healing process and new emerging technologies for skin wound care and regeneration," *Pharmaceutics*, vol. 12, no. 8, pp. 1–30, Aug. 2020, doi: 10.3390/PHARMACEUTICS12080735/PHARMACEUTICS_12_00735_PDF.PDF.
7. A. G. Tabriz and D. Douroumis, "Recent advances in 3D printing for wound healing: A systematic review," *J Drug Deliv Sci Technol*, vol. 74. 2022. doi: 10.1016/j.jddst.2022.103564.
8. T. Velnar, T. Bailey, and V. Smrkolj, "The wound healing process: An overview of the cellular and molecular mechanisms," *J Int Med Res*, vol. 37, no. 5, pp. 1528–1542, 2009, doi: 10.1177/147323000903700531.
9. K. Ghoshal and M. Bhattacharyya, "Overview of platelet physiology: Its hemostatic and nonhemostatic role in disease pathogenesis," *Sci World J*, vol. 2014, 2014, doi: 10.1155/2014/781857.
10. K. Hally, S. Fauteux-Daniel, H. Hamzeh-Cognasse, P. Larsen, and F. Cognasse, "Revisiting platelets and toll-like receptors (TLRS): At the interface of vascular immunity and thrombosis," *Int J Mol Sci*, vol. 21, no. 17, pp. 1–28, Sep. 2020, doi: 10.3390/IJMS21176150.
11. H. N. Wilkinson and M. J. Hardman, "Wound healing: Cellular mechanisms and pathological outcomes," *Adv Surg Med Spec*, pp. 341–370, Jun. 2023, doi: 10.1098/RSOB.200223.
12. G. Broughton, J. E. Janis, and C. E. Attinger, "The basic science of wound healing," *Plast Reconstr Surg*, vol. 117, no. 7 Suppl., Jun. 2006, doi: 10.1097/01.PRS.0000225430.42531.C2.
13. M. H. Norahan, S. C. Pedroza-González, M. G. Sánchez-Salazar, M. M. Álvarez, and G. Trujillo de Santiago, "Structural and biological engineering of 3D hydrogels for wound healing," *Bioact Mater*, vol. 24, pp. 197–235, Jun. 2023, doi: 10.1016/J.BIOACTMAT.2022.11.019/STRUCTURAL_AND_BIOLOGICAL_ENGINEERING_OF_3D_HYDROGELS_FOR_WOUND_HEALING.PDF.
14. K. Pfisterer, L. E. Shaw, D. Symmank, and W. Weninger, "The extracellular matrix in skin inflammation and infection," *Front Cell Dev Biol*, vol. 9, Jul. 2021, doi: 10.3389/FCELL.2021.682414/PDF.
15. R. B. Diller and A. J. Tabor, "The role of the extracellular matrix (ECM) in wound healing: A review," *Biomimetics*, vol. 7, no. 3, Sep. 2022, doi: 10.3390/BIOMIMETICS7030087.
16. P. Monika, M. N. Chandraprabha, A. Rangarajan, P. V. Waiker, and K. N. Chidambara Murthy, "Challenges in healing wound: role of complementary and alternative medicine," *Front Nutr*, vol. 8, Jan. 2022, doi: 10.3389/FNUT.2021.791899/PDF.
17. T. Ashwini *et al.*, "Transforming wound management: Nanomaterials and their clinical impact," *Pharmaceutics*, vol. 15, no. 5, May 2023, doi: 10.3390/PHARMACEUTICS15051560.
18. A. G. Tabriz and D. Douroumis, "Recent advances in 3D printing for wound healing: A systematic review," *J Drug Deliv Sci Technol*, vol. 74, Aug. 2022, doi: 10.1016/J.JDDST.2022.103564/RECENT_ADVANCES_IN_3D_PRINTING_FOR_WOUND_HEALING_A_SYSTEMATIC_REVIEW.PDF.

19. Y. Liang, J. He, and B. Guo, "Functional hydrogels as wound dressing to enhance wound healing," *ACS Nano*, vol. 15, no. 8, pp. 12687–12722, Aug. 2021, doi: 10.1021/ACSNANO.1C04206.
20. A. Smandri, A. Nordin, N. M. Hwei, K. Y. Chin, I. Abd Aziz, and M. B. Fauzi, "Natural 3D-printed bioinks for skin regeneration and wound healing: A systematic review," *Polymers*, vol. 12, no. 8. MDPI AG, Aug. 01, 2020. doi: 10.3390/polym12081782.
21. G. Han and R. Ceilley, "Chronic wound healing: A review of current management and treatments," *Adv Ther*, vol. 34, no. 3, pp. 599–610, Mar. 2017, doi: 10.1007/S12325-017-0478-Y/FIGURES/2.
22. D. T. Uchida and M. L. Bruschi, "3D Printing as a technological strategy for the personalized treatment of wound healing," *AAPS PharmSciTech*, vol. 24, no. 1, Jan. 2023, doi: 10.1208/S12249-023-02503-0.
23. S.-Y. Zheng *et al.*, "Therapeutic role of growth factors in treating diabetic wound," *World J Diabetes*, vol. 14, no. 4, p. 364, Apr. 2023, doi: 10.4239/WJD.V14.I4.364.
24. D. T. Uchida and M. L. Bruschi, "3D Printing as a technological strategy for the personalized treatment of wound healing," *AAPS PharmSciTech*, vol. 24, no. 1, Jan. 2023, doi: 10.1208/S12249-023-02503-0.
25. N. Johnson and Y. Wang, "Drug delivery systems for wound healing," *Curr Pharm Biotechnol*, vol. 16, no. 7, pp. 621–629, May 2015, doi: 10.2174/1389201016666150206113720.
26. M. T. Matter, S. Probst, S. Läuchli, and I. K. Herrmann, "Uniting drug and delivery: Metal oxide hybrid nanotherapeutics for skin wound care," *Pharmaceutics*, vol. 12, no. 8, pp. 1–17, Aug. 2020, doi: 10.3390/PHARMACEUTICS12080780/PHARMACEUTICS_12_00780_PDF.PDF.
27. T. Liu, Y. Lu, R. Zhan, W. Qian, and G. Luo, "Nanomaterials and nanomaterials-based drug delivery to promote cutaneous wound healing," *Adv Drug Deliv Rev*, vol. 193, Feb. 2023, doi: 10.1016/J.ADDR.2022.114670.
28. T. C. Ezike *et al.*, "Advances in drug delivery systems, challenges and future directions," *Heliyon*, vol. 9, no. 6, Jun. 2023, doi: 10.1016/J.HELIYON.2023.E17488/ADVANCES_IN_DRUG_DELIVERY_SYSTEMS_CHALLENGES_AND_FUTURE_DIRECTIONS.PDF.
29. J. Nandhini, E. Karthikeyan, and S. Rajeshkumar, "Nanomaterials for wound healing: Current status and futuristic frontier," *Biomed Technol*, vol. 6, pp. 26–45, Jun. 2024, doi: 10.1016/J.BMT.2023.10.001/NANOMATERIALS_FOR_WOUND_HEALING_CURRENT_STATUS_AND_FUTURISTIC_FRONTIER.PDF.
30. A. Qadir *et al.*, "Lipid engineered nanoparticle therapy for burn wound treatment," *Curr Pharm Biotechnol*, vol. 23, no. 12, pp. 1449–1459, Aug. 2021, doi: 10.2174/1389201022666210823110532.
31. T. S. Alster and P. M. Graham, "Microneedling: A review and practical guide," *Dermatol Surg*, vol. 44, no. 3, pp. 397–404, Mar. 2018, doi: 10.1097/DSS.0000000000001248.
32. S. Shi *et al.*, "A skin-like nanostructured membrane for advanced wound dressing," *Compos B Eng*, vol. 250, Feb. 2023, doi: 10.1016/J.COMPOSITESB.2022.110438.
33. H. N. Kim, Y. Hong, M. S. Kim, S. M. Kim, and K. Y. Suh, "Effect of orientation and density of nanotopography in dermal wound healing," *Biomaterials*, vol. 33, no. 34, pp. 8782–8792, Dec. 2012, doi: 10.1016/J.BIOMATERIALS.2012.08.038.
34. M. Wang *et al.*, "Nanomaterials applied in wound healing: Mechanisms, limitations and perspectives," *J Control Release*, vol. 337, pp. 236–247, Sep. 2021, doi: 10.1016/J.JCONREL.2021.07.017.
35. J. Yoon, H. Han, and J. Jang, "Nanomaterials-incorporated hydrogels for 3D bioprinting technology," *Nano Converg*, vol. 10, no. 1, Dec. 2023, doi: 10.1186/S40580-023-00402-5.
36. A. Zarrabi, I. Selestin Raja, B. Kim, and D.-W. Han, "Nanofibrous material-reinforced printable ink for enhanced cell proliferation and tissue regeneration," *Bioengineering*, vol. 11, no. 4, p. 363, Apr. 2024, doi: 10.3390/BIOENGINEERING11040363.
37. D.-G. Ahn, "Directed energy deposition (DED) process: State of the art," *Int J Precis Eng Manuf Green Technol*, vol. 8, no. 2, pp. 703–742, Mar. 2021, doi: 10.1007/S40684-020-00302-7).
38. I. Gibson, D. Rosen, B. Stucker, and M. Khorasani, "Additive manufacturing technologies," *Addit Manuf Technol*, pp. 1–675, Nov. 2020, doi: 10.1007/978-3-030-56127-7/COVER.
39. T. D. Ngo, A. Kashani, G. Imbalzano, K. T. Q. Nguyen, and D. Hui, "Additive manufacturing (3D printing): A review of materials, methods, applications and challenges," *Compos B Eng*, vol. 143, pp. 172–196, Jun. 2018, doi: 10.1016/J.COMPOSITESB.2018.02.012.
40. M. Koltsaki and M. Mavri, "A comprehensive overview of additive manufacturing processes through a time-based classification model," *3D Print Addit Manuf*, vol. 11, no. 1, pp. 363–382, Feb. 2024, doi: 10.1089/3DP.2022.0167.
41. K. Rane and M. Strano, "A comprehensive review of extrusion-based additive manufacturing processes for rapid production of metallic and ceramic parts," *Adv Manuf*, vol. 7, no. 2, pp. 155–173, Jun. 2019, doi: 10.1007/S40436-019-00253-6.

42. A. Elkaseer, K. J. Chen, J. C. Janhsen, O. Refle, V. Hagenmeyer, and S. G. Scholz, "Material jetting for advanced applications: A state-of-the-art review, gaps and future directions," *Addit Manuf*, vol. 60, Dec. 2022, doi: 10.1016/J.ADDMA.2022.103270/MATERIAL_JETTING_FOR_ADVANCED_APPLICATIONS_A_STATE_OF_THE_ART_REVIEW_GAPS_AND_FUTURE_DIRECTIONS.PDF.
43. J. Persad and S. Rocke, "Multi-material 3D printed electronic assemblies: A review," *Results Eng*, vol. 16, Dec. 2022, doi: 10.1016/J.RINENG.2022.100730/MULTI_MATERIAL_3D_PRINTED_ELECTRONIC_ASSEMBLIES_A_REVIEW.PDF.
44. L. Ladani and M. Sadeghilaridjani, "Review of powder bed fusion additive manufacturing for metals," *Metals (Basel)*, vol. 11, no. 9, Sep. 2021, doi: 10.3390/MET11091391.
45. M. Haddad, K. B. Nixon, and S. Wolff, "Sheet Lamination," *Springer Handbooks*, vol. Part F1592, pp. 407–423, 2023, doi: 10.1007/978-3-031-20752-5_25/FIGURES/21.
46. M. Pagac *et al.*, "Polymers A review of vat photopolymerization technology: Materials, applications, challenges, and future trends of 3D printing," 2021, doi: 10.3390/polym13040598.
47. H. Ben Hsieh *et al.*, "Ultra-high-throughput microarray generation and liquid dispensing using multiple disposable piezoelectric ejectors," *J Biomol Screen*, vol. 9, no. 2, pp. 85–94, Mar. 2004, doi: 10.1177/1087057103260943.
48. M. Hayat, A. Nawaz, S. Chinnam, M. Muzammal, M. S. Latif, M. Yasin, S. Ashique, G. Zengin, and A. Farid. "Formulation development and optimization of herbo synthetic gel: In vitro biological evaluation and in vivo wound healing studies," *Process Biochem*, vol. 130, pp. 116–126, Jul. 1, 2023.
49. P. Fisch, M. Holub, and M. Zenobi-Wong, "Improved accuracy and precision of bioprinting through progressive cavity pump-controlled extrusion," *Biofabrication*, vol. 13, no. 1, Jan. 2021, doi: 10.1088/1758-5090/ABC39B.
50. P. S. Gungor-Ozkerim, I. Inci, Y. S. Zhang, A. Khademhosseini, and M. R. Dokmeci, "Bioinks for 3D bioprinting: An overview," *Biomater Sci*, vol. 6, no. 5, pp. 915–946, May 2018, doi: 10.1039/C7BM00765E.
51. A. Schwab, R. Levato, M. D'Este, S. Piluso, D. Eglin, and J. Malda, "Printability and shape fidelity of bioinks in 3d bioprinting," *Chem Rev*, vol. 120, no. 19, pp. 11028–11055, Oct. 2020, doi: 10.1021/ACS.CHEMREV.0C00084.
52. J. Groll *et al.*, "A definition of bioinks and their distinction from biomaterial inks," *Biofabrication*, vol. 11, no. 1, Jan. 2019, doi: 10.1088/1758-5090/AAEC52.
53. A. Veiga, I. V. Silva, M. M. Duarte, and A. L. Oliveira, "Current trends on protein driven bioinks for 3d printing," *Pharmaceutics*, vol. 13, no. 9, Sep. 2021, doi: 10.3390/PHARMACEUTICS13091444.
54. A. Blaeser, D. F. Duarte Campos, U. Puster, W. Richtering, M. M. Stevens, and H. Fischer, "Controlling shear stress in 3D bioprinting is a key factor to balance printing resolution and stem cell integrity," *Adv Healthc Mater*, vol. 5, no. 3, pp. 326–333, Feb. 2016, doi: 10.1002/ADHM.201500677.
55. L. Ouyang, R. Yao, Y. Zhao, and W. Sun, "Effect of bioink properties on printability and cell viability for 3D bioplotting of embryonic stem cells," *Biofabrication*, vol. 8, no. 3, Sep. 2016, doi: 10.1088/1758-5090/8/3/035020.
56. R. Levato *et al.*, "Light-based vat-polymerization bioprinting," *Nat Rev Methods Primers*, vol. 3, no. 1, Dec. 2023, doi: 10.1038/S43586-023-00231-0.
57. N. Hong, G. H. Yang, J. H. Lee, and G. H. Kim, "3D bioprinting and its in vivo applications," *J Biomed Mater Res B Appl Biomater*, vol. 106, no. 1, pp. 444–459, Jan. 2018, doi: 10.1002/JBM.B.33826.
58. S. Zhang *et al.*, "Convergence of 3D bioprinting and nanotechnology in tissue engineering scaffolds," *Biomimetics*, vol. 8, no. 1, Mar. 2023, doi: 10.3390/BIOMIMETICS8010094.
59. X. Zhang *et al.*, "3D Bioprinting of biomimetic bilayered scaffold consisting of decellularized extracellular matrix and silk fibroin for osteochondral repair," *Int J Bioprint*, vol. 7, no. 4, pp. 1–14, 2021, doi: 10.18063/IJB.V7I4.401.
60. P. S. Gungor-Ozkerim, I. Inci, Y. S. Zhang, A. Khademhosseini, and M. R. Dokmeci, "Bioinks for 3D bioprinting: An overview," *Biomater Sci*, vol. 6, no. 5, pp. 915–946, May 2018, doi: 10.1039/C7BM00765E.
61. A. Sadeghianmaryan *et al.*, "Advancements in 3D-printable polysaccharides, proteins, and synthetic polymers for wound dressing and skin scaffolding – A review," *Int J Biol Macromol*, vol. 266, May 2024, doi: 10.1016/J.IJBIOMAC.2024.131207.
62. A. T. Yayehrad, E. A. Siraj, M. Matsabisa, and G. Birhanu, "3D printed drug loaded nanomaterials for wound healing applications," *Regen Ther*, vol. 24, pp. 361–376, Dec. 2023, doi: 10.1016/J.RETH.2023.08.007/3D_PRINTED_DRUG_LOADED_NANOMATERIALS_FOR_WOUND_HEALING_APPLICATIONS.PDF.
63. O. Kammona, E. Tsanaktsidou, and C. Kiparissides, "Recent developments in 3D-(Bio)printed hydrogels as wound dressings," *Gels*, vol. 10, no. 2, Feb. 2024, doi: 10.3390/GELS10020147.

64. J. D. Tanfani, J. D. Monpara, and S. Jonnalagadda, "3D Bioprinting and its role in a wound healing renaissance," *Adv Mater Technol*, vol. 8, no. 17, Sep. 2023, doi: 10.1002/ADMT.202300411.
65. D. Han *et al.*, "Rapid bacteria trapping and killing of metal-organic frameworks strengthened photoresponsive hydrogel for rapid tissue repair of bacterial infected wounds," *Chem Eng J*, vol. 396, p. 125194, Sep. 2020, doi: 10.1016/J.CEJ.2020.125194.
66. C. Benwood *et al.*, "Natural biomaterials and their use as bioinks for printing tissues," *Bioengineering*, vol. 8, no. 2, pp. 1–19, Feb. 2021, doi: 10.3390/BIOENGINEERING8020027/BIOENGINEERING_08_00027_PDF.PDF.
67. M. Hassan, K. Dave, R. Chandrawati, F. Dehghani, and V. G. Gomes, "3D printing of biopolymer nanocomposites for tissue engineering: Nanomaterials, processing and structure-function relation," *Eur Polym J*, vol. 121, Dec. 2019, doi: 10.1016/J.EURPOLYMJ.2019.109340.
68. Y. Cai, S. Y. Chang, S. W. Gan, S. Ma, W. F. Lu, and C. C. Yen, "Nanocomposite bioinks for 3D bioprinting," *Acta Biomater*, vol. 151, pp. 45–69, Oct. 2022, doi: 10.1016/J.ACTBIO.2022.08.014.
69. M. Mobaraki, M. Ghaffari, A. Yazdanpanah, Y. Luo, and D. K. Mills, "Bioinks and bioprinting: A focused review," *Bioprinting*, vol. 18, Jun. 2020, doi: 10.1016/J.BPRINT.2020.E00080.
70. G. Kannayiram, S. Sendilvelan, and M. P. R, "Importance of nanocomposites in 3D bioprinting: An overview," *Bioprinting*, vol. 32, Jul. 2023, doi: 10.1016/J.BPRINT.2023.E00280.
71. D. K. Chandra, R. L. Reis, S. C. Kundu, A. Kumar, and C. Mahapatra, "Nanomaterials-based hybrid bioink platforms in advancing 3D bioprinting technologies for regenerative medicine," *ACS Biomater Sci Eng*, Jun. 2024, doi: 10.1021/ACSBIOMATERIALS.4C00166.
72. Y. H. Kim *et al.*, "Biofabrication of nanocomposite-based scaffolds containing human bone extracellular matrix for the differentiation of skeletal stem and progenitor cells," *Biodes Manuf*, vol. 7, no. 2, pp. 121–136, Mar. 2024, doi: 10.1007/S42242-023-00265-Z.
73. G. Cidonio *et al.*, "Clay nanofiller enhances and stabilises a new injectable human bone extracellular matrix scaffold for skeletal regeneration," *Mater Today Commun*, vol. 39, Jun. 2024, doi: 10.1016/J.MTCOMM.2024.109082.
74. A. Mousavi *et al.*, "Development of photocrosslinkable bioinks with improved electromechanical properties for 3D bioprinting of cardiac BioRings," *Appl Mater Today*, vol. 36, Feb. 2024, doi: 10.1016/J.APMT.2023.102035.
75. T. Chen, K. Hou, Q. Ren, G. Chen, P. Wei, and M. Zhu, "Nanoparticle–polymer synergies in nanocomposite hydrogels: From design to application," *Macromol Rapid Commun*, vol. 39, no. 21, Nov. 2018, doi: 10.1002/MARC.201800337.
76. C. C. Piras and D. K. Smith, "Multicomponent polysaccharide alginate-based bioinks," *J Mater Chem B*, vol. 8, no. 36, pp. 8171–8188, Sep. 2020, doi: 10.1039/D0TB01005G.
77. J. Jia *et al.*, "Engineering alginate as bioink for bioprinting," *Acta Biomater*, vol. 10, no. 10, pp. 4323–4331, Oct. 2014, doi: 10.1016/J.ACTBIO.2014.06.034.
78. X. Wang, Q. Wang, and C. Xu, "Nanocellulose-based inks for 3d bioprinting: Key aspects in research development and challenging perspectives in applications—a mini review," *Bioengineering*, vol. 7, no. 2, Jun. 2020, doi: 10.3390/BIOENGINEERING7020040/BIOENGINEERING_07_00040_PDF.PDF.
79. K. Khoshmaram, F. Yazdian, Z. Pazhouhnia, and N. Lotfibakhshaiesh, "Preparation and characterization of 3D bioprinted gelatin methacrylate hydrogel incorporated with curcumin loaded chitosan nanoparticles for in vivo wound healing application," *Biomater Adv*, vol. 156, Jan. 2024, doi: 10.1016/J.BIOADV.2023.213677.
80. M. R. Ceccarini *et al.*, "Biomaterial inks from peptide-functionalized silk fibers for 3D printing of futuristic wound-healing and sensing materials," *Int J Mol Sci*, vol. 24, no. 2, Jan. 2023, doi: 10.3390/IJMS24020947.
81. H. Chen, H. Xue, H. Zeng, M. Dai, C. Tang, and L. Liu, "3D printed scaffolds based on hyaluronic acid bioinks for tissue engineering: A review," *Biomater Res*, vol. 27, no. 1, Dec. 2023, doi: 10.1186/S40824-023-00460-0.
82. T. Jiao *et al.*, "Properties of collagen/sodium alginate hydrogels for bioprinting of skin models," *J Bionic Eng*, vol. 20, no. 1, pp. 105–118, Jan. 2023, doi: 10.1007/S42235-022-00251-8.
83. S. S. Mathew-Steiner, S. Roy, and C. K. Sen, "Collagen in wound healing," *Bioengineering*, vol. 8, no. 5, May 2021, doi: 10.3390/BIOENGINEERING8050063.
84. S. Nie, Z. Li, Y. Yao, and Y. Jin, "Progress in synthesis of conductive polymer poly (3,4-Ethylenedioxythiophene)," *Front Chem*, vol. 9, Dec. 2021, doi: 10.3389/FCHEM.2021.803509/PDF.
85. A. Gelmi, M. K. Ljunggren, M. Rafat, and E. W. H. Jager, "Influence of conductive polymer doping on the viability of cardiac progenitor cells," *J Mater Chem B*, vol. 2, no. 24, pp. 3860–3867, Jun. 2014, doi: 10.1039/C4TB00142G.

86. R. Balint, N. J. Cassidy, and S. H. Cartmell, "Conductive polymers: Towards a smart biomaterial for tissue engineering," *Acta Biomater*, vol. 10, no. 6, pp. 2341–2353, 2014, doi: 10.1016/J.ACTBIO.2014.02.015/CONDUCTIVE_POLYMERS_TOWARDS_A_SMART_BIOMATERIAL_FOR_TISSUE_ENGINEERING.PDF.
87. M. Saha, "3D Printing of nanoceramics: Present status and future perspectives," *SSRN Electron J*, p. undefined-undefined, 2023, doi: 10.2139/SSRN.4388983.
88. N. Khoshnood and A. Zamanian, "Development of novel alginate-polyethyleneimine cell-laden bioink designed for 3D bioprinting of cutaneous wound healing scaffolds," *J Appl Polym Sci*, vol. 139, no. 21, Jun. 2022, doi: 10.1002/APP.52227.
89. X. Lin, H. Zhang, H. Zhang, Z. Zhang, G. Chen, and Y. Zhao, "Bio-printed hydrogel textiles based on fish skin decellularized extracellular matrix for wound healing," *Engineering*, vol. 25, pp. 120–127, Jun. 2023, doi: 10.1016/J.ENG.2022.05.022/BIO_PRINTED_HYDROGEL_TEXTILES_BASED_ON_FISH_SKIN_DECELLULARIZED_EXTRACELLULAR_MATRIX_FOR_WOUND_HEALING.PDF.
90. M. C. Ceballos-Santa *et al.*, "Aloe vera-based biomaterial ink for 3D bioprinting of wound dressing constructs," *J Biomed Mater Res B Appl Biomater*, vol. 112, no. 2, Feb. 2024, doi: 10.1002/JBM.B.35379.
91. Z. Güney, F. Karacaoglu, | Emre Barıs, | K Gökalp Gezer, and | M Murat Akkaya, "The relationship of peri-implant soft tissue wound healing with implant cover screw design: Cross-sectional study," 2023, doi: 10.1111/cid.13272.
92. S. P. Kochar, A. Reche, and P. Paul, "The etiology and management of dental implant failure: A review," *Cureus*, p. undefined-undefined, Oct. 2022, doi: 10.7759/CUREUS.30455.
93. "InterLatin and colibrí introducing the next generation to 3D printing - 3DPrint.com | The voice of 3D printing / additive manufacturing." Accessed: May 18, 2024. Available from: https://3dprint.com/94572/colibri-next-generation/.
94. A. Yusuf, A. R. Z. Almotairy, H. Henidi, O. Y. Alshehri, and M. S. Aldughaim, "Nanoparticles as drug delivery systems: A review of the implication of nanoparticles' physicochemical properties on responses in biological systems," *Polymers*, vol. 15, no. 7, p. 1596, Mar. 2023, doi: 10.3390/POLYM15071596.
95. R. M. Pearson, V. V. Juettner, and S. Hong, "Biomolecular corona on nanoparticles: A survey of recent literature and its implications in targeted drug delivery," *Front Chem*, vol. 2, Nov, 2014, doi: 10.3389/FCHEM.2014.00108.
96. N. Baig, I. Kammakakam, W. Falath, and I. Kammakakam, "Nanomaterials: A review of synthesis methods, properties, recent progress, and challenges," *Mater Adv*, vol. 2, no. 6, pp. 1821–1871, Mar. 2021, doi: 10.1039/D0MA00807A.
97. M. Ferrari, "Nanogeometry: Beyond drug delivery," *Nat Nanotechnol*, vol. 3, no. 3, pp. 131–132, 2008, doi: 10.1038/NNANO.2008.46.
98. E. Blanco, H. Shen, and M. Ferrari, "Principles of nanoparticle design for overcoming biological barriers to drug delivery," *Nat Biotechnol*, vol. 33, no. 9, pp. 941–951, Sep. 2015, doi: 10.1038/NBT.3330.
99. M. M. Mihai, M. B. Dima, B. Dima, and A. M. Holban, "Nanomaterials for wound healing and infection control," *Materials*, vol. 12, no. 13, Jul. 2019, doi: 10.3390/MA12132176.
100. A. Pormohammad, N. K. Monych, S. Ghosh, D. L. Turner, and R. J. Turner, "Nanomaterials in wound healing and infection control," *Antibiotics*, vol. 10, no. 5, May 2021, doi: 10.3390/ANTIBIOTICS10050473.
101. T. Sangnim, V. Puri, D. Dheer, D. N. Venkatesh, K. Huanbutta, and A. Sharma, "Nanomaterials in the wound healing process: New insights and advancements," *Pharmaceutics*, vol. 16, no. 3, Mar. 2024, doi: 10.3390/PHARMACEUTICS16030300.
102. S. Hamdan *et al.*, "Nanotechnology-driven therapeutic interventions in wound healing: Potential uses and applications," *ACS Cent Sci*, vol. 3, no. 3, pp. 163–175, Mar. 2017, doi: 10.1021/ACSCENTSCI.6B00371.
103. Y. Zhong, X. T. Zheng, S. Zhao, X. Su, and X. J. Loh, "Stimuli-activable metal-bearing nanomaterials and precise on-demand antibacterial strategies," *ACS Nano*, vol. 16, no. 12, pp. 19840–19872, Dec. 2022, doi: 10.1021/ACSNANO.2C08262.
104. A. Fritschen, M. Acedo Mestre, S. Scholpp, and A. Blaeser, "Influence of the physico-chemical bioink composition on the printability and cell biological properties in 3D-bioprinting of a liver tumor cell line," *Front Bioeng Biotechnol*, vol. 11, 2023, doi: 10.3389/FBIOE.2023.1093101.
105. S. Breslin and L. O'Driscoll, "Three-dimensional cell culture: The missing link in drug discovery," *Drug Discov Today*, vol. 18, no. 5–6, pp. 240–249, Mar. 2013, doi: 10.1016/J.DRUDIS.2012.10.003.

7 Natural Bioactive Compound-Loaded Nanomaterials for Wound Healing

Younis Ahmad Hajam,[1*] *Neelam*[1], *Rajesh Kumar*[2], *and Seema Rai*[3]

[1] Sant Baba Bhag Singh University, Punjab, India
[2] Himachal Pradesh University, Himachal Pradesh, India
[3] Guru Ghasidas Vishwavidayalaya (A Central University), Chhattisgarh, India
*Corresponding Author

7.1 OVERVIEW OF WOUND HEALING PROCESS

The skin is the largest organ of the body and serves as a vital protective barrier, facilitating various functions such as moisture regulation, sensory perception, and defense against external pathogens (Dabrowska et al., 2018). Despite its resilience to numerous external stimuli, injuries like burns, cuts, and abrasions can compromise its integrity, leading to ailments. While the body's self-healing ability typically resolves wounds within three months, uncontrolled infections can prolong the healing process, resulting in chronic conditions lasting for months or even years. In recent years, there has been a notable increase in chronic ailments like vascular dysfunction, obesity, and diabetes, contributing to a surge in patients with chronic wounds. For diabetic patients, the risk of chronic abscesses is particularly high. Additionally, certain communicable skin diseases and physical skin conditions can exacerbate the vulnerability to chronic wounds (Wang et al., 2021). The prolonged exposure of hypodermic tissue in chronic wounds increases the risk of complications such as bleeding and osteomyelitis, significantly impacting patients' quality of life, financial burden, and mental well-being (Kushwaha et al., 2022).

The healthcare system faces a substantial burden due to the treatment costs associated with bone and skin imperfections, amounting to billions of dollars annually in the United States alone. As wound healing becomes increasingly complex, clinicians are in constant need of novel materials and approaches. The advent of nanotechnology, particularly advancements in nanochemistry and nanomanufacturing, has revolutionized the pharmaceutical and biotechnology industries. Nanomaterials, characterized by at least one dimension below 100 nm, offer unique physicochemical properties, including enhanced adsorption capacity, antimicrobial properties, and efficient drug delivery, making them promising candidates for wound healing applications.

Traditional wound dressings, such as gauze and bandages, have been used for years, but the demand for dressings that mimic the extracellular matrix (ECM) and promote cell proliferation and angiogenesis has driven the growth of nanomaterial-based dressings (Han and Ceilley, 2017). Novel nanomaterial-based bandages, including hydrogels, nanofibers, and films, are increasingly being utilized and are projected to dominate the global market by 2021. Despite the growing interest in nanomaterials for wound healing, a comprehensive understanding of their mechanisms and

DOI: 10.1201/9781003504948-7

potential toxicity is lacking. Therefore, this chapter aims to explore the recent therapeutic role of natural products formulated as nanomaterials in wound healing, elucidate their mechanisms, address potential toxicity concerns, and propose innovative research directions to enhance their efficacy and safety in clinical practice.

The wound healing process is a complex series of biochemical and cellular events crucial for restoring skin integrity. It progresses through three overlapping stages: inflammation, proliferation (involving neo-angiogenesis, granulation, and re-epithelialization), and maturation (involving extracellular matrix remodeling) (Groeber et al., 2011; Pereira and Bártolo, 2016; Reinke and Sorg, 2012). Effective wound management relies heavily on the materials used in dressing (Singh et al., 2010), which significantly influence healing efficacy. While traditional therapies have been extensively studied, medicinal plants offer promising avenues for improving wound healing due to their diverse active components like flavonoids, essential oils, and alkaloids (Bahramsoltani et al., 2014). These natural remedies are favored for their affordability, minimal side effects, bioavailability, and effectiveness. Moreover, enhancing the efficacy of medicinal plants can be achieved through nanosizing or incorporating them into nanostructures. Nanomaterials, characterized by their nanoscale size and high surface area-to-volume ratio, can alter the physical and chemical properties of these plants (Alvarez-Román et al., 2004; Mordorski and Prow, 2016). This nanostructuring process enhances their therapeutic potential for wound healing. Natural compounds derived from plants can serve as both wound healing agents themselves and as carriers for delivering other therapeutics (Li et al., 2018). Given the advantageous properties of nanostructured medicinal plants, it prompts authors to undertake a comprehensive review of plant-based nanomaterials produced through various methods and their potential therapeutic targets in regulating the wound healing process. Injuries disrupt tissue structure and function, triggering processes such as hemorrhage, blood vessel constriction through coagulation, and inflammation. Chronic wounds pose significant challenges, contributing to high rates of morbidity and mortality, particularly among the elderly population. As these wounds become more prevalent, their impact on society and the economy is expected to grow substantially (Hajialyani et al., 2018).

7.2 PHYSIOLOGY OF WOUND HEALING

The intricate process of skin restoration involves the orchestrated interplay of various cell types, chemokines, and growth factors in a sequential manner. Traditionally, wound healing is divided into four distinct phases: hemostasis, inflammation, proliferation, and remodeling, each playing a crucial role in the overall regeneration process (Figure 7.1).

The initial phase of wound healing, known as hemostasis, promptly responds to injury by halting bleeding and reducing hemorrhaging resulting from vascular damage. Hemostasis progresses through three stages: vasoconstriction, primary hemostasis, and secondary hemostasis (Figure 7.1), with each stage intricately linked and executed in a rapid and coordinated manner (Matter et al., 2020; Rodrigues et al., 2019; Periayah et al., 2017; Hayat et al., 2023). These stages involve a cascade of biomolecules such as platelet-derived growth factor (PDGF), transforming growth factor (TGF) family, vascular endothelial growth factor (VEGF), epidermal growth factor (EGF) family (TGF and EGF), insulin-like growth factor (IGF) family, fibroblast growth factor (FGF) family (FGF and keratinocyte growth factor [KGF]), interleukin, and tumor necrosis factor (TNF-α), which not only fulfill their functions in hemostasis but also contribute to the subsequent phases of wound healing. These molecules play roles in stimulating epithelial cells, facilitating collagen deposition by fibroblasts, and promoting the restoration of damaged tissue (Sharifi et al., 2021; Rumbaut and Thiagarajan, 2010).

The inflammation phase initiates promptly following injury and typically persists for up to three days. Following the initial vasoconstriction during hemostasis, platelet degranulation triggers a cascade of complement activation, generating potent complement peptides (Figure 7.1). This cascade leads to the release of histamine, serotonin, proteases, and other cellular mediators from

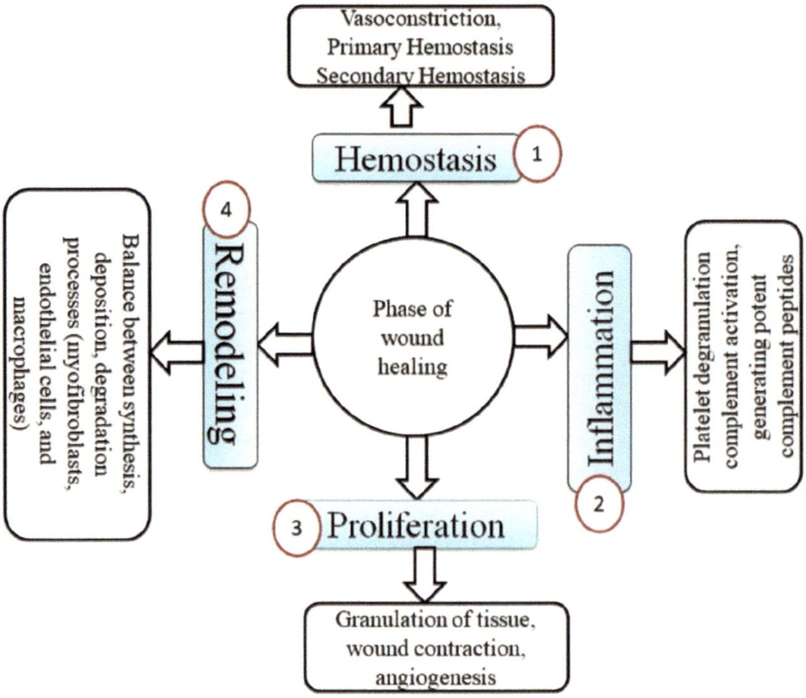

FIGURE 7.1 The process of wound healing and its steps.

basophils and mast cells. Consequently, vasodilation and increased vascular permeability occur (Cañedo-Dorantes and Cañedo-Ayala, 2019; Young and McNaught, 2011; Kratofil et al., 2017), enhancing blood flow and promoting inflammation and heat generation. Around three to four days into wound development, macrophages undertake the removal of spent neutrophils through efferocytosis (Rodrigues et al., 2019; Velnar et al., 2009). This process serves to prevent nonspecific tissue breakdown and ensures the resolution of inflammation.

Four days following the onset of wound development, a proliferation stage commences, which typically extends for approximately 21 days in cases of chronic wounds. However, the duration of this stage is also influenced by factors such as wound size and the overall health status of the patient. The proliferation stage is chiefly characterized by the granulation of tissue, wound contraction, and angiogenesis (Landén et al., 2016).

During the process of granulation, angiogenesis plays a crucial role, with tissue growth being initiated during the formation of the hemostatic plug due to the release of transforming growth factor-β (TGF-β), PDGF, and FGF by platelets (Young and McNaught, 2011). Additionally, the granulation tissue formed during the inflammation phase serves as primary tissue and eventually replaces the clot at the injury site (Figure 7.1). The re-epithelialization process is facilitated by the presence of granulation tissue, where numerous epithelial cells migrate across the newly formed tissues, resulting in the establishment of barriers between the wounded area and the external environment. Granulation tissue fosters cellular–extracellular matrix interactions, leading to the release of biomolecules along with growth factors and mechanosensory signals. These signals stimulate the differentiation of fibroblasts into myofibroblasts, aiding in wound contraction and the drawing together of wound boundaries (Sharifi et al., 2021).

The final stage of wound healing is maturation and remodeling, which typically begins approximately three weeks after the onset of wound development. The duration of this phase can vary, ranging from one year to even longer, depending on the nature of the wound. During this phase, natural epithelial growth occurs, and scar tissue undergoes maturation (Kushwaha et al., 2022).

Maturation and remodeling involve a delicate balance between synthesis, deposition, and degradation processes. Various cells, including myofibroblasts, endothelial cells, and macrophages, contribute to wound remodeling (Figure 7.1). Fibroblasts play a crucial role in crosslinking collagen, leading to the realignment of collagen fibers into organized grids that enhance the tensile strength of the tissue, achieving approximately 80% of the strength of unwounded skin (Xue and Jackson, 2015). While growth factors are important for re-epithelialization, their absence does not impede wound healing significantly. Ultimately, the result is a fully matured scar with reduced vascularity. Therefore, wound restoration is a complex process that requires precise coordination of stages; any disruption in the synchrony of these stages can lead to the development of chronic wounds (Sharifi et al., 2021).

7.3 PATHOPHYSIOLOGY OF WOUND HEALING

Skin possesses a remarkable capacity for regeneration, governed by precisely orchestrated cellular and molecular processes. However, any disruption in the normal healing sequence can completely halt wound healing, leading to the development of chronic wounds such as arterial ulcers, nonhealing surgical wounds, foot ulcers, and pressure sores (Sharifi et al., 2021). Abnormal wound healing conditions involve various interconnected factors including cell senescence/apoptosis, prolonged inflammation, increased production of matrix metalloproteinases (MMPs), matrix degradation, infection, systemic issues such as patient nutritional status, stress, and other chronic comorbidities (Guo and DiPietro, 2020; Ayuk et al., 2016).

In the acute wound microenvironment, there is a well-organized ECM and a synthesis rate of the matrix that exceeds its breakdown. Typically, normal injuries exhibit a low bacterial load, as immune cells effectively control microbial infections, allowing for prompt healing through proliferation and restoration. Imbalance in the wound healing phases leads to the formation of chronic, nonhealing wounds (Brem et al., 2007; Clinton and Carter, 2015; Piipponen et al., 2020). Chronic wounds are defined as those that fail to progress through an organized and appropriate process to achieve the usual anatomical and functional integrity (Bruce and Christie, 2016; Frykberg and Banks, 2015). It has been suggested that a time frame of six to eight weeks for wound healing, beyond which the wound should be considered chronic or nonhealing. Delayed healing can be attributed to factors such as wound infection, persistence of foreign particles or microbial proteins, chronic irritation, ischemia, and shock (Qiao et al., 2020; Chen et al., 2019; McDaniel et al., 2013). The persistent expression and overexpression of interleukins (ILs) and inflammatory cytokines disrupt the balance of essential proteases, growth factors, and cytokines, resulting in excessive degradation, proteolysis, and inadequate availability of critical receptors, growth factors, and ECM components (Brem et al., 2007). Cells involved in acute wound healing exhibit distinct functional and phenotypical differences compared to those found in chronic nonhealing wounds. The proliferation and migration of fibroblast cells are inhibited, rendering them unresponsive to growth factors. Certain immune cells responsible for bactericidal and phagocytic activity become dysfunctional, leading to the accumulation of necrotic debris at the wound periphery. Chronic wounds are also prone to biofilm formation compared to acute injuries. Impaired vascularization and insufficient oxygen and nutrient delivery to cells promote avascular necrosis, fostering bacterial growth and biofilm formation. Consequently, the presence of biofilm exacerbates inflammation, hindering ECM deposition and wound epithelialization. In chronic wounds, the wound area extends beyond the visible lesion due to pathological damage to surrounding tissues, compromising skin integrity. Tissues surrounding and beneath chronic wounds are similarly affected by the underlying abnormalities causing ulceration. Local and systemic factors influencing wound healing include inadequate blood supply, the presence of foreign particles impeding tissue repair, persistent microbial colonization and infection, wound dehiscence, compromised venous drainage, reduced macrophage activity, and elevated levels of MMPs (Ehterami et al., 2018; Hasan et al., 2019).

TABLE 7.1
Natural Bioactive Compounds in Wound Healing

Compound	Biological Activity	Importance	Reference
Curcumin	Anti-inflammatory, antioxidant, wound healing	Promotes wound healing by protecting cells, reducing ROS, and enhancing antioxidant enzymes	Phan et al., 2001; Gadekar et al., 2012; Bonte et al., 1997; Mohanty et al., 2012
Quercetin	Antioxidant, anti-inflammatory, wound healing	Accelerates wound healing by reducing oxidative stress, modulating cytokines, promoting fibroblast proliferation, and collagen deposition	Kant et al., 2017; Mi et al., 2022
Catechin	Antioxidant, wound healing	Shields wounds, reduces oxidative stress, and facilitates wound healing	Baek et al., 2020; Zhao et al., 2021
Galic acid (GA)	Antioxidant, wound healing	Protects against oxidative stress, promotes keratinocyte migration, and activates wound healing factors	Yang et al., 2016
Resveratrol	Antioxidant, wound healing	Shields against oxidative injury, accelerates wound closure, enhances skin structure, and mitigates inflammation	Zhou et al., 2021; Bilgic et al., 2022
Naringenin	Antioxidant, anti-inflammatory, wound healing	Increases antioxidant enzymes, reduces inflammatory factors, and promotes wound contraction	Al-Roujayee et al., 2017; Akrawi et al., 2020
Myricetin (Myr)	Anti-inflammatory, wound healing	Modulates inflammatory cytokines, enhances fibroblast migration, and inhibits proinflammatory cytokine production	Elshamy et al., 2020; Sklenářova et al., 2021
Pinocembrin	Antibacterial, anti-inflammatory, wound healing	Exhibits antibacterial effects, suppresses inflammation, and promotes skin fibroblast activity	Drewes and van Vuuren, 2008; Li et al., 2021
Chitosan/chitin	Hemostatic, antibacterial, wound healing	Exhibits antibacterial properties, promotes hemostasis, and facilitates wound healing	Ueno et al., 2001; Bhattarai et al., 2010
Propolis	Antibacterial, wound healing	Contains antibacterial compounds, supports wound healing processes	Afrouzan et al., 2018
Tannins	Antibacterial, wound healing	Demonstrates significant antibacterial effects and supports wound healing processes	Su et al., 2017
Allicin	Antibacterial, wound healing	Exhibits antibacterial effects and aids in wound healing	Alhashim and Lombardo, 2018
Saponins	Antibacterial, wound healing	Possess diverse biological effects including antibacterial properties	Wang et al., 2010
Cryptotanshinone	Stimulates collagen synthesis, wound healing	Accelerates wound healing by promoting collagen deposition and reducing inflammation	Song et al., 2020
Artocarpin	Stimulates collagen formation and wound healing	Enhances collagen synthesis and promotes wound healing	Yeh et al., 2017
Glucans	Immune-enhancing, wound healing	Facilitate tissue granulation, collagen deposition, and re-epithelialization	Majtan and Jesenak, 2018
Amino acids/peptides	Promotes collagen synthesis, wound healing	Stimulates collagen formation and enhances wound healing processes	Ulagesan et al., 2018

7.4 EMERGENCE OF NATURAL BIOACTIVE COMPOUND-LOADED NANOMATERIALS

Polymeric nanoparticles have been utilized to deliver various natural bioactive agents for a range of purposes, including the incorporation of bioflavonoids, a widely studied class of natural compounds known for their promising therapeutic properties and health benefits. These benefits include immune stimulation, cytotoxic effects, positive cardiovascular effects, enhanced vision, and reduction in amyloidogenic-related cell death. Furthermore, bioflavonoids exhibit efficient anticancer activity through diverse mechanisms. Their therapeutic actions have been harnessed in the treatment of various conditions such as oxidative stress, inflammation, skin disorders, cardiovascular diseases, respiratory ailments, and diabetes.

7.4.1 Curcumin

Curcumin, primarily derived from turmeric (*Curcuma longa* L.), exhibits various beneficial properties such as anti-inflammatory, antioxidant, and anticoagulant effects. Studies have consistently shown its effectiveness in promoting wound healing through its antioxidant properties. For instance, Phan et al. (2001) demonstrated that curcumin has protective effects on human dermal fibroblasts and epidermal cells when exposed to hydrogen peroxide and superoxide radicals. Similarly, Gadekar et al. (2012) evaluated the potential of curcumin in protecting keratinocytes and fibroblasts from injury induced by H_2O_2. Additionally, Bonte et al. (1997) highlighted curcumin's ability to shield human keratinocytes from damage caused by xanthine oxidase. Moreover, Mohanty et al. (2012) reported that curcumin is able to reduce reactive oxygen species (ROS) and lipid peroxidation, consequently decreasing the activation of antioxidant enzymes after wound treatment in rats. Despite its remarkable therapeutic effects, curcumin faces challenges in clinical application due to its poor water solubility, leading to issues of instability and low bioavailability.

7.4.2 Quercetin

Quercetin, a flavonoid abundant in various vegetables, fruits, and seeds such as citrus, onion, tea, and spices, is renowned for its potent antioxidant and anti-inflammatory properties (Aceituno-Medina et al., 2015). Studies, like the one conducted by Kant et al. (2017) have demonstrated that a concentration of 0.3% quercetin accelerates wound healing significantly. It achieves this by effectively ameliorating oxidative stress, modulating cytokines and growth factors, and stimulating fibroblast proliferation, vessel formation, and collagen deposition. Moreover, Mi et al. (2022) conducted an extensive investigation on the wound healing effects of quercetin extracted from *Oxytropis falcata* Bunge, a traditional Chinese legume found in Tibet. Their findings revealed that quercetin treatment not only increased collagen fiber content but also markedly reduced inflammatory factors such as TNF-α, IL-1β, and IL-6. Additionally, quercetin treatment was associated with improved levels of glutathione (GSH), a vital antioxidant and redox regulator crucial for controlling inflammation. This suggests that quercetin possesses robust antioxidant capabilities in healing skin wounds. Therefore, quercetin effectively promotes skin wound healing by enhancing fibroblast migration and proliferation while simultaneously dampening inflammation through its antioxidant actions. However, similar to many other flavonoids, quercetin's poor water solubility presents a challenge in its application for wound treatment.

7.4.3 Catechin

Catechin, a flavonoid known for its potent antioxidant activity, plays a crucial role in various physiological processes (Orhan et al., 2019; Caro et al., 2019). Baek et al. (2020) conducted a study involving the preparation of a PCL/(+)-catechin/gelatin film aimed at assessing its suitability for wound treatment. Their findings revealed that the PCL/(+)-catechin/gelatin film effectively shields wounds from external harmful factors and reduces oxidative stress, thus facilitating wound healing. Similarly, Zhao et al. investigated the efficacy of an EGCG-3-acrylamido phenyl boronic acid-acrylamide (EACPA) hydrogel in wound treatment. They found that the hydrogel not only possesses antioxidant, antibacterial, anti-inflammatory, and pro-angiogenic properties but also regulates macrophage polarity, thereby accelerating wound healing and facilitating easy dressing changes (Zhao et al., 2021). This study highlights the significant role of the antioxidant EACPA in wound healing by mitigating intracellular ROS levels in Rosup-stimulated L929 fibroblasts. Despite their remarkable activities, catechins are known to be less stable in water. To address this issue, several studies have explored various approaches, including the use of reducing agents and the formation of micro- and nanoparticles (Dube et al., 2010; Ye et al., 2019). These efforts aim to enhance the stability and efficacy of catechins for potential applications in wound treatment.

7.4.4 Galic Acid (GA)

Galic acid (GA) is ubiquitously present in plants, distributed across various parts such as fruits, leaves, and stems, boasting potent properties including antioxidant, anti-inflammatory, anticancer, and neuroprotective effects (Ng et al., 2010; Ma et al., 2003; Mansouri et al., 2014). In a study by Yang et al. (2016) studied the effect of GA on wound healing were investigated under both normal hyperglycemic conditions. The findings revealed that GA demonstrated protective effects against oxidative stress induced by H_2O_2 and ROS-induced cytotoxicity in skin cells. Moreover, GA was observed to upregulate the expression of antioxidant genes such as catalase (CAT), superoxide dismutase 2 (SOD2), and glutathione peroxidase 1 (Gpx1). Additionally, GA promoted keratinocyte migration during wound healing and activated wound healing factors such as c-Jun N-terminal kinases (JNK), focal adhesion kinases (FAK), and extracellular signal-regulated kinases (Erk). Consequently, the study suggested that GA holds promise as an antioxidant for wound treatment. However, the limited solubility of GA in organic solvents poses a challenge for its topical application on the skin.

7.4.5 Resveratrol

Resveratrol, present in over 70 plant species, boasts remarkable medicinal properties including antioxidant, anticancer, anti-inflammatory, and antibacterial effects (Burns et al., 2002; Marques et al., 2009; Malhotra et al., 2015; Means et al., 2017). Zhou et al. (2021) conducted cell and in vivo experiments to explore resveratrol's potential in wound healing. Their findings revealed that resveratrol shields against H2O2-induced injury, enhances injured cell migration, and suppresses intracellular ROS production in human umbilical vein endothelial cells (HUVECs). In vivo tests confirmed that resveratrol accelerates wound closure, enhances skin structure, and mitigates inflammation, potentially by upregulating Mn-SOD to mitigate oxidative damage. Similarly, Bilgic et al. (2022) investigated resveratrol's wound healing capabilities in Wistar albino rats. They observed higher neovascularization levels and increased glutathione peroxidase activity in resveratrol-treated wounds compared to untreated controls, indicating resveratrol's impact on wound healing through its antioxidant properties. These studies collectively underscore the promising role of resveratrol in promoting wound healing, primarily attributed to its antioxidant effects.

7.4.6 Naringenin

Naringenin, a polyphenol predominantly found in citrus fruits, exhibits remarkable biological properties including anti-inflammatory, antioxidant, cholesterol-lowering, and anticancer effects. Al-Roujayee et al. (2017) investigated naringenin's impact on inflammatory responses and oxidative stress induced by thermal burn injuries in rats. Their findings indicated that naringenin treatment led to increased activities of various antioxidant enzymes such as glutathione-S-transferase (GST), superoxide dismutase (SOD), catalase, and glutathione peroxidase (GPx), while also restoring levels of thiobarbituric acid reactive substances (TBARS) and glutathione (GSH) by day 7. Moreover, naringenin was utilized in combination with other compounds, such as chitosan, to enhance wound healing capacity. Akrawi et al. (2020) demonstrated that a nanoemulsion formulation containing naringenin and chitosan significantly promoted wound contraction in Wistar rats over a 14-day treatment period. Furthermore, naringenin exhibited anti-inflammatory and antioxidant effects in this context. Hence it might be concluded that naringenin-based drugs have potential for the treatment of burn wounds, particularly leveraging its antioxidant activities to facilitate healing.

7.4.7 Myricetin (Myr)

Myricetin (Myr), a flavonoid known for its potential in wound healing (Elshamy et al., 2020; Sklenářová et al., 2021), is naturally found in various fruits and possesses diverse biochemical properties, including antioxidant, anti-allergic, anti-inflammatory, and immunomodulatory functions (Rehman and Rather, 2019; Wang et al., 2018). Elshamy et al. (2020) conducted a study where they isolated myricetin from *Tecomaria capensis* v. *aurea* and evaluated its efficacy in wound healing in albino rats. They observed that Myr influenced inflammatory cytokines such as tumor necrosis factor-α (TNF-α) and cluster of differentiation 68 (CD68), along with interleukin-1β (IL-1β). Additionally, Myr was found to enhance the expression of serum proinflammatory cytokines (e.g., IL-1β and TNF-α) while reducing the expression of macrophage CD68, suggesting its potential therapeutic application in wound healing by modulating inflammatory cytokines and systemic reorganization. Furthermore, Sklenářova et al. (2021) investigated Myr's wound healing properties and demonstrated its ability to inhibit the production of proinflammatory cytokines (e.g., IL-6 and IL-8) in skin cells. Despite its promising effects, myricetin exhibits poor solubility in water (Hong et al., 2015), leading to challenges in its bioavailability (Dang et al., 2013). To address this limitation, it is imperative to explore strategies such as combining Myr with other compounds or biomaterials to enhance its water solubility.

7.4.8 Pinocembrin

Pinocembrin, identified in propolis, honey, and plants of the Piperaceae family, demonstrates diverse potential bioactivities beneficial for therapeutic interventions, including antibacterial, anti-inflammatory, antifibrotic, and antioxidant properties (Rasul et al., 2013). Drewes and van Vuuren (2008) highlighted its significant antibacterial efficacy against *Staphylococcus aureus* and *Pseudomonas aeruginosa*, with minimum inhibitory concentrations of 6.3 μg/mL and 45–63 μg/mL, respectively. Moreover, it exhibits anti-inflammatory effects, as evidenced by its ability to mitigate sheep red blood cell-induced delayed-type hypersensitivity reaction. Notably, pinocembrin finds application in traditional Chinese medicine for wound healing purposes. Investigations by Li et al. (2021) further elucidated its role in combating skin fibrosis, revealing its substantial reduction of bleomycin-induced fibrosis and related protein expression in xenograft mice. Mechanistically, pinocembrin suppresses TGF-β1/Smad signaling and inhibits TGF-β1-induced activation of skin fibroblasts, thereby offering potential therapeutic avenues for addressing fibrotic conditions.

7.4.9 Chitosan and Chitin

Chitosan and chitin were initially explored for their wound healing properties by Ueno et al. (2001). Chitin, characterized as poly-N-acetyl-d-glucosamine-(1–4)-poly-N-acetyl-d-glucosamine, stands out as one of the most abundant polysaccharides, primarily sourced from the exoskeleton of marine crustaceans, shrimp, crabs, insects, fungi, and yeasts, second only to cellulose (Rinaudo, 2006). Chitosan, a copolymer comprising glucosamine and N-acetylglucosamine units linked by 1–4 glucosidic bonds, represents the most significant derivative of chitin. The effectiveness of chitosan's antibacterial properties lies in its ability to be positively charged, enhancing its interaction with negatively charged molecules found in bacterial membranes, such as anionic polysaccharides, proteins, and nucleic acids (Bhattarai et al., 2010; Chung and Chen, 2008). This attribute makes chitosan particularly advantageous for wound treatment, as it boasts biocompatibility, biodegradability, nontoxicity, adsorption properties, and hemostatic qualities (Song et al., 2012; LogithKumar et al., 2016; Khoshfetrat et al., 2016). Despite these benefits, chitosan's solubility is limited in neutral and alkaline aqueous solutions with pH values exceeding pH 6.5, severely constraining its applicability. Chitosan has been incorporated into various formulations, including nanoparticles, hydrogels, micelles, hyaluronic/oleic acid-loaded systems, and glucosylation of hydrophobic molecules (Mohamed et al., 2017). These innovative approaches aim to enhance the bioavailability of chitosan and also to expand its potential applications.

7.4.10 Propolis

Propolis, used by ancient civilizations like the Egyptians, Romans, and Persians, was sourced from various origins including honey bees, tree buds, and diverse botanical sources such as poplar, willow, elm, alder, birch, and beech (Afrouzan et al., 2018; Kuropatnicki et al., 2013; Bosio et al., 2000). Comprising over 300 chemical compounds including polyphenols, phenolic aldehydes, amino acids, and steroids, propolis primarily contains flavonoids, phenylpropanoids, cinnamic acids and their esters, and glycerides, which are its most significant constituents (Huang et al., 2014; Toreti et al., 2013). These components of propolis contribute toward the antibacterial properties against gram-positive bacteria, mainly attributed to the presence of flavonoids, esters, and aromatic acids within the resin (Uzel et al., 2005).

7.4.11 Tannins

In addition to animal-derived substances possessing antibacterial properties, numerous medicinal plants used in wound healing also demonstrate significant antibacterial effects, including tannins. Su et al. (2017) documented the antibacterial activity of tannins extracted from *Entada phaseoloides* (L.) Merr., which functions by suppressing protein synthesis, altering nucleic acid metabolism, preventing changes in cell wall formation, modifying cell membrane function, and inhibiting bacterial growth.

7.4.12 Allicin

Allicin, renowned for imparting the pungent odor to garlic, serves as the active component extensively validated in various studies for its wound healing properties. In addition to its antioxidant prowess, allicin exhibits antibacterial effects, with its mechanism of action extensively investigated. Research indicates that allicin exerts its antibacterial activity against *Staphylococcus aureus* through sulfhydryl alteration of bacterial proteins (Alhashim and Lombardo, 2018; Fujisawa et al., 2009).

7.4.13 Botanical Ingredients Promoting Collagen Production

Encouraging fibroblast migration and stimulating fresh collagen deposition within the wound area are pivotal aspects of wound healing. Collagen dressing technology facilitates the creation of new tissue, fosters angiogenesis, aids in autolytic debridement, and supports re-epithelialization (Trinh et al., 2022). Therefore, compounds that effectively enhance collagen synthesis during the healing process serve as integral components in promoting optimal wound recovery.

7.4.14 Saponins

Various compound groups, categorized based on their structure, offer a wide array of biological effects (Güçlü-Üstünda et al., 2007). For example, Wang et al. (2010) identified four novel steroidal saponins, along with two known compounds, extracted from *Bletilla striata*, a popular traditional Chinese herb. These compounds exhibit diverse activities such as hemolysis (Baumann et al., 2000), antibacterial (Killeen et al., 1998; Konishi et al., 1998), antiviral (Simões et al., 1999), antioxidative (Yogeeswari and Sriram, 2005), anti-inflammatory (Just et al., 1998; Navarro et al., 2001), and collagen promotion (Yu et al., 2015). Yu et al. (2015) investigated the effects of Panax notoginseng saponins (PNS) on anterior cruciate ligament (ACL) fibroblasts, observing enhanced migration, proliferation, and expression of fibronectin, collagen I, and collagen III during ACL injury healing. This suggests that PNS may exert its effects by phosphorylating PI3K, AKT, and ERK pathways, indicating its potential therapeutic significance.

7.4.15 Cryptotanshinone

Cryptotanshinone, derived from *Salvia miltiorrhiza* Bge, acts as a natural accelerator of procollagen synthesis during wound healing. Its activity facilitates enhanced angiogenesis and collagen deposition by reducing leukocyte infiltration, promoting eNOS phosphorylation, increasing the production of VEGF and Ang-1 proteins, suppressing the expression of MMP2 and MMP9 proteins, and boosting fibroblast translation (Song et al., 2020).

7.4.16 Artocarpin

Claims have been made regarding the anti-inflammatory and anticancer properties of the prenylated flavonoid artocarpin, extracted from the plant *Artocarpus communis* (Yao et al., 2016; Lee et al., 2013; Arung et al., 2010). Yeh et al. (2017) demonstrated that artocarpin stimulates collagen formation, proliferation, and migration of human fibroblasts by activating the JNK and P38 pathways. Additionally, it enhances the proliferation of human endothelial cells through the Akt and P38 pathways, as well as human keratinocytes through the ERK and P38 pathways.

7.4.17 Glucans

Glucans, glucose polymers present in yeast, grains, and fungi, are categorized as biological response modifiers (Bohn et al., 1995). Numerous studies have illustrated the immune-enhancing properties of both particulate and soluble glucans, showcasing their effectiveness against infections, cancer, and modulation of the immune system (Browder et al., 1990; Pretus et al., 1991). In wound healing, glucans facilitate the infiltration of macrophages, promoting tissue granulation, collagen deposition, and re-epithelialization. Due to their remarkable stability and resistance to wound proteases, glucan-based wound dressings serve as optimal agents for wound healing (Majtan and Jesenak, 2018).

7.4.18 Amino Acids and Peptides

In addition to traditional medicinal plants, natural procollagen compounds containing amino acids and peptides for wound healing are also sourced from animals such as bees, mollusks, snails, and fish, as widely reported (Mellin et al., 1992). The low pH of honey may create an ideal acidic environment for fibroblasts, aiding in tasks like migration and collagen organization. Badiu et al. (2008) demonstrated that amino acids from *Rapana venosa* and *Mytilus galloprovincialis* accelerate skin wound healing by promoting dermal and epidermal neoformation. The mechanism underlying these amino acids' wound-healing effects involves the differential regulation of macrophage arginine metabolism, possibly mediated by TGF-β1. Additionally, bioactive peptides extracted from the terrestrial snail *Cryptozona bistrialis* stimulate the in vitro migration of NIH/3T3 mouse fibroblast cells and enhance wound healing in healthy and diabetic-induced Wistar albino rats (Ulagesan et al., 2018). This peptide is believed to increase wound contraction by boosting collagen content through enhanced fibroblast and epithelial cell migration to the wound site. However, the exact chemical formula of the compounds extracted from animal sources remains undisclosed.

REFERENCES

Aceituno-Medina, M.; Mendoza, S.; Rodríguez, B.A.; Lagaron, J.M.; López-Rubio, A. Improved Antioxidant Capacity of Quercetin and Ferulic Acid during in-Vitro Digestion through Encapsulation within Food-Grade Electrospun Fibers. *J. Funct. Foods* 2015, 12, 332–341.

Afrouzan, H.; Tahghighi, A.; Zakeri, S.; Es-haghi, A. Chemical Composition and Antimicrobial Activities of Iranian Propolis. *Iran. Biomed. J.* 2018, 22, 50–65.

Akrawi, S.H.; Gorain, B.; Nair, A.B.; Choudhury, H.; Pandey, M.; Shah, J.N.; Venugopala, K.N. Development and Optimization of Naringenin-Loaded Chitosan-Coated Nanoemulsion for Topical Therapy in Wound Healing. *Pharmaceutics* 2020, 12, 893.

Alhashim, M.; Lombardo, J. Mechanism of Action of Topical Garlic on Wound Healing. *Dermatologic Surg.* 2018, 44, 630–634.

Al-Roujayee, A.S. Naringenin Improves the Healing Process of Thermally-Induced Skin Damage in Rats. *J. Int. Med. Res.* 2017, 45, 570–582.

Alvarez-Román, R.; Naik, A.; Kalia, Y.N.; Guy, R.H.; Fessi, H. Skin Penetration and Distribution of Polymeric Nanoparticles. *J Control Release.* 2004, 99(1), 53–62.

Arung, E.T.; Wicaksono, B.D.; Handoko, Y.A.; Kusuma, I.W.; Shimizu, K.; Yulia, D.; Sandra, F. Cytotoxic Effect of Artocarpin onT47D Cells. *J. Nat. Med.* 2010, 64, 423–429.

Ayuk, S.M.; Abrahamse, H.; Houreld, N.N. The Role of Matrix Metalloproteinases in Diabetic Wound Healing in Relation to Photobiomodulation. *J. Diabetes Res.* 2016, 2016, 2897656.

Badiu, D. L.; Balu, A. M.; Barbes, L.; Luque, R. Nita, R.; Radu, M.; Tanase, E.; Rosoiu, N. Physico-Chemical Characterisation of Lipids from Mytilus galloprovincialis (L.) and Rapana venosa and their Healing Properties on Skin Burns. *Lipids.* 2008, 43, 829–841.

Baek, S.; Park, H.; Kim, M.; Lee, D. Preparation of PCL/(+)-Catechin/Gelatin Film for Wound Healing Using Air-Jet Spinning. *Appl. Surf. Sci.* 2020, 509, 145033.

Bahramsoltani, R.; Farzaei, M.H.; Rahimi, R. Medicinal Plants and Their Natural Components as Future Drugs for the Treatment of Burn Wounds: An Integrative Review. *Arch Dermatol Res.* 2014, 306(7), 601–617.

Baumann, E.; Stoya, G.; Völkner, A.; Richter, W.; Lemke, C.; Linss, W. Hemolysis of Human Erythrocytes with Saponin Affects the Membrane Structure. *Acta Histochem.* 2000, 102, 21–35.

Bhattarai, N.; Gunn, J.; Zhang, M. Chitosan-Based Hydrogels for Controlled, Localized Drug Delivery. *Adv. Drug Deliv. Rev.* 2010, 62, 83–99.

Bilgic, T. Partial Effect of Resveratrol on Wound Healing: Study on Wistar Albino Rats. *Eurasian J. Med. Investig.* 2022, 5, 460–468.

Bohn, J.A.; BeMiller, J.N. (1→3)-β-D-Glucans as Biological Response Modifiers: A Review of Structure-Functional Activity Relationships. *Carbohydr. Polym.* 1995, 28, 3–14.

Bonté, F.; Noel-Hudson, M.-S.;Wepierre, J.; Meybeck, A. Protective Effect of Curcuminoids on Epidermal Skin Cells under Free Oxygen Radical Stress. *Planta Med.* 1997, 63, 265–266.

Bosio, K.; Avanzini, C.; D'avolio, A.; Ozino, O.; Savoia, D.In Vitro Activity of Propolis against Streptococcus Pyogenes. *Lett. Appl. Microbiol.* 2000, 31, 174–177.

Brem, H.; Stojadinovic, O.; Diegelmann, R.F.; Entero, H.; Lee, B.; Pastar, I.; Golinko, M.S.; Rosenberg, H.; Tomic-Canic, M. Molecular Markers in Patients with Chronic Wounds to Guide Surgical Debridement. *Mol. Med.* 2007, 13, 30–39.

Browder, W.; Williams, D.; Pretus, H.; Olivero, G.; Enrichens, F.; Mao, P.; Franchello, A. Beneficial Effect of Enhanced Macrophage Function in the Trauma Patient. *Ann. Surg.* 1990, 211, 605–612, discussion 612–613.

Bruce, E.D.; Christie, M.S. Toxicological Outcomes and Pharmacological Needs in Chronic Wound Healing. *EC Pharmacol. Toxicol.* 2016, 1, 15–32.

Burns, J.; Yokota, T.; Ashihara, H.; Lean, M.E.J.; Crozier, A. Plant Foods and Herbal Sources of Resveratrol. *J. Agric. Food Chem.* 2002, 50, 3337–3340.

Cañedo-Dorantes, L.; Cañedo-Ayala, M. Skin Acute Wound Healing: A Comprehensive Review. *Int. J. Inflamm.* 2019, 2019, 3706315.

Caro, A.A.; Davis, A.; Fobare, S.; Horan, N.; Ryan, C.; Schwab, C. Antioxidant and Pro-Oxidant Mechanisms of (+) Catechin in Microsomal CYP2E1-Dependent Oxidative Stress. *Toxicol. Vitr.* 2019, 54, 1–9.

Chen, C.; Liu, Y.; Sun, L.; Chen, G.;Wu, X.; Ren, J.; Zhao, Y. Antibacterial Porous Microcarriers with a Pathological State Responsive Switch for Wound Healing. *ACS Appl. Bio Mater.* 2019, 2, 2155–2161.

Chung, Y.-C.; Chen, C.-Y. Antibacterial Characteristics and Activity of Acid-Soluble Chitosan. *Bioresour. Technol.* 2008, 99, 2806–2814.

Clinton, A.; Carter, T. Chronic Wound Biofilms: Pathogenesis and Potential Therapies. *Lab. Med.* 2015, 46, 277–284.

Da̧browska, A.; Spano, F.; Derler, S.; Adlhart, C.; Spencer, N.; Rossi, R. The Relationship Between Skin Function, Barrier Properties, and Body-Dependent Factors. *Ski. Res. Technol.* 2018, 24, 165–174.

Dang, Y.; Lin, G.; Xie, Y.; Duan, J.; Ma, P.; Li, G.; Ji, G. Quantitative Determination of Myricetin in Rat Plasma by Ultra Performance Liquid Chromatography Tandem Mass Spectrometry and Its Absolute Bioavailability. *Drug Res. (Stuttg).* 2013, 64, 516–522.

Drewes, S.E.; van Vuuren, S.F. Antimicrobial Acylphloroglucinols and Dibenzyloxy Flavonoids from Flowers of Helichrysum Gymnocomum. *Phytochemistry* 2008, 69, 1745–1749.

Dube, A.; Ng, K.; Nicolazzo, J.A.; Larson, I. Effective Use of Reducing Agents and Nanoparticle Encapsulation in Stabilizing Catechins in Alkaline Solution. *Food Chem.* 2010, 122, 662–667.

Ehterami, A.; Salehi, M.; Farzamfar, S.; Vaez, A.; Samadian, H.; Sahrapeyma, H.; Mirzaii, M.; Ghorbani, S.; Goodarzi, A. In Vitro and in Vivo Study of PCL/COLL Wound Dressing Loaded with Insulin-Chitosan Nanoparticles on Cutaneous Wound Healing in Rats Model. *Int. J. Biol. Macromol.* 2018, 117, 601–609.

Elshamy, A.I.; Ammar, N.M.; Hassan, H.A.; El-Kashak, W.A.; Al-Rejaie, S.S.; Abd-ElGawad, A.M.; Farrag A.-R.R.H. Topical Wound Healing Activity of Myricetin Isolated from Tecomaria Capensis v. Aurea. *Molecules* 2020, 25, 4870.

Frykberg, R.G.; Banks, J. Challenges in the Treatment of Chronic Wounds. *Adv. Wound Care* 2015, 4, 560–582.

Fujisawa, H.; Watanabe, K.; Suma, K.; Origuchi, K.; Matsufuji, H.; Seki, T.; Ariga, T. Antibacterial Potential of Garlic-Derived Allicin and Its Cancellation by Sulfhydryl Compounds. *Biosci. Biotechnol. Biochem.* 2009, 73, 1948–1955.

Gadekar, R.; Saurabh, M.K.; Thakur, G.S.; Saurabh, A. Study of Formulation, Characterisation andWound Healing Potential of Transdermal Patches of Curcumin. *Asian J. Pharm. Clin. Res.* 2012, 5, 225–230.

Groeber F.; Holeiter M.; Hampel M.; Hinderer S.; Schenke-Layland, K. Skin tissue engineering – in Vivo and in Vitro Applications. *Adv Drug Deliv Rev.* 2011, 63(4–5), 352–366.

Güçlü-Üstünda˘ g, Ö.; Mazza, G.; Guclu-Ustundag, Ö.; Mazza, G. Saponins: Properties, Applications and Processing. *Crit. Rev. Food Sci. Nutr.* 2007, 47, 231–258.

Guo, S.; DiPietro, L.A. Factors Affecting Wound Healing. *J. Dent. Res.* 2010, 89, 219–229.

Hajialyani, M.; Tewari, D.; Sobarzo-Sánchez, E.; Nabavi, S. M.; Farzaei, M. H.; Abdollahi, M. (2018). Natural Product-Based Nanomedicines for Wound Healing Purposes: Therapeutic Targets and Drug Delivery Systems. *Int J Nanomedicine*, 5023–5043.

Han, G.; Ceilley, R. Chronic Wound Healing: A Review of Current Management and Treatments. *Adv. Ther.* 2017, 34, 599–610.

Hasan, N.; Cao, J.; Lee, J.; Hlaing, S.P.; Oshi, M.A.; Naeem, M.; Ki, M.-H.; Lee, B.L.; Jung, Y.; Yoo, J.-W. Bacteria-Targeted Clindamycin Loaded Polymeric Nanoparticles: Effect of Surface Charge on Nanoparticle Adhesion to MRSA, Antibacterial Activity, and Wound Healing. *Pharmaceutics* 2019, 11, 236.

Hayat, M.; Nawaz, A.; Chinnam, S.; Muzammal, M.; Latif, M.S.; Yasin, M.; Ashique, S.; Zengin, G.; Farid, A. Formulation Development and Optimization of Herbo Synthetic Gel: In Vitro Biological Evaluation and in Vivo Wound Healing Studies. *Process Biochemistry.* 2023 Jul 1, 130, 116–126.

Hong, C.; Xie, Y.; Yao, Y.; Li, G.; Yuan, X.; Shen, H. A Novel Strategy for Pharmaceutical Cocrystal Generation without Knowledge of Stoichiometric Ratio: Myricetin Cocrystals and a Ternary Phase Diagram. *Pharm. Res.* 2015, 32, 47–60.

Huang, S.; Zhang, C.-P.; Wang, K.; Li, G.; Hu F.-L.Recent Advances in the Chemical Composition of Propolis. *Molecules* 2014, 19, 19610–19632.

Just, M.; Recio, M.; Giner, R.; Cuéllar, M.; Máñez, S.; Bilia, A.; Ríos, J.-L. Anti-Inflammatory Activity of Unusual Lupane Saponins from Bupleurum Fruticescens. *Planta Med.* 1998, 64, 404–407.

Kant, V.; Jangir, B.L.; Nigam, A.; Kumar, V.; Sharma, S. Dose Regulated Cutaneous Wound Healing Potential of Quercetin in Male Rats. *Wound Med.* 2017, 19, 82–87.

Khoshfetrat, A.B.; Khanmohammadi, M.; Sakai, S.; Taya, M. Enzymatically-Gellable Galactosylated Chitosan: Hydrogel Characteristics and Hepatic Cell Behavior. *Int. J. Biol. Macromol.* 2016, 92, 892–899.

Killeen, G.F.; Madigan, C.A.; Connolly, C.R.;Walsh, G.A.; Clark, C.; Hynes, M.J.; Timmins, B.F.; James, P.; Headon, D.R.; Power, R.F. Antimicrobial Saponins of Yucca Schidigera and the Implications of Their in Vitro Properties for Their in Vivo Impact. *J. Agric. Food Chem.* 1998, 46, 3178–3186.

Konishi, M.; Hano, Y.; Takayama, M.; Nomura, T.; Hamzah, A.S.; Ahmad, R.B.; Jasmani, H. Triterpenoid Saponins from Hedyotis Nudicaulis. *Phytochemistry.* 1998, 48, 525–528.

Kratofil, R.M.; Kubes, P.; Deniset, J.F. Monocyte Conversion During Inflammation and Injury. *Arter. Thromb. Vasc. Biol.* 2017, 37, 35–42.

Kuropatnicki, A.K.; Szliszka, E.; Krol, W. Historical Aspects of Propolis Research in Modern Times. *Evid. Based Complement. Altern. Med.* 2013, 2013, 964149.

Kushwaha, A.; Goswami, L.; Kim, B.S. Nanomaterial-Based Therapy for Wound Healing. *Nanomaterials* 2022, 12(4), 618.

Landén, N.X.; Li, D.; Ståhle, M. Transition from Inflammation to Proliferation: A Critical Step During Wound Healing. *Cell. Mol. Life Sci.* 2016, 73, 3861–3885.

Lee, C.-W.; Ko, H.-H.; Lin, C.-C.; Chai, C.-Y.; Chen, W.-T.; Yen, F.-L. Artocarpin Attenuates Ultraviolet B-Induced Skin Damage in Hairless Mice by Antioxidant and Anti-Inflammatory Effect. *Food Chem. Toxicol.* 2013, 60, 123–129.

Li S, Zhang T, Xu W, et al. Sarcoma-Targeting Peptide-Decorated Polypeptide Nanogel Intracellularly Delivers Shikonin for Upregulated Osteosarcoma Necroptosis and Diminished Pulmonary Metastasis. *Theranostics.* 2018, 8(5), 1361–1375.

Li, X.; Zhai, Y.; Xi, B.; Ma, W.; Zhang, J.; Ma, X.; Miao, Y.; Zhao, Y.; Ning, W.; Zhou, H.; et al. Pinocembrin Ameliorates Skin Fibrosis via Inhibiting Tgf-_1 Signaling Pathway. *Biomolecules* 2021, 11, 1240.

LogithKumar, R.; KeshavNarayan, A.; Dhivya, S.; Chawla, A.; Saravanan, S.; Selvamurugan, N.A Review of Chitosan and Its Derivatives in Bone Tissue Engineering. *Carbohydr. Polym.* 2016, 151, 172–188.

Ma, J.; Luo, X.-D.; Protiva, P.; Yang, H.; Ma, C.; Basile, M.J.; Weinstein, I.B.; Kennelly, E.J. Bioactive Novel Polyphenols from the Fruit of Manilkara Zapota (Sapodilla). *J. Nat. Prod.* 2003, 66, 983–986.

Majtan, J.; Jesenak, M. _-Glucans: Multi-Functional Modulator of Wound Healing. *Molecules* 2018, 23, 806.

Malhotra, A.; Bath, S.; Elbarbry, F. An Organ System Approach to Explore the Antioxidative, Anti-Inflammatory, and Cytoprotective Actions of Resveratrol. *Oxid. Med. Cell. Longev.* 2015, 2015, 803971.

Mansouri, M.T.; Soltani, M.; Naghizadeh, B.; Farbood, Y.; Mashak, A.; Sarkaki, A. A Possible Mechanism for the Anxiolytic-like Effect of Gallic Acid in the Rat Elevated plus Maze. *Pharmacol. Biochem. Behav.* 2014, 117, 40–46.

Marques, F.Z.; Markus, M.A.; Morris, B.J. Resveratrol: Cellular Actions of a Potent Natural Chemical That Confers a Diversity of Health Benefits. *Int. J. Biochem. Cell Biol.* 2009, 41, 2125–2128.

Matter, M.T.; Probst, S.; Läuchli, S.; Herrmann, I.K. Uniting Drug and Delivery: Metal Oxide Hybrid Nanotherapeutics for Skin Wound Care. *Pharmaceutics.* 2020, 12, 780.

McDaniel, J.C.; Roy, S.; Wilgus, T.A. Neutrophil Activity in Chronic Venous Leg Ulcers—A Target for Therapy? *Wound Repair Regen.* 2013, 21, 339–351.

Means, J.; Gerdes, B.; Koulen, P. Distinct Mechanisms Underlying Resveratrol-Mediated Protection from Types of Cellular Stress in C6 Glioma Cells. *Int. J. Mol. Sci.* 2017, 18, 1521.

Mellin, T.N.; Mennie, R.J.; Cashen, D.E.; Ronan, J.J.; Capparella, J.; James, M.L.; Disalvo, J.; Frank, J.; Linemeyer, D.; Gimenez-Gallego, G.; et al. Acidic Fibroblast Growth Factor Accelerates Dermal Wound Healing. *J Investig.Dermatol.* 1992, 7, 1–14.

Mi, Y.; Zhong, L.; Lu, S.; Hu, P.; Pan, Y.; Ma, X.; Yan, B.; Wei, Z.; Yang, G. Quercetin Promotes Cutaneous Wound Healing in Mice through Wnt/_-Catenin Signaling Pathway. *J. Ethnopharmacol.* 2022, 290, 115066.

Mohamed, R.R.; Elella, M.H.A.A.; Sabaa, M.W. Cytotoxicity and Metal Ions Removal Using Antibacterial Biodegradable Hydrogels Based on N-Quaternized Chitosan/Poly (Acrylic Acid). *Int. J. Biol. Macromol.* 2017, 98, 302–313.

Mohanty, C.; Das, M.; Sahoo, S.K. Sustained Wound Healing Activity of Curcumin Loaded Oleic Acid Based Polymeric Bandagein a Rat Model. *Mol. Pharm.* 2012, 9, 2801–2811.

Mordorski B, Prow T. Nanomaterials for Wound Healing. *Curr Dermatol Rep.* 2016, 5(4), 278–286.

Navarro, P.; Giner, R.M.; Recio, M.C.; Máñez, S.; Cerdá-Nicolás, M.; Ríos, J.-L.In Vivo Anti-Inflammatory Activity of Saponins from Bupleurum Rotundifolium. *Life Sci.* 2001, 68, 1199–1206.

Ng, T.B.; He, J.S.; Niu, S.M.; Pi, Z.F.; Shao, W.; Liu, F.; Zhao, L. A Gallic Acid Derivative and Polysaccharides with Antioxidative Activity from Rose (Rosa Rugosa) Flowers. *J. Pharm. Pharmacol.* 2010, 56, 537–545.

Orhan, I.E.; Akkol, E.K.; Suntar, I.; Yesilada, E. Assessment of Anticholinesterase and Antioxidant Properties of the Extracts and (+)-Catechin Obtained from Arceuthobium oxycedri (D.C.) M. Bieb (Dwarf Mistletoe). *South African J. Bot.* 2019, 120, 309–312.

Pereira RF, Bártolo PJ. Traditional Therapies for Skin Wound Healing. *Adv Wound Care.* 2016, 5(5), 208–229.
Periayah, M.H.; Halim, A.S.; Mat Saad, A.Z. Mechanism Action of Platelets and Crucial Blood Coagulation Pathways in Hemostasis. *Int. J. Hematol. Oncol. Stem Cell Res.* 2017, 11, 319–327.
Phan, T.-T.; See, P.; Lee, S.-T.; Chan, S.-Y. Protective Effects of Curcumin against Oxidative Damage on Skin Cells in Vitro: Its Implication for Wound Healing. *J. Trauma Inj. Infect. Crit. Care* 2001, 51, 927–931.
Piipponen, M.; Li, D.; Landén, N.X. The Immune Functions of Keratinocytes in Skin Wound Healing. *Int. J. Mol. Sci.* 2020, 21, 8790.
Pretus, H.A.; Ensley, H.E.; McNamee, R.B.; Jones, E.L.; Browder, I.W.; Williams, D.L. Isolation, Physicochemical Characterization and Preclinical Efficacy Evaluation of Soluble Scleroglucan. *J. Pharmacol. Exp. Ther.* 1991, 257, 500–510.
Qiao, Y.; He, J.; Chen, W.; Yu, Y.; Li, W.; Du, Z.; Xie, T.; Ye, Y.; Hua, S.Y.; Zhong, D.; et al. Light-Activatable Synergistic Therapy of Drug-Resistant Bacteria-Infected Cutaneous Chronic Wounds and Nonhealing Keratitis by Cupriferous Hollow Nanoshells. *ACS Nano* 2020, 14, 3299–3315.
Rasul, A.; Millimouno, F.M.; Ali Eltayb, W.; Ali, M.; Li, J.; Li, X. Pinocembrin: A Novel Natural Compound with Versatile Pharmacological and Biological Activities. *BioMed Res. Int.* 2013, 2013, 379850.
Rehman, M.U.; Rather, I.A. Myricetin Abrogates Cisplatin-Induced Oxidative Stress, Inflammatory Response, and Goblet Cell Disintegration in Colon of Wistar Rats. *Plants* 2019, 9, 28.
Reinke, J.M.; Sorg, H. Wound Repair and Regeneration. *Eur Surg Res.* 2012, 49(1), 35–43.
Rinaudo, M. Chitin and Chitosan: Properties and Applications. *Prog. Polym. Sci.* 2006, 31, 603–632.
Rodrigues, M.; Kosaric, N.; Bonham, C.A.; Gurtner, G.C. Wound Healing: A Cellular Perspective. *Physiol. Rev.* 2019, 99, 665–706.
Rumbaut, R.E.; Thiagarajan, P. Platelet-VesselWall Interactions in Hemostasis and Thrombosis. *Colloq. Ser. Integr. Syst. Physiol. Mol. Funct.* 2010, 2, 1–75.
Sharifi, S.; Hajipour, M.J.; Gould, L.; Mahmoudi, M. Nanomedicine in Healing Chronic Wounds: Opportunities and Challenges. *Mol. Pharm.* 2021, 18, 550–575.
Simões, C.M.O.; Amoros, M.; Girre, L. Mechanism of Antiviral Activity of Triterpenoid Saponins. *Phyther. Res.* 1999, 13, 323–328.
Singh, V.A.; Aditi, A.S.; Gade, W.N., et al. Nanomaterials: New Generation Therapeutics in Wound Healing and Tissue Repair. *Curr. Nanosci.* 2010, 6(6), 577–586.
Sklenářová, R.; Svrčková, M.; Hodek, P.; Ulrichová, J.; Franková, J. Effect of the Natural Flavonoids Myricetin and Dihydromyricetin on the Wound Healing Process in Vitro. *J. Appl. Biomed.* 2021, 19, 149–158.
Song, E.-H.; Shang, J.; Ratner, D.M. Polysaccharides. In *Polymer Science: A Comprehensive Reference.* Amsterdam, The Netherlands: Elsevier, 2012, pp. 137–155.
Song, M.; Chen, L.; Zhang, L.; Li, C.; Coffie, J.W.; Fang, Z.; Zhang, L.;Wang, S.; Gao, X.;Wang, H. Cryptotanshinone Enhances Wound Healing in Type 2 Diabetes with Modulatory Effects on Inflammation, Angiogenesis and Extracellular Matrix Remodelling. *Pharm. Biol.* 2020, 58, 845–853.
Su, X.; Liu, X.; Wang, S.; Li, B.; Pan, T.; Liu, D.; Wang, F.; Diao, Y.; Li, K. Wound-Healing Promoting Effect of Total Tannins from Entada phaseoloides (L.) Merr.in Rats. *Burns* 2017, 43, 830–838.
Toreti, V.C.; Sato, H.H.; Pastore, G.M.; Park, Y.K. Recent Progress of Propolis for Its Biological and Chemical Compositions and Its Botanical Origin. *Evid. Based Complement. Altern. Med.* 2013, 697390.
Trinh, X.T.; Long, N.V.; Van Anh, L.T.; Nga, P.T.; Giang, N.N.; Chien, P.N.; Heo, C.Y. A Comprehensive Review of Natural Compounds for Wound Healing: Targeting Bioactivity Perspective. *Int. J. Mol. Sci.* 2022, 23(17), 9573.
Ueno, H.; Mori, T.; Fujinaga, T. Topical Formulations and Wound Healing Applications of Chitosan. *Adv. Drug Deliv. Rev.* 2001, 52, 105–115.
Ulagesan, S.; Sankaranarayanan, K.; Kuppusamy, A. Functional Characterisation of Bioactive Peptide Derived from Terrestrial Snail Cryptozona Bistrialis and Its Wound-Healing Property in Normal and Diabetic-Induced Wistar Albino Rats. *Int. Wound J.* 2018, 15, 350–362.
Uzel, A.; Sorkun, K.K.; Önça˘ g, Ö.; Ço˘gulu, D.; Gençay, Ö.; Salih, B. Chemical Compositions and Antimicrobial Activities of Four Different Anatolian Propolis Samples. *Microbiol. Res.* 2005, 160, 189–195.
Velnar, T.; Bailey, T.; Smrkolj, V. The Wound Healing Process: An Overview of the Cellular and Molecular Mechanisms. *J. Int. Med. Res.* 2009, 37, 1528–1542.
Wang, M.; Huang, X.; Zheng, H.; Tang, Y.; Zeng, K.; Shao, L.; Li, L. Nanomaterials Applied in Wound Healing: Mechanisms, Limitations and Perspectives. *J. Control. Release* 2021, 337, 236–247.
Wang, T.; Li, Q.; Bi, K. Bioactive Flavonoids in Medicinal Plants: Structure, Activity and Biological Fate. *Asian J. Pharm. Sci.* 2018, 13, 12–23.

Wang, Y.; Wang, L.; Hu, Y.; Zhang, L.; Wang, Z. Isolation and Identification of Two Steroid Compounds from Oviductus Ranae. *Natural Product Research* 2010, 24(16), 1518–1522.

Xue, M.; Jackson, C.J. Extracellular Matrix Reorganization During Wound Healing and Its Impact on Abnormal Scarring. *Adv. Wound Care* 2015, 4, 119–136.

Yao, X.; Wu, D.; Dong, N.; Ouyang, P.; Pu, J.; Hu, Q.; Wang, J.; Lu, W.; Huang, J. Moracin C, A Phenolic Compound Isolated from Artocarpus Heterophyllus, Suppresses Lipopolysaccharide-Activated Inflammatory Responses in Murine Raw264.7 Macrophages. *Int. J. Mol. Sci.* 2016, 17, 1199.

Yang, D.J.; Moh, S.H.; Son, D.H.; You, S.; Kinyua, A.W.; Ko, C.M.; Song, M.; Yeo, J.; Choi, Y.H. Kim K.W. Gallic Acid Promotes Wound Healing in Normal and Hyperglucidic Conditions. *Molecules.* 2016, 21(7), 899.

Ye, J.-H.H.; Augustin, M.A. Nano- and Micro-Particles for Delivery of Catechins: Physical and Biological Performance. *Crit. Rev. Food Sci. Nutr.* 2019, 59, 1563–1579.

Yeh, C.-J.; Chen, C.-C.; Leu, Y.-L.; Lin, M.-W.; Chiu, M.-M.; Wang, S.-H. The Effects of Artocarpin on Wound Healing: In Vitro and in Vivo Studies. *Sci. Rep.* 2017, 7, 15599.

Yogeeswari, P.; Sriram, D. Betulinic Acid and Its Derivatives: A Review on Their Biological Properties. *Curr. Med. Chem.* 2005, 12, 657–666.

Young, A.; McNaught, C.-E.The physiology of wound healing. *Surgery* 2011, 29, 475–479.

Yu, L.; Xie, J.; Xin, N.;Wang, Z. Panax Notoginseng Saponins Promote Wound Repair of Anterior Cruciate Ligament through Phosphorylation of PI3K, AKT and ERK. *Int. J. Clin. Exp. Pathol.* 2015, 8, 441–449.

Zhao, X.; Pei, D.; Yang, Y.; Xu, K.; Yu, J.; Zhang, Y.; Zhang, Q.; He, G.; Zhang, Y.; Li, A.; et al. Green Tea Derivative Driven Smart Hydrogels with Desired Functions for Chronic Diabetic Wound Treatment. *Adv. Funct. Mater.* 2021, 31, 2009442.

Zhou, X.; Ruan, Q.; Ye, Z.; Chu, Z.; Xi, M.; Li, M.; Hu, W.; Guo, X.; Yao, P.; Xie, W. Resveratrol Accelerates Wound Healing by Attenuating Oxidative Stress-Induced Impairment of Cell Proliferation and Migration. *Burns* 2021, 47, 133–139.

8 Bioinspired Nanomaterials for Wound Healing

Alamgir Khan,[1] Arshad Farid,[1] and Gagan Preet[2]*

[1] Gomal University, D. I. Khan, Pakistan
[2] University of Aberdeen, Aberdeen, UK
[*] Corresponding Author

8.1 INTRODUCTION

The development of various methods for creating nanomaterials has been spurred by the advancements made in nanotechnology. Bioinspired or biomimetic nanoplatforms are one of these tactics that have been used more and more for various biomedical applications. During the synthesis cycle, bioinspired or biomimetic nanoparticles (NPs) are created by taking inspiration from nature or its constituent parts. Technological advances in bioengineering and bioformation science have allowed analysts to lead the development of novel varieties of these nanoplatforms (Wu et al., 2022).

Compared to current strategies, bioinspired nanomaterials demonstrate essential viability in healing wounds. Bioinspired nanomaterials can be customized to create wound dressings that endorse anti-inflammatory processes and inhibit microbial biofilm growth. Bioinspired nanomaterials' sweeping power in wound healing surpasses previous evaluations, creating new avenues for research (Arshad et al., 2023).

Wounds on the skin leave it susceptible to microbial infections, which makes wound care challenging. Contamination of wounds prompts increased formation of wound exudates, which contain protein nutrients advantageous for bacterial development. This worsens bacterial expansion, delays the healing process, and disturbs proper collagen deposition. Initially sterile, wounds quickly become populated with microscopic organisms, predominantly *Staphylococcus aeruginosa* and *Staphylococcus aureus* (Von Muller et al., 2020).

The wound healing process is intricate and can be hindered by different variables. Key controllers of the wound healing cascade include growth factors and cytokines, which coordinate several cell activities such as cell movement, enzyme production, matrix synthesis, and cell differentiation. Surprisingly, these growth factors exert their effects even in minute amounts. There are several therapy options to manage chronic wounds and biofilms. Microbial resistance is a crucial challenge in any scenario, especially when it comes to established biofilms.

Bioinspired nanomaterials have arisen as a promising pathway in this manner, produced from natural sources like polysaccharides, phytochemicals, amino acids, polyphenols, and vitamins (Ma et al., 2022). Wound dressings using bioinspired nanocarriers, including bandages, hydrogels, and membranes have shown qualities of nonadhesion alongside great mechanical strength, porosity, and advanced flexibility. Moreover, formulations consolidating bioinspired nanomaterials show a penchant for environmental manageability, marked by diminished hazards and pollution (Hazra et al., 2022).

The safety, effectiveness, patient compliance, and mucoadhesion of traditional procedures are comparatively low. Moreover, bioinspired nanoparticles have multiple uses, such as enhanced self-healing capacity, antioxidant and antimicrobial properties, hemostasis, and mucoadhesion (Hayat et al., 2023).

DOI: 10.1201/9781003504948-8

8.2 BIOINSPIRED NANOPARTICLES

8.2.1 Bioinspired Metallic Nanoparticles

8.2.1.1 Bioinspired Green Synthesis of Metallic Nanoparticles

Drawing motivation from nature, the field of bioinspired metallic nanoparticles is growing. This strategy is favored for its simplicity and cost-friendliness. Also, these nanoparticles can be readily modified with different carbohydrates, lipids, proteins, and antibodies to enhance their natural properties (Gurunathan, 2019).

Reliance on harsh synthetic compounds and stabilizers in the amalgamation of metallic nanoparticles prompts harmful impacts and high energy utilization. Industrially, it is crucial to achieve well-dispersed nanoparticles with controlled sizes. To resolve these issues, different endeavors have been made to utilize food byproducts, like banana peels (Ibrahim, 2015) and orange peels (Castro et al., 2013), in the combination cycle. These by-products can be repurposed into valuable products. Propelled by this potential, scientists orchestrated gold (Au), silver (Ag), and platinum (Pt) nanoparticles by utilizing an aqueous extract obtained from *Garcinia mangostana* skin (Pedraza-Chaverri et al., 2008). This fruit extract is known for its applications in reducing inflammation, diarrhea, and cholera. The nanoparticles were then tried for their antimicrobial viability, both autonomously and joined with different classes of anti-infection agents. The findings revealed that, when tested specifically against gram-negative bacterial strains, silver nanoparticles (AgNPs) outperformed gold nanoparticles (AuNPs) and platinum nanoparticles (PtNPs) in terms of antibacterial capabilities. Additionally, every group of nanoparticles showed synergistic effects when combined with various antibiotic classes, suggesting a relationship between antimicrobials and nanoparticles. Moreover, *Bacillus* species that had previously shown resistance to streptomycin showed vulnerability to treatment with a combination of AuNPs and antibiotics. According to Nishanthi et al. (2019), metallic nanoparticles generally made bacterial strains more susceptible to drugs.

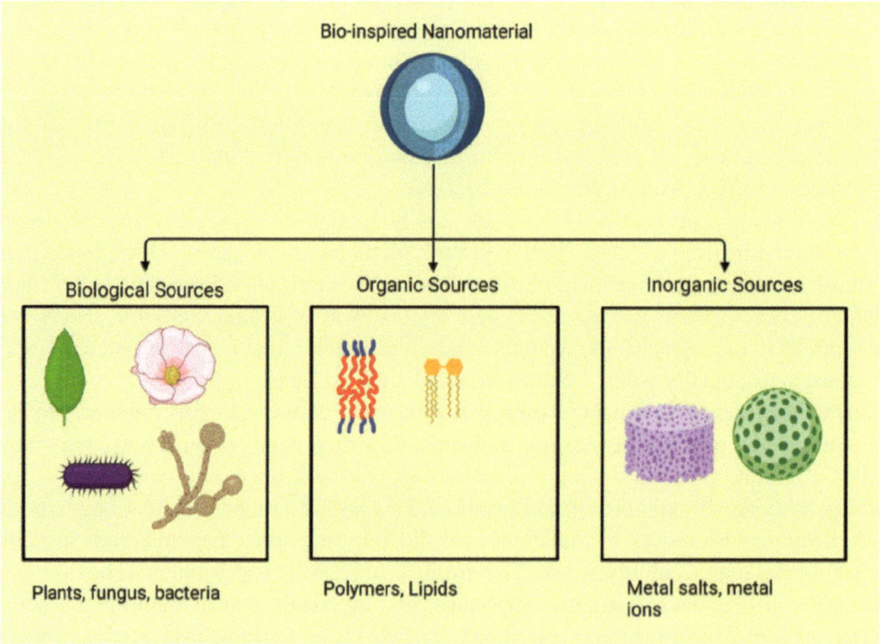

FIGURE 8.1 Synthesis of bioinspired nanoparticles.

8.2.1.2 Bioinspired Gold Nanoparticles (AuNPs)

Among the most advanced nanomaterials today, graphene has demonstrated significant scientific promise for upcoming uses in nanotechnology (Mao et al., 2013). Researchers created reduced graphene oxide-coated AuNPs. They reduced both chloroauric acid and graphene oxide (GO) using seed extract from *Syzygium cumini*. Their physicochemical properties were meticulously characterized and subjected to antibacterial tests. According to Kadiyala et al. (2018), this nanocomposite displayed substantial antibacterial action against Escherichia coli when compared to other strains, as demonstrated by the antibacterial toxicology assays.

8.2.1.3 Bioinspired Silver Nanoparticles (AgNPs)

Different fractions from extract of *Elaeagnus umbellata* were utilized to synthesize silver nanoparticles by reducing silver nitrate (Ali et al., 2020). Their size and shape were evaluated in terms of their use. Atomic force microscopy (AFM) and scanning electron microscopy (SEM) were applied to examine their method of action and evaluate their size, shape, and bactericidal activity. They were 40 nm in size and were monodisperse and nontoxic. Additionally, they showed a high capacity for killing gram-positive bacteria (*Staphylococcus aureus*) and gram-negative bacteria (*Escherichia coli*). They exhibited electrostatic interactions with the bacterial cell wall. NPs denatured proteins through a reaction with sulfhydryl groups, after entering the bacterium, which altered the function of the enzymes.

8.2.1.4 Bioinspired Iron Oxide Nanoparticles

Due to their unique properties, SPIONs, or superparamagnetic iron oxide nanoparticles, are widely used. However, their increased toxicity, decreased magnetic characteristics, and higher cost—which essentially stem from the use of harsh chemical reagents during production—restrict their use in biomedicine. SPION-stacked silica nanocapsules were created with a bimodal reactant peptide surfactant-settled nanoemulsion arrangement in order to overcome these challenges. SPIONs were incorporated into the nanoemulsion oil phase, and stability was provided by the peptide's electrostatic repellent surface. Through their interaction with biosilification, the reactant peptides promoted the formation of silica shell nanocapsules containing iron oxide. In general, the ability of silica nanocapsules for drug delivery is demonstrated by the epitomization of iron oxide within them (Wilson et al., 2021).

8.2.2 Polymeric Bioinspired Nanoparticles and Novel Formulations Based on Bioinspired Polymeric NPs

Polymeric drug delivery frameworks are arising as creative advancements that can meet various biomedical requirements. These frameworks offer multiple valuable features and adaptable changes tailored to explicit targets. Bioinspired polymers are obtained from natural sources like polysaccharides, polyphenols, phytochemicals, amino acids, and vitamins. These precursors are used to make polymers like starch, cellulose, chitosan, fucoidan, gelatin, alginate, hyaluronic acid, and agar. In the domain of drug delivery, bioinspired polymers have brought about significant advancements due to their upgraded biocompatibility, biodegradability, hemocompatibility, ease of use and assembling, and reduced cytotoxicity. They can be modified by attaching different functional groups using various strategies to make bioinspired NPs. These bioinspired NPs are highly regarded in biomedical sciences since they offer benefits over synthetically formed agents, such as reduced energy use, diminished toxicity, and simpler synthesis processes (Jeong et al., 2022).

Bioinspired polymers can be altered utilizing innovative plan advancements to make nanoparticles (Khan et al., 2019). Nanoparticles have antimicrobial properties and specifically diminish inflammation. They can be utilized for wound recuperating either straightforwardly or as scaffolds coordinated into hydrogels and nanocomposites (Bhattacharya et al., 2019). Traditional wound healing drugs are often limited by their inability to really infiltrate cell membranes. Interestingly, nanoparticles can easily overcome this hindrance. By rupturing cell membranes, changing gene expression, and preventing the production of enzymes, NPs cause cell death (Dakal et al., 2016). A treatment for wounds based on nanoparticles should be gentle, flexible, and biocompatible. Applications for wound treatment have been investigated with a variety of organically derived nanomaterials. For instance, chitosan is thought to have hemostatic effects on blood coagulation, whereas cellulose supports the healing of skin tissue (He et al., 2021). Moreover, silk fibroin invigorates collagen development, further improving wound healing (Zhou et al., 2010).

8.2.2.1 Chitosan

Biodegradability of chitosan results in the release of N-acetyl glucosamine, a monomer with notable antibacterial and antifungal characteristics. A review led by Azad et al. (2004) found that a chitosan network film significantly assisted in the granular layer's rebuilding and re-epithelialization. In order to shield the wound from further pollution, chitosan forms a protective barrier around it. Additionally, chitosan has the power of stimulating growth factors, which in turn promotes wound healing and recuperation (Negm et al., 2020).

8.2.2.2 Hyaluronic Acid

Another biomaterial that can be synthesized into a nanomaterial is hyaluronic acid (HA), which contains the monomers β-d-glucuronic acid and N-acetyl-d-glucosamine. Furthermore, HA is essential for wound inflammation, cellular migration, growth factor activation, and angiogenesis during the healing process. It has a great ability to retain water and is very biocompatible and non-immunogenic (Snetkov et al., 2019).

El-Aassar and associates (2020) created a wound dressing with hyaluronic acid and polygalacturonic acid contained in silver nanoparticles. A microbiologically active nanofibrous mat was created by electrospinning additional materials for the wound dressing (Najafiasl et al., 2020). AgNPs, gelatin, and sodium alginate were combined to create an antibacterial hydrogel by Diniz and associates (2020). They discovered that wounds treated with hydrogels loaded with AgNP were smaller than wounds left uncoated, leading to differences in the histology of the skin tissues that were healing.

8.2.2.3 Sodium Alginate

Brown seaweed is the natural source of sodium alginate, a polymer. In addition to being biocompatible and biodegradable, it can gel when it comes into contact with water. It has the ability to create a moist wound environment and absorb exudates, which is helpful in promoting healing (Obot et al., 2017).

8.2.2.4 Cellulose

Cellulose has a β-d-glucose monomer and can be utilized to create nanomaterials such as nanofibers, nanocrystals, and other compounds (Portela et al., 2019). Cellulose is a biomaterial that is biocompatible and nonallergenic, which helps absorb wound exudates. Pei et al. (2015) reported that cellulose sponge wound dressings were made using economical and environmentally friendly processes, starting with a cellulose solution in an aqueous NaOH/urea medium. They discovered that using the newly created material to treat wounds sped up the healing process by seven days when compared to using conventional gauze.

8.2.2.5 Silk Fibroin

A distinctive protein isolated from silkworm silk called silk fibroin is thought to provide a number of anticipated advantages for wound healing. It supports cell division and proliferation, facilitates the migration and adhesion of cells involved in tissue repair such as fibroblasts and keratinocytes, and stimulates the synthesis of growth factors essential for wound healing (Grabska-Zielinska and

Sionkowska, 2021). Silk fibroin may be shaped to create silk nanoparticles and nanofilms, and it also promotes cell migration and expansion during wound healing due to its disulfide bond connections (Patil et al., 2020). Chitosan and silk fibroin (CS/SF) mixing layers crosslinked to alginate dialdehyde (ADA) were prepared as wound dressings, facilitating cell connections and multiplication, according to a review (Gu et al., 2013).

8.2.2.6 Pectin

Another naturally occurring dynamic polymer with hydrophilic qualities is pectin, which is filled in as a monomer by d-galacturonic acid. Pectin not only helps wound exudates to be expelled, but it also creates an acidic environment around the wound, making it resistant to bacteria. Additionally, pectin exhibits soothing qualities that aid in the wound bed's reduction of inflammation and encourage healing (Ghai et al., 2012). Pectin's antithrombogenic properties and abundance in galacturonic acid and esters enhance its anti-inflammatory properties and ability to form nanocomposites and nanofilms (Munarin et al., 2012). For instance, a hydrogel with pectin was developed to treat skin wounds, which resulted in a 25% reduction in the overall wound healing time (Kodoth et al., 2019). According to Rajapaksha et al. (2020), a review was conducted on the use of pectin nanoparticles derived from *Spirulina maxima* for wound recovery purposes. The results indicated a 48.9% better wound healing viability when compared to vehicle-treated groups.

8.2.2.7 Fucoidan

Fucoidan is a biopolymer with anticoagulant, antibacterial, and anti-inflammatory characteristics that is derived from α-L-fucose monomers. It is capable of being changed into various compounds, such as nanoflakes and nanospikes. Furthermore, by activating growth factors, fucoidan contributes to the skin's collagen framework being rebuilt. Alginate, chitosan, and fucoidan were combined to create biomultifunctional composite wipes through electrostatic interactions, which also included $Ca2+$ crosslinking and lyophilization. This led to enhanced wound closure as a result of the wipes' improved antibacterial, anti-inflammatory, and hemostatic effects (Hao et al., 2020).

8.2.2.8 Gelatin

Gelatin has biodegradable and highly retentive sugar and amino acid monomers that facilitate wound healing by encouraging cell adhesion and signal transduction (Galdoporpora et al., 2019). Gelatin has anti-inflammatory and antioxidant qualities that aid in reducing wound bed irritation and promoting healing. It also promotes skin repair (Tyuftin and Kerry, 2021). Electrospun chitosan/gelatin nanofibrous nanocomposite frameworks reinforced with graphene were applied for wound healing and antibacterial activities (Ali et al., 2020). In this review, bacterial growth was inhibited by up to 80% when supported by nanofibers coated with graphene nanosheets. Similarly, gelatin and chitosan-based nanocomposite films were developed for wound healing; the addition of coordinated nanoparticles significantly increased the films' antibacterial viability and antioxidant activities (Hajji et al., 2022).

8.2.2.9 Collagen

Because collagen is a biopolymer made of amino acids joined by amide bonds, it is anti-inflammatory and biodegradable. It typically stimulates tissue growth and represses matrix metalloproteinase. A hybrid scaffold that combined collagen/chitosan and metallic nanosilver particles demonstrated promise in accelerating wound healing by increasing α-smooth muscle actin fluxes and accelerating the mobility of fibroblasts (You et al., 2017).

8.2.2.10 Alginate Nanoparticles

Natural and synthetic nanoparticles have been widely synthesized and reported in the field of nanotechnology for the aim of drug delivery. Alginate NPs improve the bioavailability and viability of different drugs by addressing a broad array of polymeric drug delivery challenges

(Soppimath et al., 2001). Due to its cationic nature, alginate, a distinctive sugar polysaccharide, exhibits mucoadhesive qualities (Malafaya et al., 2007) that allow it to adhere to the plasma membrane.

Various techniques have been developed periodically to produce alginate nanoparticles. However, conventional approaches need precise control over the size and dispersion of NPs, often involving the use of strong natural solvents that may be toxic to living things. A typical oddity in aquatic environments, bubble bursting generates nanosized and tiny estimated particles (Fitzgerald, 1991). The main mechanisms causing this cycle include wave breaking, which consists of jetting and bubble film disintegration (Spiel, 1998). Using the microbubble-bursting method, alginate NPs were combined to produce particles of a size range 80–200 nm. The research team produced microbubbles of varying sizes in a highly regulated manner using a T intersection microfluidic device. The thickness of the alginate arrangement employed in this cycle and the size of the delivered NPs were directly correlated (Elsayed et al., 2015).

8.2.3 Bioinspired Carbon Dots

Highly valued for their size-dependent optical characteristics, quantum dots are potent transporters that are widely used in biomedicine. Either way, these dots' preparation methods are often laborious, multifaceted, and have poor repeatability. Among these methods, the hydrothermal methodology stands out as a green chemistry approach for delivering fluorescent carbon quantum dots (C-dots) for a wide range of applications (Kasibabu et al., 2015). Carbon was introduced into the system using waste materials and common assets like orange juice, sugarcane, and ginger. Watermelon peel (Zhou et al., 2012), milk (Wang and Zhou, 2014), lignin, sugarcane juice (Mehta et al., 2014), coffee beans, chicken eggs, food waste, banana, hair (Liu et al., 2014), ginger (Li et al., 2014), onion waste (Bandi et al., 2016), honey, bread, candle residue, chitosan, and gelatin are some of the additional carbon sources that have also been studied.

Asiya F. Shaikh used the unique properties of natural resources to combine the hydrothermal method and *Citrus limetta* juice to create a rapid, extremely luminous C-dots synthesis procedure. This juice's high sugar content, which is a source of carbohydrates, supplies the carbon needed to start the synthesis of C-dots (Shaikh et al., 2019).

8.2.4 Bioinspired Dressing Mat for Wound Healing

Treatment for wounds is difficult since certain illnesses, like diabetes and cardiovascular conditions, exacerbate their chronic nature (Qu et al., 2018). According to Chhabra et al. (2016), wound dressings mimic the extracellular matrix (ECM), adhere to the wound, and eventually migrate there, all of which help the healing and regeneration of the skin. Due to its excellent biocompatibility, biodegradability, and accessibility, polycaprolactone (PCL) has been utilized extensively in wound healing (Ravichandran et al., 2019). Its mechanical characteristics and capacity for tissue regeneration can be improved by blending it with other polymers (Sawadkar et al., 2020). In order to do this, ECM-like electrospun mats of environmentally safe and biodegradable polyhydroxybutyrate/poly-3-caprolactone (PHB/PCL) were developed to provide structural and biochemical support for tissue regeneration. Using melanin–TiO2 nanostructures, which are inspired by the natural component melanin, which is frequently employed as a defense against microbial infection, the previously made mats were modified. Strong antibacterial activity was shown by these coated mats against both gram-positive and gram-negative bacterial strains. They showed good hydrophilicity, water retention ability, and in vitro activity during their contact and adhesion (Avossa et al., 2021).

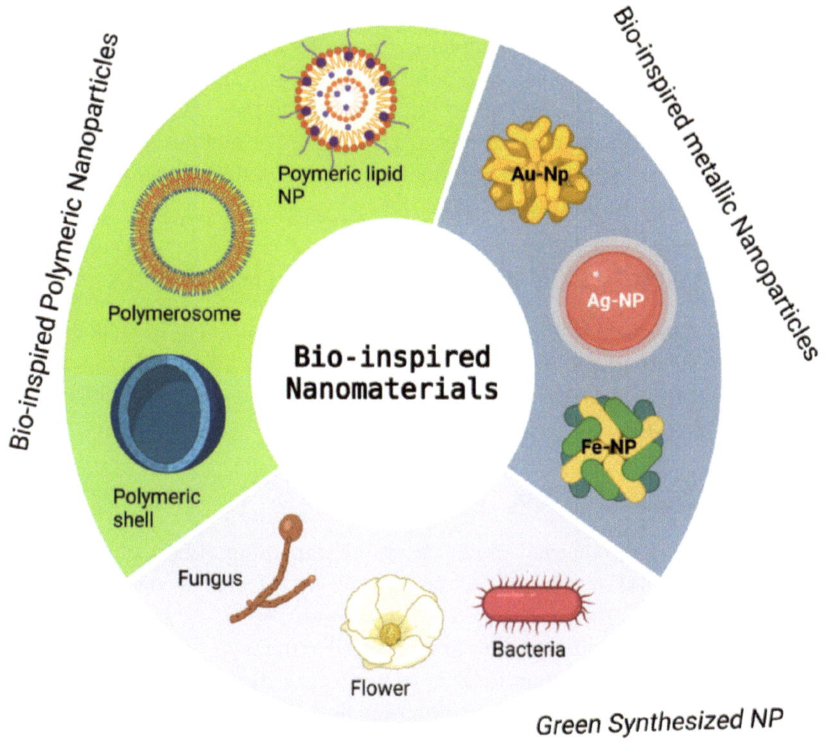

FIGURE 8.2 Bioinspired nanomaterials.

8.3 ANTIMICROBIAL AND ANTIBIOFILM ACTIONS OF BIOINSPIRED NANOMATERIALS

Pathogens have developed resistance within biofilms as a result of antimicrobial drug resistance (Singh et al., 2021). This development is caused by microbial communities made up of bacteria, yeast, or fungi growing in infected wounds. Incorporating bacteria into an extracellular polymeric matrix, biofilms cling to surfaces. Due to the existence of resistant genes, they are immune to human immunity and antibacterial medications (Sauer et al., 2022). They are a frequent cause of patient morbidity and mortality in healthcare settings.

Promising paths for the development of new antibacterial drugs that work with bioinspired materials to stop bacterial colonization and biofilm formation and speed up wound healing are provided by nanoscience and nanotechnology. Broad spectrum antimicrobials with low toxicity can be encapsulated in bioinspired nanomaterials, which also have innate antibacterial capabilities (Raie et al., 2017). Because of their inherent antibacterial activity, nanomaterials can aid in the immobilization, impregnation, and prolonged release of antimicrobials, resulting in a synergistic effect (Obeizi et al., 2020).

High surface area nanoparticles with important physicochemical characteristics prevent the growth of harmful microorganisms and the creation of biofilms while accelerating the healing process. They cause oxidative damage and genetic changes in bacteria by inducing different antibiofilm activities (Liu et al., 2021). Conventional antibiotics may eventually be replaced with nanomaterials-based coatings on medical devices, kits for healthcare, and implants that could prevent biofilm-associated illnesses and the deaths that accompany them.

With the help of green nanotechnology, it is feasible to produce a potent, stable antibacterial action that may be recycled without causing harm to healthy cells. Hassiba et al. have created novel hybrid nanomaterials that can be used as wound dressings and have antibacterial properties. The metal cations dissolve and are released when the NP in which they are embedded comes into contact with the membrane of pathogenic bacteria, preventing the synthesis of ATP and respiratory enzymes. Reactive oxygen species (ROS) production also jeopardizes other transport pathways and the integrity of the membrane (Hassiba et al., 2017).

8.4 TARGETED DELIVERY OF BIOINSPIRED NANOMATERIALS

Multifunctionalized nanomaterial-based solutions hold great promise in treating and preventing wound infections when compared to conventional therapeutic techniques. In this sense, substantial research on wound-healing applications has been made possible by the repeatable physicochemical properties of nanomaterials.

8.4.1 Phytoengineered Gold NPs

Skin shields and protects the internal organs from the damaging effects of the external environment. Any disturbance of skin homeostasis brought on by infected wounds might be fatal if left untreated (Kohlmeyer et al., 2021).

Kabeerdass and colleagues attempted to assess herbal extracts' potential to get beyond treatment obstacles posed by multidrug-resistant (MDR) bacteria. The disadvantages of treatments that just use AuNPs can be significantly reduced by combining herbal extract with AuNPs. Because *Limonia acidissima* essence (LA) is an environmentally friendly product with enhanced antioxidant and anti-inflammatory properties, it was used in this study. The plant extract was used to biosynthesize nanoparticles, which were then coated onto fabrics made of mixed cotton and investigated in vivo, ex vivo, and in vitro. The MTT assay was also used to assess the migration and survival of L929 cell lines in the presence of biosynthesized LA–AuNPs. The produced LA–AuNPs demonstrated the widest inhibitory zone in wound infections and performed effectively as antibacterial agents. Moreover, the green synthesis of LA–AuNPs from *L. acidissima* leaf extracts is effective and eco-friendly (Kabeerdass et al., 2022).

Boomi and associates investigated the creation of phytoengineered gold nanoparticles as a secure and efficient therapeutic approach for mending complex wounds. *Acalypha indica*, a naturally occurring plant, was used to increase the green synthesis of AuNPs, which led to advancements in wound healing treatments. Following the synthesis of 2.5 ml of *A. indica* leaf extract,

TABLE 8.1
Different Nanoparticles and Their Key Features

Nanoparticles	Features
Phytoengineered AuNPs	Maximum inhibition zones
NPs loaded with *Vitis vinifera*	Antidermatophytic
Injectable methyl cellulose/hyaluronic acid	Skin recovery and remodeling
Thiolated bandage based on ZnONPs	Wound exudate absorption
Polymeric NPs loaded with lignin	Wound management in rectal and anal surgery
SNEDDS based on poly-L-lysine	Intestinal wound healing and biofilm inhibition
NPs based on PVA/ZnO	Re-epithelialization and collagen activation

Source: Arshad et al. (2023).

cotton was pretreated and sterilized. Cotton fabric that had been pretreated was treated with 100 ml of *A. indica* leaf extract and AuNPs. The NPs underwent histological analyses, antioxidant activity assessments, in vitro antibacterial evaluations, and instrumental setups. It was determined that green AuNPs' potential as future therapies is based on their antibacterial, antioxidant, free radical scavenging, and wound healing properties (Boomi et al., 2020).

8.4.2 Clotrimazole and *Vitis vinifera* Extract-Loaded Chitosan NPs

A common medical condition that has a significant impact on morbidity and quality of life is fungal infections of wounds, which result in damage to skin, nails, and hair (Cimolai, 2021). These infections are typically brought on by keratinophilic fungi and dermatophytosis. The development of a resistance mechanism, low absorption, and high toxicity of the current antifungal regimen are the main problems with it. As a result, researchers have looked into cutting-edge medications that use plant extracts to reduce adverse effects.

Regarding this, nanocarriers based on natural biomaterials were the focus of research by Esraa E. Elshaer and colleagues. Grape (*Vitis vinifera*) juice extract (VJ) possesses antioxidant, antiviral, antibacterial, and anticancer qualities. In this fascinating work, VJ extract was loaded onto chitosan NPs, which are able to transport medicines through several endocytic pathways and pass through a variety of biological barriers. The chitosan NPs were prepared by the ionic gelation technique, and VJ (20 mg) was subsequently encapsulated within the nanocarriers (NCs) via an incorporation process. It was then time to add clotrimazole to the NC–VJ reaction mixture. Ultimately, an NC–VJ–clotrimazole ointment was created, and both ex vivo and in vivo testing were done on it.

Antifungal activity, antioxidant activity, skin retention, skin penetration, histology, hemolysis, and hemocompatibility were a few of the attributes for the assessment of the treatment's effectiveness. It was shown that the formulation resulted in a prolonged drug-release pattern, tissue remodeling, and complete wound healing in just seven days. This ointment based on NC–VJ–clotrimazole NPs has strong wound-healing properties, which make it a potential innovative anti-dermatophyte agent (Elshaer et al., 2022).

8.4.3 Injectable Methylcellulose and Hyaluronic Acid-Based Silver Nanoparticles (AgNPs)

Transdermal hydrogels are preferred for treating wounds. AgNPs containing methylcellulose and hyaluronic acid were produced as an enhanced wound dressing for surgical bleeding and wounds (Qiu et al., 2022). In tissue engineering, the injectable biodegradable polymer can be utilized to entrap medications as cell carriers, enabling selective internalization. Carbohydrate polymers include cellulose and methylcellulose, but because it is directly related to the formation of collagen, another vital carbohydrate polymer called HA is most crucial for the healing of wounds. The antibacterial action of NPs against *E. coli* and *S. aureus* in wounds, as well as their ability to heal wounds, were evaluated using X-ray diffraction (XRD), Fourier transform infrared spectrum (FTIR) profiling, UV spectroscopy, transmission electron microscopy (TEM), and scanning electron microscopy (SEM). With a diameter of about 35 nm, NPs successfully coupled with the polymers, improving swelling ability, degradation, and stability over the course of gelation. Improved antibacterial activity was also shown by the data, with a recovery rate of 99.6%. According to research, MC-HA/AgNP hydrogels are very helpful for children who have suffered serious burns in terms of skin remodeling and regeneration.

The most frequent side effects of a *Salmonella typhi* infection are intestinal perforations or wounds, which are followed by higher morbidity and death. The self-nanoemulsifying drug delivery system (SNEDDS) has become a competitive alternative in recent years for controlling drug parameters. Initially, poly-L-lysine (PLL)-based SNEDDS (PLL-SNEDDS) was developed by Arshad et al. (2021) to jointly improve ciprofloxacin's ability to heal wounds. Enhanced anti-inflammatory and antibiofilm activity, together with improved spleen and intestinal healing, were the outcomes of this experiment.

8.4.4 Thiolated Chitosan and Alginate-Loaded Zinc Oxide NPs

Because of the high microbial load and significant bleeding associated with post-surgical site infections, healing might be difficult; thus, it's critical to use the best topical dressings in this situation. For mitigating the side effects of conventional dosing regimens, Arshad et al. developed a bandage coated with thiolated polymer chitosan and natural polymer alginate in conjunction with zinc oxide (ZnO) NPs for the treatment of post-operative surgical site infections. The production of ROS, which is connected to the demise of bacterial strains, is attributed to ZnO NPs. An anionic natural polymer of polysaccharides, alginate is obtained from brown algae. Because it contains divalent cations, it is hemostatic, biodegradable, hydrophilic, and antibacterial. It can even form crosslinks in mild environments.

Thiolated bandages were evaluated physicochemically in vivo, ex vivo, and in vitro to determine their safety and effectiveness. The effects of thiolated and nonthiolated bandages on hemostatic, mucoadhesive, antibacterial, medication release, swelling, degradation, wound healing, and porosity were compared. ZnO NPs that are biodegradable and have a favorable polydispersity index (PDI) were successfully produced with particle sizes ranging from 300 to 400 nm. These were combined with an affordable, biocompatible polymer matrix to create two different bandages that were tested.

Comparing chitosan–alginate thiolated bandages to nonthiolated ones in vitro showed improved hemostatic action, efficient biodegradation, elevated swelling index, and higher porosity. Studies on controlled release showed that ZnO NPs may be reasonably managed; about 70% were released after 72 hours. Using thiolated bandages, in vivo research showed improved skin healing and re-epithelialization at the surgical site. By extending skin contact, effectively absorbing wound exudate, inhibiting microbial growth, and encouraging tissue regeneration, thiolated bandages promoted overall wound healing and restored normal skin (Arshad et al., 2019).

8.4.5 Lignin-Conjugated Polymeric NPs

Gram-negative bacteria like *E. coli* and *Klebsiella* spp. are well-known for their ability to quickly build colonies and cause frequent diseases. Because these pathogens can pierce cell membranes, microbial management becomes more difficult during anorectal surgery. As a potential remedy, Dai and associates looked into dextran/glycol (DEX/GLY) nanoparticles loaded with lignin. These NPs were characterized using animal modeling, swelling index, antibacterial characteristics, and in vitro biodegradation. Later on, a particular type of wound dressing was created, composed of cotton fabric coupled with nanoparticles loaded with lignin. Notably, there was no harm to the tissue after removing this wound wrap from the wound site. According to the study, lignin-loaded DEX/GLY NPs have potential as biomaterials for wound care after anorectal surgery, reducing scarring, and avoiding tissue injury (Dai et al., 2022).

8.5 CONCLUSION

Wound management and healing are a complex process that can be impeded by different factors. Wounds on the skin leave it susceptible to microbial infections, which makes wound care challenging. When a wound becomes contaminated, more wound exudates are produced. These exudates contain protein ingredients that are favorable for the growth of bacteria. Produced from natural sources, bioinspired nanomaterials have emerged as a viable avenue in this regard. Advanced flexibility, high mechanical strength, porosity, and nonadhesion have all been demonstrated by wound dressings made with bioinspired nanocarriers. Compared to current methodologies, bioinspired nanomaterials have demonstrated essential viability in wound healing. They can be customized to create wound dressings that enhance anti-inflammatory processes and inhibit biofilm formation and growth. Bioinspired nanoparticles' sweeping power in wound healing surpasses previous evaluations, providing new avenues for research.

REFERENCES

Ali, S., Perveen, S., Ali, M., Jiao, T., Sharma, A. S., Hassan, H., ... Chen, Q. (2020). Bioinspired morphology-controlled silver nanoparticles for antimicrobial application. *Materials Science and Engineering: C, 108*, 110421.

Arshad, R., Razlansari, M., Hosseinikhah, S. M., Tiwari, A., Ajalli, N., Manicum, A. L. E., ... Tabish, T. A. (2023). Antimicrobial and anti-biofilm activities of bio-inspired nanomaterials for wound healing applications. *Drug Discovery Today*, 103673.

Arshad, R., Sohail, M. F., Sarwar, H. S., Saeed, H., Ali, I., Akhtar, S., ... Shahnaz, G. (2019). ZnO-NPs embedded biodegradable thiolated bandage for postoperative surgical site infection: In vitro and in vivo evaluation. *PLoS One, 14*(6), e0217079.

Arshad, R., Tabish, T. A., Naseem, A. A., ul Hassan, M. R., Hussain, I., Hussain, S. S., & Shahnaz, G. (2021). Development of poly-L-lysine multi-functionalized muco-penetrating self-emulsifying drug delivery system (SEDDS) for improved solubilization and targeted delivery of ciprofloxacin against intracellular Salmonella typhi. *Journal of Molecular Liquids, 333*, 115972.

Avossa, J., Pota, G., Vitiello, G., Macagnano, A., Zanfardino, A., Di Napoli, M., ... Luciani, G. (2021). Multifunctional mats by antimicrobial nanoparticles decoration for bioinspired smart wound dressing solutions. *Materials Science and Engineering: C, 123*, 111954.

Azad, A. K., Sermsintham, N., Chandrkrachang, S., & Stevens, W. F. (2004). Chitosan membrane as a wound-healing dressing: characterization and clinical application. *Journal of Biomedical Materials Research Part B: Applied Biomaterials: An Official Journal of The Society for Biomaterials, The Japanese Society for Biomaterials, and The Australian Society for Biomaterials and the Korean Society for Biomaterials, 69*(2), 216–222.

Bandi, R., Gangapuram, B. R., Dadigala, R., Eslavath, R., Singh, S. S., & Guttena, V. (2016). Facile and green synthesis of fluorescent carbon dots from onion waste and their potential applications as sensor and multicolour imaging agents. *RSC advances, 6*(34), 28633–28639.

Bhattacharya, D., Ghosh, B., & Mukhopadhyay, M. (2019). Development of nanotechnology for advancement and application in wound healing: A review. *IET Nanobiotechnology, 13*(8), 778–785.

Boomi, P., Ganesan, R., Prabu Poorani, G., Jegatheeswaran, S., Balakumar, C., Gurumallesh Prabu, H., ... Saravanan, M. (2020). Phyto-engineered gold nanoparticles (AuNPs) with potential antibacterial, antioxidant, and wound healing activities under in vitro and in vivo conditions. *International Journal of Nanomedicine*, 7553–7568.

Castro, L., Blázquez, M. L., González, F., Muñoz, J. A., & Ballester, A. (2013). Gold, silver and platinum nanoparticles biosynthesized using orange peel extract. *Advanced Materials Research, 825*, 556–559.

Chhabra, H., Deshpande, R., Kanitkar, M., Jaiswal, A., Kale, V. P., & Bellare, J. R. (2016). A nano zinc oxide doped electrospun scaffold improves wound healing in a rodent model. *RSC Advances, 6*(2), 1428–1439.

Cimolai, N. (2021). The complexity of co-infections in the era of COVID-19. *SN Comprehensive Clinical Medicine, 3*(7), 1502–1514.

Dai, M., Zhang, J., Liu, N., & Zhang, X. H. (2022). Precise engineering of lignin incorporated dextran/glycol nanomaterials for wound dressings for the care of anorectal surgery. *Journal of Polymers and the Environment, 30*(1), 206–216.

Dakal, T. C., Kumar, A., Majumdar, R. S., & Yadav, V. (2016). Mechanistic basis of antimicrobial actions of silver nanoparticles. *Frontiers in Microbiology, 7*, 1831.

Diniz, F. R., Maia, R. C. A., de Andrade, L. R. M., Andrade, L. N., Vinicius Chaud, M., da Silva, C. F., ... Severino, P. (2020). Silver nanoparticles-composing alginate/gelatine hydrogel improves wound healing in vivo. *Nanomaterials, 10*(2), 390.

El-Aassar, M. R., Ibrahim, O. M., Fouda, M. M., El-Beheri, N. G., & Agwa, M. M. (2020). Wound healing of nanofiber comprising Polygalacturonic/Hyaluronic acid embedded silver nanoparticles: In-vitro and in-vivo studies. *Carbohydrate polymers, 238*, 116175.

Elsayed, M., Huang, J., & Edirisinghe, M. (2015). Bioinspired preparation of alginate nanoparticles using microbubble bursting. *Materials Science and Engineering: C, 46*, 132–139.

Elshaer, E. E., Elwakil, B. H., Eskandrani, A., Elshewemi, S. S., & Olama, Z. A. (2022). Novel Clotrimazole and Vitis vinifera loaded chitosan nanoparticles: Antifungal and wound healing efficiencies. *Saudi Journal of Biological Sciences*, *29*(3), 1832–1841.

Fitzgerald, J. W. (1991). Marine aerosols: A review. *Atmospheric Environment. Part A. General Topics*, *25*(3–4), 533–545.

Galdopórpora, J. M., Morcillo, M. F., Ibar, A., Perez, C. J., Tuttolomondo, M. V., & Desimone, M. F. (2019). Development of silver nanoparticles/gelatin thermoresponsive nanocomposites: Characterization and antimicrobial activity. *Current Pharmaceutical Design*, *25*(38), 4121–4129.

Ghai, K., Gupta, A. K., & Gupta, P. K. (2012). Pectin: A versatile biopolymer with numerous health benefits and medical uses. *Journal of Biologically Active Products from Nature*, *2*(4), 250–255.

Grabska-Zielińska, S., & Sionkowska, A. (2021). How to improve physico-chemical properties of silk fibroin materials for biomedical applications?—Blending and cross-linking of silk fibroin—A review. *Materials*, *14*(6), 1510.

Gu, Z., Xie, H., Huang, C., Li, L., & Yu, X. (2013). Preparation of chitosan/silk fibroin blending membrane fixed with alginate dialdehyde for wound dressing. *International journal of Biological Macromolecules*, *58*, 121–126.

Gurunathan, S. (2019). Rapid biological synthesis of silver nanoparticles and their enhanced antibacterial effects against Escherichia fergusonii and Streptococcus mutans. *Arabian Journal of Chemistry*, *12*(2), 168–180.

Hajji, S., Ktari, N., Ben Salah, R., Boufi, S., Debeaufort, F., & Nasri, M. (2022). Development of nanocomposite films based on chitosan and gelatin loaded with chitosan-tripolyphosphate nanoparticles: antioxidant potentials and applications in wound healing. *Journal of Polymers and the Environment*, 1–22.

Hao, Y., Zhao, W., Zhang, L., Zeng, X., Sun, Z., Zhang, D., ... Zhou, Q. (2020). Bio-multifunctional alginate/chitosan/fucoidan sponges with enhanced angiogenesis and hair follicle regeneration for promoting full-thickness wound healing. *Materials & Design*, *193*, 108863.

Hassiba, A. J., El Zowalaty, M. E., Webster, T. J., Abdullah, A. M., Nasrallah, G. K., Khalil, K. A., ... Elzatahry, A. A. (2017). Synthesis, characterization, and antimicrobial properties of novel double layer nanocomposite electrospun fibers for wound dressing applications. *International Journal of Nanomedicine*, 2205–2213.

Hayat, M., Nawaz, A., Chinnam, S., Muzammal, M., Latif, M. S., Yasin, M., Ashique, S., Zengin, G., & Farid, A. (2023, July 1). Formulation development and optimization of herbo synthetic gel: In vitro biological evaluation and in vivo wound healing studies. *Process Biochemistry*, *130*, 116–126.

Hazra, R. S., Hasan Khan, M. R., Kale, N., Tanha, T., Khandare, J., Ganai, S., & Quadir, M. (2022). Bioinspired materials for wearable devices and point-of-care testing of cancer. *ACS Biomaterials Science & Engineering*, *9*(5), 2103–2128.

He, W., Wu, J., Xu, J., Mosselhy, D. A., Zheng, Y., & Yang, S. (2021). Bacterial cellulose: functional modification and wound healing applications. *Advances in Wound Care*, *10*(11), 623–640.

Ibrahim, H. M. (2015). Green synthesis and characterization of silver nanoparticles using banana peel extract and their antimicrobial activity against representative microorganisms. *Journal of Radiation Research and Applied Sciences*, *8*(3), 265–275.

Jeong, G. J., Khan, S., Tabassum, N., Khan, F., & Kim, Y. M. (2022). Marine-bioinspired nanoparticles as potential drugs for multiple biological roles. *Marine Drugs*, *20*(8), 527.

Kabeerdass, N., Kandasamy, S., Albasher, G., Alamri, O., Alsultan, N., Thangaswamy, S., & Mathanmohun, M. (2022). Limonia acidissima leaf mediated gold nanoparticles synthesis and their antimicrobial and wound healing properties. *Materials Letters*, *314*, 131893.

Kadiyala, N. K., Mandal, B. K., Ranjan, S., & Dasgupta, N. (2018). Bioinspired gold nanoparticles decorated reduced graphene oxide nanocomposite using Syzygium cumini seed extract: Evaluation of its biological applications. *Materials Science and Engineering: C*, *93*, 191–205.

Kasibabu, B. S. B., D'souza, S. L., Jha, S., Singhal, R. K., Basu, H., & Kailasa, S. K. (2015). One-step synthesis of fluorescent carbon dots for imaging bacterial and fungal cells. *Analytical Methods*, *7*(6), 2373–2378.

Khan, I., Saeed, K., & Khan, I. (2019). Nanoparticles: Properties, applications and toxicities. *Arabian Journal of Chemistry*, *12*(7), 908–931.

Kodoth, A. K., Ghate, V. M., Lewis, S. A., Prakash, B., & Badalamoole, V. (2019). Pectin-based silver nanocomposite film for transdermal delivery of Donepezil. *International Journal of Biological Macromolecules*, *134*, 269–279.

Kohlmeyer, K. J., Toman, C., & Pruet, J. (2021). *New ways of synthesizing potential pterin anti-fungal drugs.* Summer Interdisciplinary Research Symposium. https://scholar.valpo.edu/sires/94

Li, C. L., Ou, C. M., Huang, C. C., Wu, W. C., Chen, Y. P., Lin, T. E., ... Chang, H. T. (2014). Carbon dots prepared from ginger exhibiting efficient inhibition of human hepatocellular carcinoma cells. *Journal of Materials Chemistry B*, *2*(28), 4564–4571.

Liu, S. S., Wang, C. F., Li, C. X., Wang, J., Mao, L. H., & Chen, S. (2014). Hair-derived carbon dots toward versatile multidimensional fluorescent materials. *Journal of Materials Chemistry C*, *2*(32), 6477–6483.

Liu, Y., Huang, Y., Kim, D., Ren, Z., Oh, M. J., Cormode, D. P., ... Koo, H. (2021). Ferumoxytol nanoparticles target biofilms causing tooth decay in the human mouth. *Nano Letters*, *21*(22), 9442–9449.

Ma, N., Li, B., Shu, P., You, G., Zhou, H., Wang, Q., & Zhao, L. (2022). Bio-inspired polydopamine coated hemoglobin as potential oxygen carriers for wound healing. https://doi.org/10.21203/rs.3.rs-1218595/v1

Malafaya, P. B., Silva, G. A., & Reis, R. L. (2007). Natural–origin polymers as carriers and scaffolds for biomolecules and cell delivery in tissue engineering applications. *Advanced Drug Delivery Reviews*, *59*(4–5), 207–233.

Mao, H. Y., Laurent, S., Chen, W., Akhavan, O., Imani, M., Ashkarran, A. A., & Mahmoudi, M. (2013). Graphene: Promises, facts, opportunities, and challenges in nanomedicine. *Chemical Reviews*, *113*(5), 3407–3424.

Mehta, V. N., Jha, S., & Kailasa, S. K. (2014). One-pot green synthesis of carbon dots by using Saccharum officinarum juice for fluorescent imaging of bacteria (Escherichia coli) and yeast (Saccharomyces cerevisiae) cells. *Materials Science and Engineering: C*, *38*, 20–27.

Munarin, F. A. B. I. O. L. A., Tanzi, M. C., & Petrini, P. A. O. L. A. (2012). Advances in biomedical applications of pectin gels. *International Journal of Biological Macromolecules*, *51*(4), 681–689.

Najafiasl, M., Osfouri, S., Azin, R., & Zaeri, S. (2020). Alginate-based electrospun core/shell nanofibers containing dexpanthenol: A good candidate for wound dressing. *Journal of Drug Delivery Science and Technology*, *57*, 101708.

Naskar, A., & Kim, K. S. (2020). Recent advances in nanomaterial-based wound-healing therapeutics. *Pharmaceutics*, *12*(6), 499.

Negm, N. A., Hefni, H. H., Abd-Elaal, A. A., Badr, E. A., & Abou Kana, M. T. (2020). Advancement on modification of chitosan biopolymer and its potential applications. *International Journal of Biological Macromolecules*, *152*, 681–702.

Nishanthi, R., Malathi, S., & Palani, P. (2019). Green synthesis and characterization of bioinspired silver, gold and platinum nanoparticles and evaluation of their synergistic antibacterial activity after combining with different classes of antibiotics. *Materials Science and Engineering: C*, *96*, 693–707.

Obeizi, Z., Benbouzid, H., Ouchenane, S., Yılmaz, D., Culha, M., & Bououdina, M. (2020). Biosynthesis of Zinc oxide nanoparticles from essential oil of Eucalyptus globulus with antimicrobial and anti-biofilm activities. *Materials Today Communications*, *25*, 101553.

Obot, I. B., Onyeachu, I. B., & Kumar, A. M. (2017). Sodium alginate: A promising biopolymer for corrosion protection of API X60 high strength carbon steel in saline medium. *Carbohydrate Polymers*, *178*, 200–208.

Patil, P. P., Reagan, M. R., & Bohara, R. A. (2020). Silk fibroin and silk-based biomaterial derivatives for ideal wound dressings. *International Journal of Biological Macromolecules*, *164*, 4613–4627.

Pedraza-Chaverri, J., Cárdenas-Rodríguez, N., Orozco-Ibarra, M., & Pérez-Rojas, J. M. (2008). Medicinal properties of mangosteen (Garcinia mangostana). *Food and Chemical Toxicology*, *46*(10), 3227–3239.

Pei, Y., Ye, D., Zhao, Q., Wang, X., Zhang, C., Huang, W., ... Zhang, L. (2015). Effectively promoting wound healing with cellulose/gelatin sponges constructed directly from a cellulose solution. *Journal of Materials Chemistry B*, *3*(38), 7518–7528.

Portela, R., Leal, C. R., Almeida, P. L., & Sobral, R. G. (2019). Bacterial cellulose: a versatile biopolymer for wound dressing applications. *Microbial Biotechnology*, *12*(4), 586–610.

Qiu, Y., Sun, X., Lin, X., Yi, W., & Jiang, J. (2022). An injectable metal nanoparticle containing cellulose derivative-based hydrogels: Evaluation of antibacterial and in vitro-vivo wound healing activity in children with burn injuries. *International Wound Journal*, *19*(3), 666–678.

Qu, X., Liu, H., Zhang, C., Lei, Y., Lei, M., Xu, M., ... Liu, C. (2018). Electrofabrication of functional materials: Chloramine-based antimicrobial film for infectious wound treatment. *Acta Biomaterialia*, *73*, 190–203.

Raie, D. S., Mhatre, E., Thiele, M., Labena, A., El-Ghannam, G., Farahat, L. A., ... Kovács, Á. T. (2017). Application of quercetin and its bio-inspired nanoparticles as anti-adhesive agents against Bacillus subtilis attachment to surface. *Materials Science and Engineering: C*, *70*, 753–762.

Rajapaksha, D. C., Edirisinghe, S. L., Nikapitiya, C., Dananjaya, S. H. S., Kwun, H. J., Kim, C. H., ... De Zoysa, M. (2020). Spirulina maxima derived pectin nanoparticles enhance the immunomodulation, stress tolerance, and wound healing in zebrafish. *Marine Drugs*, *18*(11), 556.

Ravichandran, S., Radhakrishnan, J., Jayabal, P., & Venkatasubbu, G. D. (2019). Antibacterial screening studies of electrospun Polycaprolactone nano fibrous mat containing Clerodendrum phlomidis leaves extract. *Applied Surface Science*, *484*, 676–687.

Sauer, K., Stoodley, P., Goeres, D. M., Hall-Stoodley, L., Burmølle, M., Stewart, P. S., & Bjarnsholt, T. (2022). The biofilm life cycle: expanding the conceptual model of biofilm formation. *Nature Reviews Microbiology*, *20*(10), 608–620.

Sawadkar, P., Mohanakrishnan, J., Rajasekar, P., Rahmani, B., Kohli, N., Bozec, L., & Garcia-Gareta, E. (2020). A synergistic relationship between polycaprolactone and natural polymers enhances the physical properties and biological activity of scaffolds. *ACS Applied Materials & Interfaces*, *12*(12), 13587–13597.

Shaikh, A. F., Tamboli, M. S., Patil, R. H., Bhan, A., Ambekar, J. D., & Kale, B. B. (2019). Bioinspired carbon quantum dots: an antibiofilm agents. *Journal of Nanoscience and Nanotechnology*, *19*(4), 2339–2345.

Singh, K. R., Nayak, V., & Singh, R. P. (2021). Introduction to bionanomaterials: an overview. *Bionanomaterials: Fundamentals and Biomedical Applications*, 1–1.

Snetkov, P., Morozkina, S., Uspenskaya, M., & Olekhnovich, R. (2019). Hyaluronan-based nanofibers: Fabrication, characterization and application. *Polymers*, *11*(12), 2036.

Soppimath, K. S., Aminabhavi, T. M., Kulkarni, A. R., & Rudzinski, W. E. (2001). Biodegradable polymeric nanoparticles as drug delivery devices. *Journal of Controlled Release*, *70*(1–2), 1–20.

Spiel, D. E. (1998). On the births of film drops from bubbles bursting on seawater surfaces. *Journal of Geophysical Research: Oceans*, *103*(C11), 24907–24918.

Tyuftin, A. A., & Kerry, J. P. (2021). Gelatin films: Study review of barrier properties and implications for future studies employing biopolymer films. *Food Packaging and Shelf Life*, *29*, 100688.

von Müller, C., Bulman, F., Wagner, L., Rosenberger, D., Marolda, A., Kurzai, O., ... Vylkova, S. (2020). Active neutrophil responses counteract Candida albicans burn wound infection of ex vivo human skin explants. *Scientific Reports*, *10*(1), 21818.

Wang, L., & Zhou, H. S. (2014). Green synthesis of luminescent nitrogen-doped carbon dots from milk and its imaging application. *Analytical Chemistry*, *86*(18), 8902–8905.

Wilson, R. J., Hui, Y., Whittaker, A. K., & Zhao, C. X. (2021). Facile bioinspired synthesis of iron oxide encapsulating silica nanocapsules. *Journal of Colloid and Interface Science*, *601*, 78–84.

Wu, G., Hui, X., Hu, L., Bai, Y., Rahaman, A., Yang, X. F., & Chen, C. (2022). Recent advancement of bioinspired nanomaterials and their applications: A review. *Frontiers in Bioengineering and Biotechnology*, *10*, 952523.

You, C., Li, Q., Wang, X., Wu, P., Ho, J. K., Jin, R., ... Han, C. (2017). Silver nanoparticle loaded collagen/chitosan scaffolds promote wound healing via regulating fibroblast migration and macrophage activation. *Scientific reports*, *7*(1), 10489.

Zhou, J., Cao, C., Ma, X., & Lin, J. (2010). Electrospinning of silk fibroin and collagen for vascular tissue engineering. *International journal of Biological Macromolecules*, *47*(4), 514–519.

Zhou, J., Sheng, Z., Han, H., Zou, M., & Li, C. (2012). Facile synthesis of fluorescent carbon dots using watermelon peel as a carbon source. *Materials Letters*, *66*(1), 222–224.

9 Wound Dressing and Tissue Regeneration

Role of Nanomaterials from Diagnosis to Treatment Strategy

Nishant Kumar[1], Prince Ahad Mir[1], Sameena Ramzan[2], Apporva Chawla[1], Murtaza Ahmad Khanday[2], Sumel Ashique[3,4], Roohi Mohi-ud-din[5], and Reyaz Hassan Mir[2*]*

[1] Khalsa College of Pharmacy, Punjab, India
[2] University of Kashmir, Jammu and Kashmir, India
[3] Bengal College of Pharmaceutical Sciences and Research, West Bengal, India
[4] Lovely Professional University, Punjab, India
[5] Sher-I-Kashmir Institute of Medical Sciences (SKIMS), Jammu and Kashmir, India
[*] Corresponding Authors

9.1 INTRODUCTION

The softer tissues that make up the skin constitute around 15% of the entire weight of the human body [1, 2]. Thus, the biggest organ in the human organism may be regarded as the skin. The primary function of the skin in everyday existence is to separate one's internal and external surroundings to prevent bacterial, chemical, and physical intrusions. Moreover, its roles include controlling metabolism and body heat [2]. However, a range of extrinsic stimuli, including heat, mechanical, and chemical agents, can make the skin vulnerable to various sorts of lesions. The skin's capacity for repairing itself is severely limited when extensive burns or wounds occur, and these difficult-to-heal wounds carry an elevated probability of amputation or mortality. Attempting to cater to the absence of barrier defense at locations of wounds and to speed up the recovery process, a variety of dressings for injuries have been produced. These include passive supplies like gauze, cotton wool, and plasters; interactive substances like hydrogels, polymeric films, and foamed covers; and biologically active compounds like alginate, collagens, and substances containing growth factors [3, 4].

Thus, maintaining the integrity of the skin is crucial for general well-being. Wounds resulting from burns, trauma, chronic illnesses, and surgery can provide significant challenges to worldwide medical frameworks, including disability and anguish. According to the outcomes and underlying factors, skin wounds are categorized as acute or chronic. Small, acute wounds can usually be repaired and cured by the skin's natural repairing process. Growth factors, extracellular matrix remodeling, and various interactively coordinated reactions within cells facilitate the wound healing process [5, 6]. The skin's architectural and operational integrity are restored throughout the repair phase. Extensive injury to the skin and long-term wounds, on the other hand, heal slowly and

DOI: 10.1201/9781003504948-9

are challenging to repair because they are prone to pathogens and losing fluids. However, extensive and prolonged skin lesions compromise the functional and structural integrity of the skin, impacting millions of individuals (about 8.2 million) and presenting significant social and economic issues to healthcare providers worldwide [7, 8]. For instance, according to five-year fatality figures [9], the death rate from chronic diabetic foot ulcers (DFUs) is 30.5%, which is similar to the death incidence from cancer (31%). According to previous statistics, the yearly cost of managing chronic wounds is above USD 12 billion, and by 2023, it is predicted to reach USD 35 billion [10–12]. The "silent epidemic" of chronic wounds is becoming worse by aging populations, diabetes, and obesity. Furthermore, aging, diabetes, or improper care can cause a variety of chronic wounds, including ulcers caused by pressure, diabetic ulcers, and vascular ulcers, that recover slowly and are typically linked to higher rates of death and disability [13]. In light of the challenges in assessing and managing wound care, cutaneous wound healing, big wound care, and chronic wound care remain unresolved treatment areas despite extensive research and breakthroughs [13, 14]. Thus, the creation of new and enhanced skin wound healing techniques is of great clinical significance on a worldwide scale in the field of healthcare.

Throughout the past 20 years, there has been notable growth in the use of nanomaterials, specifically nanoparticles (NPs), in wound recovery and avoiding infections [15–17]. A while ago, the focused method that has attracted a lot of attention is the incorporation of NPs into scaffolds for wound repair [17, 18]. Antimicrobial nanoparticles come in a range of forms these days, ranging from polymers like lipids (liposomes) [19], cellulose [20], or mesoporous silica [21], which operate as antimicrobial drug transporters.

Metal/metalloid-based nanoparticles are a different kind of antimicrobial nanomedicines that have been shown in multiple investigations to facilitate wound healing [18, 22, 23]. Their outstanding antibacterial efficacy may or may not be closely linked to this recovery process [23–25]. While the concept of nanoantibiotics, also known as nanobots, has been known for at least ten years [26], and there have been several evaluations of the application of nanomaterials as antiseptics [27] or antimicrobials [26, 28–31], there hasn't been as much focus on whether or not they truly enhance wound repair results. A few research studies have been released on the issue of nanoparticles specifically utilized to enhance wound repair, despite the fact that several articles have been issued in recent years investigating the antibacterial efficiency of different nanomaterials [23, 32]. In recent times, nanomaterials in wound dressings were reviewed in a great study that was given by Stoica et al. [16]. To assess nanomaterials' full potential and transition from being an academic curiosity to a frequent fixture in the therapeutic context, a report of this kind is necessary.

9.2 SKIN ARCHITECTURE AND FUNCTIONS

The epidermis, dermis, and hypodermis are the three separate layers of the skin, which is the biggest organ in the entire body. An anisotropic collagenous extracellular matrix (ECM) contains around 50 distinct kinds of cells that are scattered throughout the skin [33]. Appropriate shielding against extreme elements (UV light, bacteria, chemicals, heat, etc.), stopping the human organism from losing water, controlling body temperature, and aiding in the process of metabolism (vitamin D production) are among the primary roles of the skin. The basement membrane connects the skin's two primary anatomical sections, the dermis and epidermis. The adipose tissues and the third layer, the hypodermis, are subcutaneous layers that sit underneath these layers (Figure 9.1). The topmost waterproof part of the skin is called the epidermis, and it is made up of avascular keratinized stratified squamous epithelium. It is essential for regulating the amount of moisture that enters the human body [34]. The passage of interstitial liquids from the dermal vasculature typically provides nourishment for the epidermal layer [35].

Antigen-presenting dendritic Langerhans cells, keratinocytes, Merkel cells, melanocytes, and other specialized cells make up the epidermal layer [36]. Around 90% of all epidermal cells are keratinocytes, which are the major epidermal cells found in the outermost layer of skin. Water

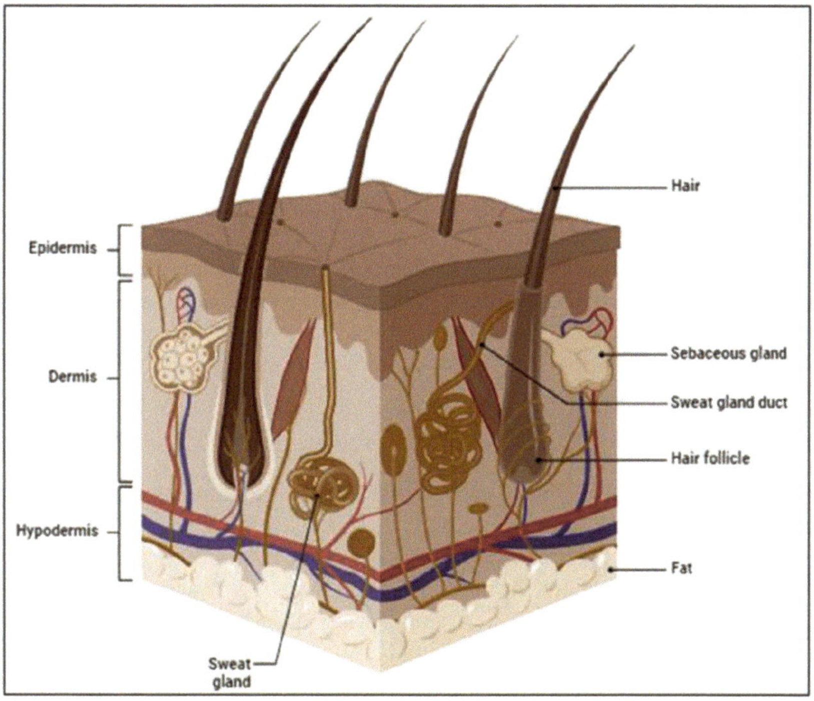

FIGURE 9.1 Structure of skin.

transport and epidermal regeneration are the two principal functions of keratinocytes, the basic cell type found in the superficial skin layer. Keratinocytes release a range of now-recognized growth factors, chemokines, and cytokines to regulate immunological and inflammatory processes in pathogenic skin diseases [37]. The Merkel cells, which are nonepithelial cells, transmit contact sensory data to the brain, including tension and texture. Melanocytes release melanin, a dye found in the skin, which keratinocytes then absorb. Antimicrobial peptides secreted by Langerhans cells assist with adaptive as well as innate immunity [38].

The stratum corneum, stratum granulosum, stratum spinosum, stratum lucidum, and stratum basale are other components of the epidermis. In essence, stem cells found in the skin's baseline give rise to keratinocytes, which can develop for up to six weeks before changing into corneocytes and finally forming the stratum corneum layer [39]. Many protein molecules, including keratin, which is essential to the stratum corneum's structure, are produced by keratinocytes. Apart from proteins, the stratum corneum's complex fats and cellular envelopes act as an essential defense against microorganisms and dehydration. The movement of fully developed cuboidal basal keratinocytes with large nuclei, phospholipid membranes, and cytoplasm starts from the baseline after about every month [40].

An increased build-up of lipids and keratin takes place throughout this turnover phase, and it eventually experiences terminal differentiation to form the stratum corneum. The dermoepidermal junction, the underlying layer that divides the epidermal and dermal layers, is the point at which hemidesmosomes, which are components of cell adhesion, are joined. Keratinocytes from the basal layer, ECM elements, basal lamina, filaments, and anchoring fibrils make up the multifaceted dermoepidermal interface arrangement [41]. As a way to prevent the possibility of the epidermis separating from the dermis layer during the phase of wound repair, it is also necessary to restore the dermoepidermal interface [42]. Roughly 90% of the total mass of skin is made up of the dermis, the deeper layer of skin located atop the hypodermis and underneath the epidermal layer [43].

Connective tissue that is rich in extracellular matrix, vascular endothelial cells, fibroblasts, Schwann cell-type, mast cells, perspiration glands, arteries, veins, hair roots, and unbound and encased nerve terminals make up the dermal layer [42, 44]. The superficial (papillary dermis) and deeper (reticular dermis) sections of the dermis are made up of dense collagen fibers and fibroblasts, as well [45]. The primary dermis cells, fibroblasts, produce glycosaminoglycans and collagen along with elastin, which give the skin its fluidity and toughness [45]. The outermost layer of the skin receives physical and metabolic nourishment from the dermal layer, which houses the skin's neuronal, circulatory, and lymphatic systems, thanks to its architectural characteristics. The hypodermis, the covering of subcutaneous tissue underneath the dermis, is what links the skin's dermal coat to other tissues like musculature or skeleton [46]. The porous connective tissue, fatty tissue cells, liquids that absorb proteoglycan and glycosaminoglycan, blood vessels, and fibers make up the hypodermis zone [6]. The hypodermal membrane is primarily responsible for controlling body heat and storing fat.

9.3 WOUNDS AND WOUND HEALING PROCESS

According to definitions, wounds are any injuries or illnesses that impair the structural integrity of the skin and are brought on by medical issues like diabetes or vascular diseases, burns, scrapes, or strains, as well as by surgical procedures [47]. Based on the length of time they take to recover and how they do so, wounds are divided into two different groups: acute and chronic. Acute wounds are a broad category of lesions that can result from a variety of factors, including radiation, rapid fluctuations in temperature, and chemical compounds. Acute wounds can be further categorized into three categories—complete thickness, deep dermis, or superficial injuries—based on their depth and dimension [48].

With an organized, conventional wound recuperation procedure, acute wounds typically recover in 4–12 weeks, restoring physiological and structural skin strength [49]. On the other hand, chronic wounds, which result from underlying medical disorders including diabetes, immunological illnesses, or venous stasis, are more challenging to repair. Chronic wounds can also result from previous infections, inflammations, tumors, or other physical causes. Limited mitogenic action, increased cytokine and protease levels, limited growth factor release, excessive generation of free radicals (reactive oxygen species) and matrix metalloproteinases (MMPs), repressed angiogenesis, fibrosis, and ECM degradation are some of the hallmarks of chronic lesions [50, 51]. In contrast to acute wounds, chronic wounds take longer to heal—more than 12 weeks—which increases the risk of spreading infection [52]. The optimum and functional reconstitution of skin integrity is further hindered by the prolonged recovery period associated with chronic wounds. Therefore, in order to speed up the healing of wounds, especially those that are chronic, it is imperative that novel techniques and tactics be developed.

Wound reconstruction is a complicated and dynamic biological process that is categorized into four interconnected and contiguous stages: homeostasis, inflammation, proliferation, and tissue remodeling [53]. Following an acute injury, a meticulously orchestrated series of occurrences sets in motion the wound repairing procedure, facilitating the advancement of recovery via the aforementioned overlapped but discrete stages of healing [54]. The several wound repair episodes are characterized by a sequence of closely controlled processes that include the systemic discharge of different growth factors (GFs) and cytokines, chemotaxis, phagocytosis, collagen remodeling with neocollagen production, and collagen breakdown [55].

Adults recover more slowly than newborns do, and the wound healing process frequently leads to scarring (fibrosis healing) instead of rejuvenation. As a result, novel approaches to high-quality, scarless wound repair are additionally required. In the event of chronic or nonhealing injuries, the recovery process often halts and is unable to precisely restore skin tissue because of the huge wound's diameter or depth, which causes keloid development and unsightly scars [55]. Hemostasis, which occurs as soon as an accident occurs, is the first stage of wound recuperation during which the body tries to halt bleeding by constricting blood vessels and forming a scab by aggregating platelets [42, 56]. When the platelets reach the area of injury, they aid in the clotting of blood, avoiding interference with regular blood flow. By temporarily closing the injury, this blood clotting helps to prevent needless blood loss. Additionally, this procedure prevents the post-injury transmission of bacteria [32, 57]. Numerous chemotactic substances are released by injured epithelium and endothelial cells, as well as blood coagulation, which draws inflammatory cells and starts the next stage of healing. Following hemostasis comes the stage of inflammatory responses, during which immune responses travel to the wound area in an effort to disrupt signaling molecules to avoid bacterial infection. These cells include neutrophils, macrophages, monocytes, and lymphocytes. Multiple growth regulators and cytokines, including reactive oxygen species (ROS), tumor necrosis factor (TNF), platelet-derived growth factor (PDGF), fibroblast growth factor (FGF), transforming growth factor (TGF), epidermal growth factor (EGF), and interleukins (IL-1, IL-6, and IL-17) are released during this process [58]. Diverse white blood cells migrate through the circulation during the inflammatory phase and arrive at the area of injury where they will continue to absorb and break down microbiological pollutants [59]. White blood cells migrate nearer to the wound zone through the dilatation and greater permeation of the capillaries.

Commencing the wound repairing procedure, neutrophils—which have an antibacterial function—arrive at the location of the wound initially and remain there for 24 hours. Then, in order to continue breaking down bacteria and leftover debris for skin rebuilding, additional immune cells like macrophages and lymphocytes are drawn in. This process is facilitated by growth regulators released by the neutrophils (vascular endothelial growth factor) along with neutrophil cytokines (IL-17) [58, 60]. The inflammatory period of wound repair lasts between one and four days after the damage and is crucial for phagocytosis and debris elimination, which equip the damaged area for rejuvenation [57, 61].

Pro-inflammatory (M1 phagocytic) to anti-inflammatory (M2 pro-regenerative) macrophage morphologies change in the final phases of inflammation, coinciding with neutrophil death [62, 63]. To start the process of proliferation, the immune cells' growth factors as well as cytokines in the inflammation stage attract a variety of cells, including fibroblasts, endothelial cells, keratinocytes, and stem/progenitor cells [64]. In this stage, fibroblasts (producing extracellular matrix to produce granulation tissue), endothelial migration and development of cells (vascularization), and wound re-epithelialization (restoring the epidermal barrier) begin the process of skin tissue repair [32, 65]. The duration and characteristics of each phase that occur during the wound healing process are listed in Table 9.1.

TABLE 9.1
Duration and Characteristic of Each Stage of Wound Healing

Stage	Duration	Characteristics
Haemostasis	1–3 days	Aggregation of platelets. The formation of a clot to cease bleeding.
Inflammation	3–20 days	The wound is shielded from microbial infection by neutrophils and reactive oxygen species. The wound warms up, swells, and turns red in color. Dead cells are replaced by epithelial cells.
Proliferation	7–40 days	Multiple cells merge together to generate new tissue. Granulation tissue is produced by ECM to replace the initial clot formation.
Remodeling	40 days–2 years	The wound heals completely. Fibroblasts enclose the wound, like fresh skin. The formation of a scar.

Source: [67].

By generating extracellular matrix components such as proteoglycans, hyaluronic acid, procollagen, and elastin, which serve as an appropriate substrate for the development of new blood vessels, fibroblasts multiply and create granulation tissue. Endothelial cells facilitate the formation of new blood arteries by supplying the cells with oxygen and nutrients. The granulation tissue's chemicals have the tendency to break down the blood clot that has already formed, and the fibroblasts' production of collagen gives the wound its mechanical strength. Re-epithelialization results from the development and proliferation of keratinocytes, giving the lesion a superficial epidermal covering [66]. Additionally, extracellular matrix accumulation, angiogenesis/revascularization, keratinization, and re-epithelialization are all facilitated by cellular interaction among the various skin cells [57].

The remodeling or maturation phase that follows this stage replaces the granulation tissue with collagen type 1 production and results in the formation of different skin appendages (sweat glands, sebaceous glands, and hair follicles) to return the injured area to its pre-injury functioning condition [32]. In order to provide the skin with more tensile strength, collagen type III is converted to collagen type I during this stage, and fibroblasts develop into myofibroblasts [68]. Remodeling the extracellular matrix is crucial because dysfunctional wound healing resulting in hypertrophic scars is caused by excessive accumulation of collagen and matrix metalloproteinase activities regulation [69, 70]. For acute wounds, wound healing occurs when each of these stages is finished. On the other hand, persistent wounds hinder the healing process by not recovering through all four stages. Chronic wounds are characterized by hypoxia and changed cellular phenotypes, and not all wound regions are healing at the same rate [71, 72]. Chronic wounds are characterized by ongoing inflammation, which typically does not advance to further stages of wound repair because of molecular variations in their etiology from acute wounds. This results in an extended inflammatory phase during chronic wound repair due to the overabundance of inflammatory cell recruitment and the dispersion of multiple growth variables into the area of the wound [54]. Figure 9.2 shows the four stages of wound repair.

Furthermore, inadequate circulation and growth regulators contribute to the slower recovery of chronic wounds. Although the pro-inflammatory cytokine cascade is activated for a longer amount of time in chronic wounds than in acute wound repair, wound patches in chronic wounds show higher levels of proteases. When combined, an organized strategy is necessary for the evaluation and management of chronic wounds.

9.4 CONVENTIONAL APPROACHES EMPLOYED FOR WOUND HEALING

Acute wounds often recover according to the standard wound recuperation protocol, culminating in predictable and well-organized skin tissue restoration with the fulfillment of all wound repairing stages [74]. Nevertheless, since they require a great deal of healing time, chronic or hard-to-heal wounds are treated according to the "TIME" guidelines, which stand for tissue (T) evaluation and debridement of nonviable tissues, infection/inflammation (I) control and etiology review, equilibrium of moisture (M), and epithelial edge (E) progression [75]. In recent years, the area around the wound, the edge, and periwound skin are all included in a therapy plan according to the modern triangular wound assessment (TWA) technique [76]. Furthermore, irrespective of the cause of the wounds, treating chronic wounds requires a methodical strategy. For the restoration of skin wounds, traditional wound care techniques encompass a range of surgical techniques, nonsurgical treatments, and medication-based treatment plans [77]. Debridement, skin grafts, and skin flaps are examples of operative wound recovery treatments; dermal compositions, wound dressings, and skin replacements are examples of nonsurgical remedies. The debridement procedure begins following the first evaluation of the wound size and wound cleanliness [78]. In an effort to prevent wound repair compromise and to expedite extracellular matrix remodeling, surgical debridement of devitalized tissue conditions the wound bed [79]. Because surgical debridement is so successful, it continues to be the gold standard for wound care [80]. This quick and efficient surgery is carried out under anesthesia; yet, there are hazards involved with the technique, including the greater

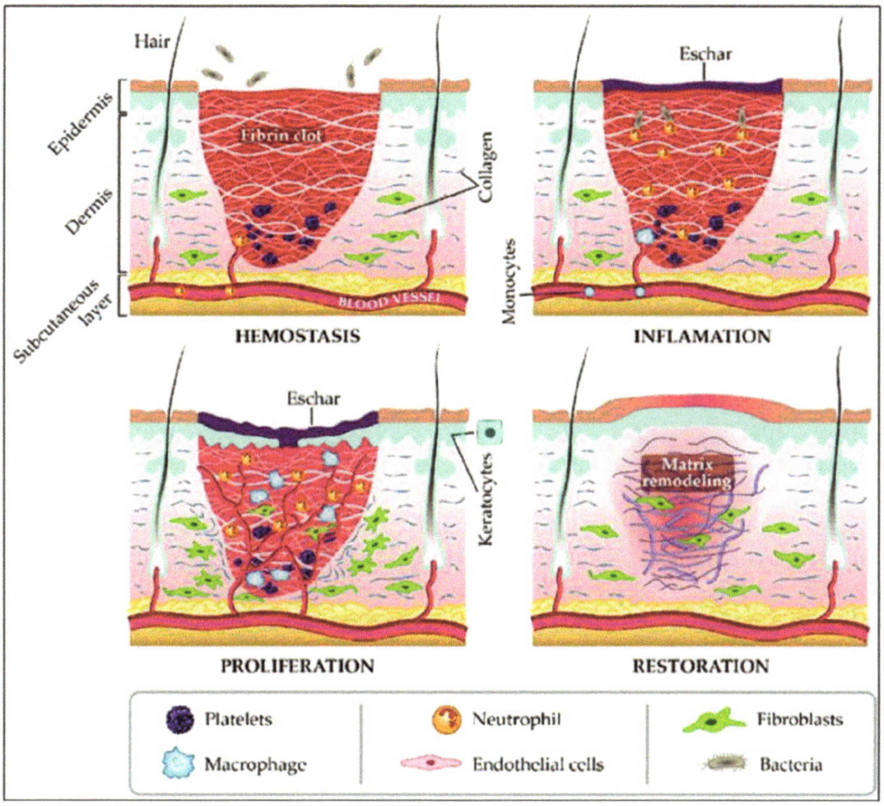

FIGURE 9.2 Wound healing in four phases [73].

likelihood of general anesthesia and potential tissue damage. Skin transplants that are customized for the kind and etiology of wounds are among the additional surgical therapy options for wound care. The most frequently performed surgical option of wound therapy is skin grafting, which is used for wounds with significant tissue destruction and persistent wounds.

Split-thickness and full-thickness skin transplants are the two categories into which they are divided depending on thickness. Skin grafts with split-thickness are considered the gold standard in surgery for treating tiny lesions with only minor epidermal loss. During this procedure, the injured skin region is covered with mostly the epidermis and a small amount of dermis taken from normal skin. Patients whose wounds encompass greater than 30% of their bodies, nevertheless, should not employ this transplantation technique. The sensation of pain, itching, and greater contraction throughout the recuperation phase that results in scarring are further drawbacks. On the other hand, because full-thickness skin transplants address both dermal and epidermal healing, they are beneficial for larger, deeper lesions.

Because there is less constriction and the treated tissue is scarless, these transplants promote a quicker recovery. Nevertheless, the limited availability of donor skin and the need for a vascularized wound surface constrain the use of full-thickness transplants. Additionally, depending on the origins of the donors, skin transplants are divided into three categories: autografts, allografts, and xenografts [81]. Since healthy skin is removed from the patient and transplanted to the wounded region to restore skin integrity, autografts continue to be the surgical gold standard for healing the skin [82].

Immune rejection is avoided, and local vasculature and epidermal integrity are restored when autologous tissue is used in autografts. Nevertheless, there are a number of drawbacks to using autografts, including an excruciating repair procedure, a lack of typical skin locations, and resulting scarring. Another choice to autografts is an allograft, in which a separate individual provides the donor skin. The instantaneous accessibility of donor skin and extended preservation for potential use in the years to come are benefits of allografts. Xenografts, on the contrary, are skin grafts that are taken from other animals and placed inside of humans. Porcine skin grafts, for instance, have been used to treat burn wounds with burn percentages <30%. Restrictions related to allografts and xenografts include immunological rejection, the possibility of infection transmission, difficult recovery, and scarring.

Various kinds of wound dressings, topical treatments, scaffold/hydrogel-based skin grafts, and skin replacements are examples of traditional nonsurgical wound remedies. These nonsurgical treatments regulate constriction and re-epithelialization, appropriately govern the spread of infection and inflammation, and debride the wound bed. The most popular method of administering medication to the skin is through topical products, which include gels, emulsions, pastes, creams, foams, lotions, sprays, and other products. Topical antibacterial medications are useful for treating a variety of bacterial infections. They come in the form of semisolid pharmaceutical preparations that include ingredients like neomycin, povidone-iodine, and silver sulfadiazine. Antibiotics applied to the skin work better during the inflammatory stage of wound reconstruction, when infections are far more likely to occur. Topical antibiotic administration ought to be stopped subsequently to prevent adverse effects and hypersensitivity. Different ointment preparations that contain collagenase and dexpanthenol have been used to speed up re-epithelialization and extracellular matrix remodeling while also controlling the growth of fibroblast [83]. Because of its advantageous effects, topical applications that incorporate growth regulators have been investigated in wound treatment in the past few years. However, the use of growth regulators is restricted by their poor bioavailability as a result of their quick evacuation from the area of injury. Novel methods that apply growth regulators topically to the wound site that are encapsulated in nanoparticles and have improved stability and bioavailability aim to address this drawback. Improved collagen production and angiogenesis were observed in recent research with the topical use of microspheres expressing fibroblast growth factor (FGF-10) [84]. Wound dressings are another nonsurgical treatment that can help in the healing process. High accessibility to oxygen and the maintenance of humid conditions are provided by wound dressings, both of which are necessary for effective wound treatment. The ideal characteristics of a wound dressing are shown in Figure 9.3. For the medical management of acute wounds, dressings containing multiple antimicrobial compounds are among the many commonly used types. There are now over 3,000 different kinds of wound dressings on the market, which are broken down into

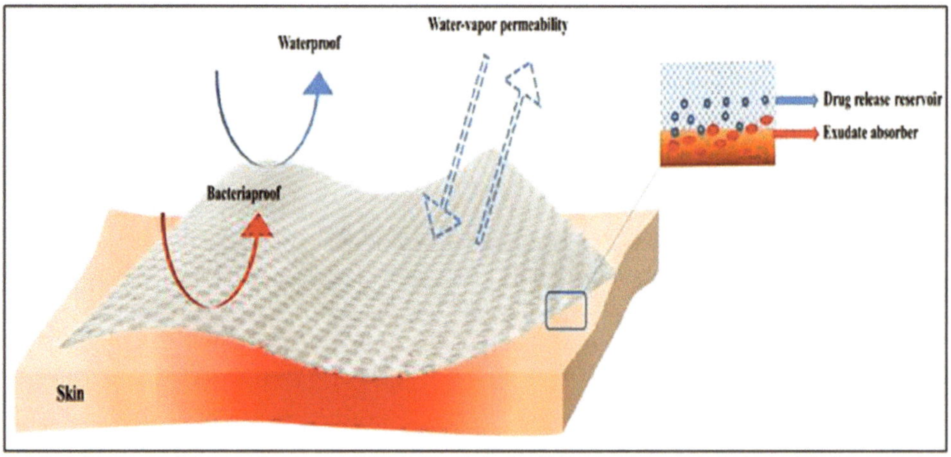

FIGURE 9.3 The ideal characteristics of a wound dressing [86].

three categories: biological, artificial, and conventional wound dressing treatments based on the type and origin of the wound [85].

Film-based, hydrogel-based, and polymeric wound dressings are the most widely used types of wound dressings for wound repair purposes. Over the past 30 years, polymeric film-based dressings have been used, with a steady advancement in both usage and design. Transparent and semipermeable, film-based dressings readily adapt to the surface of the wound. Film-based dressings have the benefit of a high vapor transfer percentage, bacterial impermeability, and the capacity to automatically remove scabs. Polymeric films, originating from a variety of natural and artificial polymers, have been employed as wound dressings for burns, abrasions, topical wounds, surgical incisions, and infections. Film-based wound dressings still require development due to their adhesiveness, inability to consume large volumes of wound exudates, and need for frequent modifications. Presently, there are several film-based coverings approved by the Food and Drug Administration on the market, with different permeability, adhesiveness, and conformability, such as Biooclusive™, Tegaderm™, Hydrofilm™, and Opsite™ [87].

Since they have the capacity to retain moisture, absorb a high amount of wound exudate, and encourage the spread and growth of fibroblasts and keratinocytes, moist wound dressings utilizing hydrogels have attracted a lot of attention for use in wound healing procedures over the past ten years. These hydrogel-based wound dressings offer cooling properties, have a high absorbent capacity, and supply the essential moisture while reducing discomfort. Furthermore, burns, deep wounds, surface wounds, necrotic wounds, dry wounds, and other conditions can all be treated using hydrogel-based bandages. The development of biomaterials and design techniques has led to the availability of injectable hydrogels, self-healing hydrogels, and hydrogels with the ability to discharge drugs, growth factors, and antibacterial agents for better wound healing [88–90].

The majority of wound dressings are made of artificial and organic biomaterials, combined with growth factors, medications, and other kinds of bioactive substances. Natural biomaterial-based wound dressings, such as silk fibroin, alginate, hyaluronic acid, and others, imitate the extracellular matrix that surrounds the skin and hasten the repair process. The management of various wounds has also employed composite wound dressings made from natural, synthetic, organic, or inorganic biomolecules. Wound dressings using nanoparticles, growth regulators, antibiotics, and bioactive substances have become more bioactive due to the development of nanotechnology. Nowadays, 3D-printed/bioprinted wound dressings and stimuli-responsive (pH, temperature, oxygen) wound dressings are widely used for wound repair purposes. This is an enhanced technique.Bioengineered skin replacements are among the other most popular enhanced traditional wound healing techniques. Tissue-engineered skin replacements, which imitate natural skin and promote skin regeneration by forming a semipermeable barrier surrounding the wound site, are made up of three main components: biomaterials, cells, and growth regulators. For the creation of skin replacements, a range of natural biomaterials, growth regulators, and cellular structures are being investigated. Unlike previous traditional surgical treatments, skin replacements lessen the risk of death and disability related to chronic wounds without requiring further surgical operations. The efficiency and security of skin replacements are additional benefits.In accordance with the percentage of harm that occurs to the skin, there are three categories of skin replacements: dermal skin substitutes, epidermal skin alternatives, and epidermal–dermal skin replacements. The most commonly employed dermal and epidermal skin replacements on the market include products like Matriderm, Integra, Dermagraft, Alloderm, and Transcyte. These commercialized skin replacements are widely used for the management of burns, chronic ulcers, and partial and complete thickness wounds [91]. Keratinocytes, growth regulators, and a biomaterial scaffold for maintenance make up epidermal skin replacements. On the other hand, fibroblasts, growth regulators, and biomaterial matrix aid in wound repair are found in dermal skin replacements. Nevertheless, for certain cutaneous lesions in which both layers are injured, epidermal and dermal skin replacements separately are insufficient. As a result, advanced biological skin replacements that replicate the dermal and epidermal sections and supply cytokines, growth regulators, and extracellular matrix elements to aid in wound repair are needed. Some commercialized dermoepidermal cell replacements, such as Appligraf and OrCel, are being utilized to treat chronic wounds, diabetic foot ulcers, and partial- and complete-thickness burns [92].

Nevertheless, certain drawbacks still exist, such as the possibility of tissue rejection and the expensive nature of epidermal–dermal cellular skin replacements, which makes more recent research investigations and technological developments necessary to address these issues. Furthermore, to promote wound repair, the most popular traditional wound care techniques—wound dressings and grafts—have been widely used in conjunction with more recent technologies, like bioengineered skin replacements. Antibacterial nanoparticles combined with conventional cotton gauze provide another method for better wound recovery. Furthermore, superficial therapies, including both nanoparticles and nanostructured carriers, have enhanced wound repairing therapy options. Other alternatives to conventional techniques include the use of stem cell treatment, vacuum-assisted wound closure, bioengineered skin replacements, and growth factor/cytokine therapy [93]. In large-scale research, the majority of the aforementioned techniques still lack efficacy and evaluation. Consequently, in order for cutaneous wound repair to meet patient demands with regard to efficacy, affordability, and aesthetic appearance, it is imperative that traditional wound treatment be improved through the investigation of new and creative ways. Numerous cutting-edge and novel techniques, including 3D bioprinting, cold plasma therapy, platelet-rich plasma-based therapy, and ECM-based therapies, provide enormous promise for better wound repair efficiency in light of the latest developments in science, technology, and medical precision.

9.5 INNOVATIVE STRATEGIES FOR WOUND HEALING

Due to variations in the biological and molecular processes of various wound types, there is currently no therapeutic method that is adequate for full wound recovery. Novel modern technological improvements have significantly increased our understanding of evaluation and treatment options in wound healing research. The transition from dry dressings to moist wound dressings, wound dressings that respond to stimuli, treatment based on growth factors, skin manufactured via tissue engineering, human skin replacements created by bioengineering, gene therapy, nanotherapeutics, and stem cell therapy have brought about a revolutionary breakthrough in this sector. Furthermore, progress in cutting-edge technologies like platelet-rich plasma (PRP), 3D bioprinting, and ECM-based methods have facilitated the development of personalized wound healing techniques. However, the evaluation and management of persistent wounds continue to be a matter of worry. Hence, it is essential to develop novel and advanced therapeutic approaches for wound healing that consider efficacy, risk–benefit ratio, and economics. The following sections will outline and analyze many emerging and creative approaches that are currently being used and investigated for wound healing, including their benefits and limitations.

9.5.1 Nanotherapeutics-Based Strategies

Due to its intricate etiology and the resulting local and systemic challenges, the treatment of chronic wounds is much more complicated than that of acute wounds. Prior methods used the systemic delivery of antibiotics, antimicrobial agents, and topical application of medications to enhance the recovery process of challenging wounds. Nevertheless, these methods have drawbacks, including inadequate medication absorption into deeper layers of the epidermis and resistance development due to chronic utilization of antibiotics [94, 95]. Hence, it is essential to implement advanced treatment methods that may eliminate the dangers of widespread bacterial infection and enhance the transportation of medication to the underlying tissues in long-lasting wounds to achieve effective wound recovery. The developing treatment possibilities comprise physiotherapy, stem cell therapy, and nanotherapeutics, as well as other biological approaches including nitric oxide and ROS generators and microbiome therapy. Out of the several new methods for treatment, tactics involving nanotherapeutics have shown remarkable effectiveness in improving wound recovery. These tactics include the use of a variety of nanoparticles [22, 96]. Nanotherapeutics involves the use of various types of nanomaterials (such as nanoemulsions, nanofibers, nanoparticles, and nanogels) stuffed with antibacterial agents, interferons, growth factors, antimicrobial peptides, and other substances. These tiny molecules are delivered to the injury site to address various kinds of wounds. Furthermore, nanotechnology-based methods have shown promise in addressing the challenges often associated with traditional wound recovery techniques, such as infection, inadequate absorption into deeper layers of epidermal tissue, and prolonged wound healing. In the past 20 years, nanotechnology has experienced significant progress, providing new opportunities for delivering medications, antibacterial agents, antibiotics, biomacromolecules (such as peptides and proteins), growth factors, DNA/RNA, and other therapeutic substances for the purpose of healing chronic wounds. Nanotechnology plays a crucial role in regulating bacterial infection, inflammatory processes, and hemostasis, as well as encouraging cell replication [97, 98]. In addition, the use of various tiny carriers in nanoparticulate drug delivery systems allows for the sustained and controlled release of a diverse range of medications and medicinal products. This method ensures that medications are delivered in an ideal amount, with a longer half-life, higher bioavailability, enhanced pharmacokinetic profiles, and reduced intervals for drug administration [99, 100]. The medicinal function of nanocarriers in wounds is due to their tiny dimensions (10 to 100 nm) and physicochemical features, which enable the distribution of drugs into cells, the preservation of wet conditions with improved permeability, and durability in deterioration [101, 102]. Moreover, the enhanced encapsulation efficiency of various medicines and biomolecules inside nanocarriers significantly improves the efficacy of delivering them for wound recovery and skin rejuvenation purposes [103]. Nanotherapeutic approaches offer multiple benefits in managing infectious microbes in chronic wounds. The benefits include enhanced medication stability and absorption, increased drug absorption into microbial biofilms and tissue barriers, better drug-microbe interactions, and the capability to augment drug levels at infected locations [104]. There are two primary classifications of nanoparticles that are now being used with specific characteristics and effectiveness for the treatment of long-lasting wounds [96]. The first group consists of organic nanoparticles, such as solid lipid nanoparticles, nanosheets, ethosomes, liposomes, nanoemulsions, nanogels, polymeric nanomaterials, and nanofibers. Another primary classification of nanomaterials consists of inorganic nanomaterials, including magnetic nanoparticles, carbon nanotubes, metallic nanomaterials, and quantum dots [105].

Moreover, nanomaterials used in treatments for wounds may be divided into two distinct categories: (1) nanoparticles with inherent qualities that enhance the speed of wound recovery, and (2) nanoparticles used as carriers for delivering drugs. Furthermore, nanoparticles may be genetically modified to serve as either a therapeutic tool or as carriers for regulated and prolonged release of different medicinal compounds, thereby enhancing the mechanism of repairing wounds. Due to their many benefits, metal oxide nanoparticles, metal nanoparticles, polymeric nanomaterials (nanoemulsions, nanoparticles, and nanofibers), and other nanotherapeutics are extensively used for treating chronic wounds. Nanostructures composed of metals like zinc, gold, silver, gallium, and metal oxides, including oxides of zinc, aluminum, iron, titanium, and copper, have remarkable antibacterial capabilities that promote wound regeneration [106–108]. The antimicrobial characteristics are due to the generation of ROS, inhibitory enzymes, damaging impact on bacterial DNA, and the potential to disrupt biofilms, hence inhibiting bacterial development [109]. Furthermore, chronic wound care has made use of many materials like carbon-based nanoparticles, nitric oxide nanoparticles, cerium, and bioactive glass with inherent bioactivity [110]. Prior consideration should be given to the toxicity generated by metal-based nanoparticles, as it may limit their in vivo applicability. To address this issue, optimizing the dimension and characteristics of metallic nanotherapeutics might offer advantages by mitigating toxicities. In addition to metal-based nanotherapeutics, polymeric nanoparticles that possess intrinsic biocompatibility as well as biodegradability are advancing the field of wound recovery applications. Polymers derived from polysaccharides, proteins, cationic polymers (chitosan), polyamides, polyester, and other similar substances are used in polymeric nanotechnology. For skin rejuvenation and wound recovery, polymeric nanoparticles come in a variety of forms, including nanosheets, nanofibers, nanogels, nanoemulsions, liposomes, and polymeric micelles [111]. Encapsulating growth factors and antibiotics inside polymeric nanoparticles is another important technique for employing polymeric nanotherapeutics to reduce the time it takes for wounds to heal by preventing infection [112]. Table 9.2 and Table 9.3 summarize the functions of different inorganic and organic nanoparticles in wound healing.

9.5.2 Stem Cell Therapy-Based Strategies

Regenerative therapy has recently developed as a groundbreaking discipline that offers new treatment approaches that preserve the natural structure of the skin and enhance the recovery of wounds [131, 132]. Stem cell-based treatment in regenerative healthcare has attracted significant attention due to its characteristics, such as long-term self-renewal capability and the ability to differentiate into various cell types [133]. Stem cells and progenitor cells have garnered significant interest as cell treatments for wound recovery. The surviving stem cells located around the site of the wounds contribute to the recovery process because of their ability to change and adapt (plasticity). In addition, stem cell-based therapy has shown significant promise in the treatment of persistent injuries that are unresponsive to traditional treatments. The treatment for persistent wound repair using

TABLE 9.2
Functions of Inorganic NPs in wound Healing

Nanoparticle	Function	Reference
Silver NPs	Promote angiogenesis Provide antimicrobial effects Minimize inflammation via modulating the immune system responses	[113, 114]
Gold NPs	Stimulate the wound electrically Utilized in photothermal therapy (PTT) to foster the regeneration of tissue Employed as nanocarriers for antimicrobial drugs	[115, 116]
Copper NPs	Aid in the closure of wounds by managing cytokines, growth factors, and cells Involved in the healing of wounds Enhance the synthesis of collagen, epithelialization, and keratinocyte proliferation	[117, 118]
Silica NPs	Silicic acid promotes the movement of skin fibroblasts to aid in topical wound healing	[119, 120]
Zinc oxide NPs	Exhibit both bacteriostatic as well as bactericidal qualities Exhibit increased permeability via lipid and skin membranes	[121, 122]
TiO_2 NPs	Have antibacterial and anti-inflammatory benefits Increase biocompatibility	[123, 124]
Carbon nanotubes	Impair the integrity of the microbial cell membrane	[125]

TABLE 9.3
Function of Organic NPs in Wound Healing

Nanoparticle	Function	Reference
Lipid NPs	Targeted administration of peptides, proteins, and medications	[126]
Polymeric NPs	Prevents degradation of drugs by proteases	[127]
Nanobiotics	Treat lesions produced by bacteria resistant to multiple drugs	[128]
Micelles	Encase polymeric NPs, water-loving, and drugs with hydrophobic nature	[129]
Cyclodextrin	Improve the stability and solubility of drugs	[130]

stem cells involves the use of growth factors, management of inflammatory reactions, and activation of immunological processes to accelerate re-epithelialization and vascularization [134]. Several experimental and medical investigations conducted in recent decades have shown significant effects on the improvement of healing wounds employing stem cells [135]. The therapeutic efficacy of stem cell-based wound therapy is mostly due to its capacity to release growth factors and pro-regenerative cytokines, which facilitate the rejuvenation of skin in the management of chronic injuries [132]. Furthermore, autologous stem cells possess a remarkable capacity for division, promote the formation of new blood vessels, and are well accepted by patients with few negative responses. Stem cells possess an extraordinary capacity to differentiate into many cell types, making them an ideal option for facilitating the innate healing mechanism, stimulating cell growth, and expediting the closing of wounds. Prior studies have shown that stem cells have a favorable impact on wound restoration by directly and indirectly stimulating the cells in cutaneous tissue, releasing active chemicals, regulating inflammatory processes, and restructuring the extracellular matrix [136]. Due to the positive impacts of these processes, the use of stem cells may lead to quicker wound healing, avoidance of wound contraction and scar formation, and prompt repair and renewal of the wound. However, the primary objective of progenitor cell-based treatment in clinical practice is to enhance the standard of wound recovery for wound management. Induced pluripotent stem cells (iPSCs), adult mesenchymal stem cells (MSCs), and embryonic stem cells (ESCs) are the main forms of stem cells used for healing wounds and skin rejuvenation [137]. Endothelial colony-forming cells have been underutilized as a potential source of stem cells for wound repair, mostly owing to the ethical problems connected with their use. MSCs are the most compelling source of stem cells for healing wounds due to their capacity to regulate inflammatory processes, stimulate the growth of new blood vessels and tissue, exert antibacterial properties, minimize scarring, and promote the proliferation of fibroblasts [138, 139]. MSCs are derived from several sources including amniotic fluid, umbilical cord, bone marrow, blood, adipose tissue, and Wharton's jelly stem cells. MSCs have a role in all four stages of wound recovery by aiding migration to the wound site, stimulating vascular development, releasing growth factors/cytokines, and promoting re-epithelialization. The first human trial for treating severe skin burn wounds employed mesenchymal stem cells generated from bone marrow. Following grafting of the skin, the study demonstrated enhanced neovascularization and alleviation of pain [140]. Another study that dealt with burn injuries used acellular dermal supporting matrix to wrap the wounds after transplanting bone marrow stem cells from patients. The patients had burns that covered 80% of their body surface area and had hypertrophic scarring [141]. Contraction of wounds was decreased, the ECM was modulated, and revascularization was promoted after stem cell implantation. Additionally, a different clinical trial was conducted to treat diabetic ulcers, which are chronic nonhealing wounds, employing a biodegradable collagen membrane (Coladerm) in conjunction with autologous MSCs generated from skin fibroblasts and bone marrow. The findings showed that after 29 days of combination therapy, the wound's dimension was reduced and circulation was improved [142]. One study found that epidermal wound repair was enhanced after MSC transplantation via paracrine signaling of vascular endothelial growth factor [143]. There are also some limits to consider when using MSCs in cell therapy-based wound management techniques, such as concerns about their long-term safety and limited survivability after transplantation. The post-implantation survival of transplanted MSCs has been improved by the use of many techniques [144]. Combining stem cells with scaffold-based medicinal approaches is one approach. Here, a 3D scaffold matrix made from polymers that were either synthetic or natural was used to seed the stem cells that would be transplanted. Afterward, the scaffold was transported to the location of the injury for transplantation. Stem cells with a decellularized silk fibroin scaffold significantly enhanced wound recovery, cellular activity conservation, and stem cell survival [145, 146]. To transport MSCs to the area of a wound, Kamolz *et al.* [147] used Matrigel in conjunction with Matriderm. In addition, this method was expanded to include growth factors in the scaffolds to enhance blood circulation and wound healing [148]. These studies showed that stem cell and scaffold-based delivery methods, when combined with growth factors, improve healing after injury.

In addition, there have been a number of promising clinical trials employing scaffold-based treatment techniques to expedite wound restoration in chronic injuries [149].

Skin also contains a variety of additional stem cells that may speed up the healing process, in addition to MSCs [150]. Various types of stem cells, including those found in the sebaceous gland, interfollicular, and bulging areas, are housed in the outermost layer of skin [151]. In contrast, the epidermal layer is home to two separate types of stem cells: those found in the dermal papilla, which are located in hair follicles, and those found in the perivascular space. Reportedly, these stem cells may hasten the recovery of wounds with little risk of immunological rejection and ethical problems [152]. Further investigation into human clinical studies is necessary to validate their effects, nevertheless, before we can draw any definite conclusions about their application in healing injuries. The field of regenerative therapy has recently unveiled iPSCs as a novel form of stem cells to address the limitations of MSCs and ESCs [150, 153]. Unlike ESCs, which raise ethical problems and may cause immunological rejection, iPSCs seem very similar to ESCs [154]. Potentially game-changing for transplant treatment, iPSCs are transformed cells produced from somatic cells that include reprogramming factors (e.g., Sox2, Klf4, Oct3/4, and c-Myc). According to Bilousova *et al.* [155], iPSCs were able to differentiate into skin, glands that produce sebum, and hair follicles. Reconstitution of hair follicles was shown in another work using iPSC-derived epithelial stem cells [156]. Recovery from wounds may be aided by induced iPSCs or exosomes produced by these cells, according to many studies [156, 157]. Recent research have shown remarkable advancements in the bioengineering of skin replacements using iPSCs derived from various somatic cells, including folliculogenic human epithelial stem cells, fibroblasts and keratinocytes [137]. In addition, Yan *et al.* [158] used microvesicles produced from iPSCs that included microRNA (miR-16-5p) to improve burn injury re-epithelialization and collagen deposition. Through the integration of genome-editing techniques such as CRISPR, TALEN, and ZFN, the iPSCs-based stem cells strategy can provide long-term, effective treatment for chronic wounds. As a whole, stem cell-based treatment and other methods outperformed more traditional methods of improving wound healing. To understand the actual clinical scenario in rejuvenating wound recovery, however, more research into stem cell sources, processing techniques, and delivery routes is required.

9.5.3 Bioprinting-Based Strategies

Over the past 20 years, multiple treatments have been widely used for managing both acute and chronic wounds. Nevertheless, the majority of the available medical alternatives relied on manual methods, which were impeded by the need for a considerable amount of time to heal extensive wounds or burns. In order to surpass this constraint, the field of three-dimensional (3D) bioprinting has evolved as an efficient and automated technique to tackle the obstacles in regenerative therapies, such as wound recovery. 3D bioprinting is a method of additive fabrication that allows for the creation of bio-acceptable artificial skin. This is achieved by depositing live cells, biomolecules, biomaterials, and growth factors in a precise layer-by-layer manner. This robotic technology is a versatile instrument that excels in clinical applications due to its precision and versatility. The benefits of 3D-bioprinted skin entities for healing wounds and skin rejuvenation are as follows: (1) automated and quicker production, resulting in reduced time and cost; (2) the power to incorporate various cells and biomolecules through the procedure to enhance sensation, coloration, and blood circulation; (3) accurate deposition of multiple biomaterials and cells in different locations; and (4) the capability for large-scale manufacturing with excellent elasticity and flexibility [159–161]. The bioprinted skin counterparts accurately replicate the natural structure and diversity of the skin by precisely placing numerous cells and biomaterials [162]. The bioprinted skin constructions must possess a range of characteristics in terms of their functional and compositional aspects. First and foremost, bioprinted skin replacements must possess the capability to efficiently transfer nourishment and wound exudates. Furthermore, it is crucial for bioprinted counterparts to precisely place many types of skin cells, including Langerhans cells, melanocytes, fibroblasts, adipocytes, and keratinocytes, at certain layers and positions. Furthermore, it is crucial for the permeation, deterioration, and mechanical attributes of the bioprinted structure to closely resemble those of the natural skin structure. Additionally, the bioprinted form should possess robustness, biocompatibility, and biodegradability, enabling it to withstand exterior pressures and stresses that occur in living organisms [163, 164]. In addition, the current progress in 3D bioprinting procedures may be used to create an in vitro 3D disease simulation and a high throughput platform for drug screening. Furthermore, the development of personalized therapeutics for various kinds of wounds utilizing autologous cells necessitates the implementation of a multifaceted strategy that involves the use of application-specific bioprinting methods.

Four distinct bioprinting methods are primarily used for the purpose of topical wound recovery and skin tissue rejuvenation. The various bioprinting processes are categorized according to their native prototypes, comprising droplet-on-demand (inkjet)-based, extrusion-based, laser-based, and stereolithography-based bioprinting technologies [165–168]. Extrusion-based bioprinting (EBB) strategies are widely used for skin bioprinting due to their numerous advantages. These include accessibility, cost-effectiveness, excessive printing speed, the potential to mimic complicated tissues, and the ability to print a wide range of viscous biomaterials [169]. In the field ofEBB, the process of depositing pre-designed, geometrically precise 3D complex structures onto the skin involves using pneumatic pressure, screw-based, piston, or microfluidic mechanisms to push bioinks through a nozzle in a continuous manner [170]. Nevertheless, the EBB approach is hindered by a recurring issue of blocked nozzles across several nozzle types. Similarly, the aforementioned bioprinting processes have their own advantages and disadvantages that need enhancements regarding equipment designs and procedures. To address the constraints of bioprinting innovation, potential approaches for skin bioprinting include the development of advanced bioprinting technologies and the integration of hybrid bioprinting systems that combine two separate current bioprinting methods. Some examples of enhanced versions of current bioprinting methods involve continuous liquid interface production (CLIP) [171], multimaterial multinozzle 3D printing (MM3D) [172], and computed axial lithography (CAL) [173]. CLIP-based bioprinting surpasses traditional stereolithography (SLA) printing technologies due to its much quicker printing speed, which is 100 times faster. Additionally, it establishes a "no polymerization zone" by using an oxygen-permeable window underneath the UV projection. This ensures a continuous SLA process [171]. CLIP may be expanded to enable the printing of multilayer skin constructions on a larger scale by using various photo-initiators, methods based on natural light, and photopolymers that are compatible with living organisms. Bioprinting with MM3D technology uses a voxel-by-voxel method, as opposed to the traditional layer-by-layer technique. This involves a specialized printer that operates at high speed and has the capability to switch among up to eight different substances [172]. This bioprinting technique offers advantages for printing skin tissue by enabling the preservation of variations in stiffness, not only across different layers but also within the length and width of each individual layer. This bioprinting technology, together with computer-aided design and computer-aided manufacturing (CAL), was created in 2019, much later than the development of CLIP in 2015. The purpose of this approach is to be used specifically for bioprinting applications related to skin. Hence, more investigation is necessary to fully explore the potential of bioprinting methods for the rejuvenation of skin tissue. Another innovative bioprinting technique is CAL, which employs volumetric building instead of the traditional layer-by-layer approach for 3D bioprinting [173]. CAL has many benefits, including using natural light rather than UV light or laser light, and the ability to print on an existing solid object with more versatility. In addition to the development of novel or enhanced bioprinting processes, another strategy to enhance skin bioprinting utilizes hybrid bioprinting systems, which combine two or more current bioprinting approaches. For instance, Kim *et al.* have shown the creation of epidermal and dermal skin layers utilizing inkjet and extrusion bioprinting methods, respectively [174]. Furthermore, hybrid bioprinting methods show significant promise for the field of skin bioprinting and need further investigation to fully unlock their potential. Another crucial factor for achieving effective wound recovery and skin engineering using 3D bioprinting methods is the use of biomaterial ink or bioink (Figure 9.4) [175, 176]. Bioinks consist of biological materials, cells, and growth factors/biomolecules, while biomaterial ink has all components of bioink except for cells. Various types of biomaterial hydrogels, such as hyaluronic acid, collagen, chitosan, gelatin, silk fibroin, alginate, and cellulose, have been employed for cutaneous bioprinting. Artificial biopolymers are also being used to enhance the mechanical characteristics of the bioprinted structures [177, 178]. All biological inks, whether made from natural or manufactured biopolymers, must exhibit certain essential qualities. The features included in this list are biological compatibility, abundant availability, biodegradability, excellent printing capability, mechanical stability, and the capacity to retain high form accuracy after the bioprinting procedure. Furthermore, bioinks use live cells that might influence the immunological response post-implantation. Hence, the choice of cells should be carefully deliberated, taking into account this pivotal aspect. When it comes to skin bioprinting, it is preferable to use primary skin cells such as melanocytes, stem cells, keratinocytes, or fibroblasts to create skin structures while ensuring that all biological functions are preserved following the bioprinting process [179]. Viscoelastic behavior is a significant characteristic of bioink that has an impact on the capacity to bioprint skin and also effects cell movement, growth, and remodeling of the extracellular matrix [180]. Skin bioprinting for wound treatment applications involves two tactics: ex vivo and in situ bioprinting. Despite the specific bioinks or bioprinting processes used, both approaches are used. Ex vivo bioprinting involves the use of extrusion, inkjet, or laser-based methods to print the superficial and subcutaneous layers of the skin. If needed, these printed layers may then be matured in a controlled laboratory environment. Once the bioprinted skin constructions have fully developed, they are transplanted into the patient's wounds. Alternatively, bioinks, which consist of skin cells in a hydrogel biomaterial, are applied directly to the wounded area with a portable extrusion bioprinter in an in situ bioprinting method. This process aims to replicate the architecture of the skin and may include crosslinking the bioinks if necessary

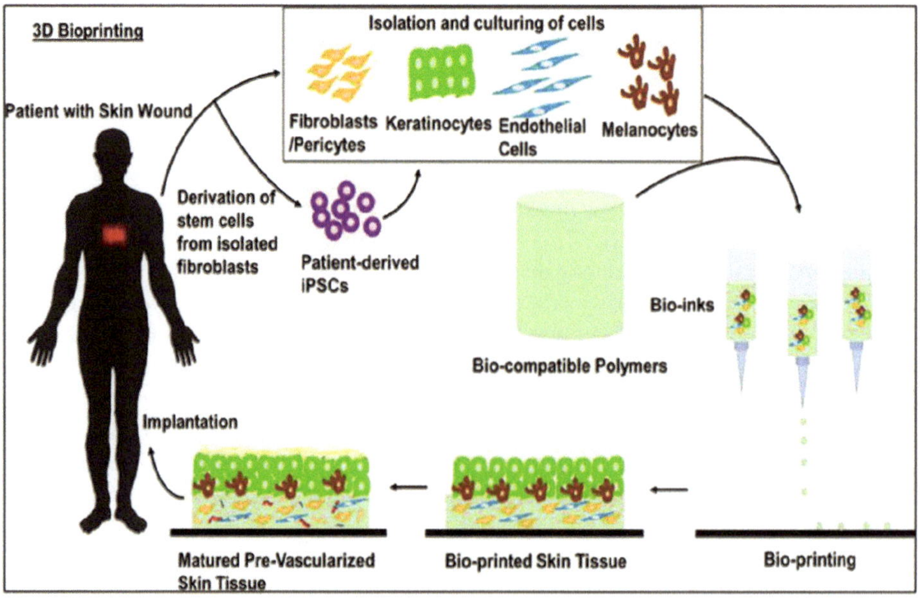

FIGURE 9.4 An illustration of how tailored skin tissue may be printed in three dimensions. After being separated from the patient's biopsy, the cells are cultivated in vitro until they are multiplied sufficiently. It is also possible to reprogram isolated fibroblasts into induced pluripotent stem cells (iPSCs), which can develop into the required cell types and quantities more quickly. Typically, bioinks are made of living cells enclosed in a biopolymer. Skin tissues are bioprinted via 3D technology in a conformationally specified way, then the tissue is allowed to develop before being implanted into the patient [197].

[181]. In 2009, Lee *et al.* introduced the first instance of bioprinted skin, which included the use of collagen hydrogel and a bioink made from fibroblasts from humans [182]. In addition, Koch *et al.* created skin analogues by using a collagen hydrogel-based bioink that consisted of fibroblasts and keratinocytes [183]. Binder *et al.* used a 3D inkjet bioprinter to create skin replacements for wound healing. They used fibroblasts from humans and keratinocytes in their fabrication process [184]. Subsequently, notable advancements were documented in the field of skin bioprinting, employing a diverse array of bioinks and bioprinting methodologies. 3D-bioprinted skin alternatives are a substitute for conventional skin grafts in the process of regenerating skin tissue makeup with appendages. They address the concern of needing donors and surgery [185]. In addition, this developing skin bioprinting technique has been used to treat a range of persistent and nonhealing infections, including burn injuries, diabetic foot ulcers, venous ulcers, and pressure ulcers. Research was conducted to create a portable 3D bioprinting device that may be employed for treating deep burn wounds in pigs. The device was designed to minimize the development of scars [186]. The portable bioprinter enables the on-site printing of skin tissue sheets, while preserving many cell types and compositional differences seen in various levels of the skin. This technology employs superficial and dermal skin cells together with various biopolymers, including alginate, collagen, and hyaluronic acid hydrogels, for the purpose of bioprinting skin. Another research project included creating an actual human skin that closely resembled native skin in terms of structure and biomechanics. This skin analogue also had a distinctive undulated pattern at the dermal–epidermal interface. Significant advancements have occurred in the area of skin bioprinting during the last ten years [187–189] employing 3D bioprinting techniques, researchers were able to achieve the printing of dermis initially [190], followed by the printing of bilayered dermal and epidermal layers, and ultimately the printing of skin trilayers consisting of epidermis, dermis, and hypodermis [191, 192]. The significant advancements in skin bioprinting can be attributed to the investigation of more advanced and enhanced bioink compositions, the availability of a diverse selection of bioprinters with varying capabilities and characteristics, remarkable progress in bioprinting technologies, and the increasing understanding of cell structure and metabolic processes at all stages of the bioprinting process.

The construction of 3D-bioprinted skin structures included incorporating pro-angiogenic peptides into hydrogel patches as part of a novel skin bioprinting approach. The enhanced bioinks demonstrated exceptional printability and vascularization capabilities, contributing positively to the process of repairing skin wounds [193, 194]. Recently conducted studies successfully created a 3D-bioprinted peptide patch with angiogenic properties to enhance the recovery process of skin wounds. The peptide-containing patch produced using bioprinting exhibited enhanced biocompatibility, angiogenesis, and wound healing in both laboratory and live organism settings [195]. Another investigation employed a bioprinted tissue coculture platform utilizing a combination of plasma-derived fibrinogen containing macrophages, fibronectin, factor XIII, and thrombin (referred to as an FPM "bioink") for the purpose of promoting wound repair and re-epithelialization in a murine superficial wound model by the process of extrusion bioprinting [196]. Currently, there is no existing bioprinted skin equivalent that has all the components of natural skin. However, this subject is still in its early stages and more research is needed to explore bioprinting methods and bioinks for the purpose of regenerating skin and healing wounds.

9.5.4 Extracellular Matrix (ECM)-Based Strategies

In the course of the wound regeneration process, the outermost layer's matrix typically guides the various stages of recovery that occur following skin injury. Persistent injuries have delayed recovery because the epithelial matrix lacks a functioning ECM, which is crucial for promoting repair in conjunction with exterior variables [198]. Long-term injuries experience degradation of the ECM due to an abundance of active matrix metalloproteinases, a higher concentration of inactive fibroblasts, and a continuous aggressive phase. This degradation hinders proper recovery [199]. Restoring the functioning ECM in long-term injuries helps to promote, guide, and organize the recovery process, ultimately improving the closure of the injury. Hence, to enhance the recovery procedure in long-term injuries, it is essential to have a thorough comprehension of the ECM elements and their interactions with cells. The ECM is a complex and intricate 3D molecular system that plays a crucial role in maintaining cellular structure and providing mechanical assistance to various tissues, such as the skin. The ECM is composed of many macromolecules, including fibrous proteins (laminin, vitronectin, collagen, elastin, and fibronectin), proteoglycans, and glycosaminoglycans [200]. The ECM is essential for guiding the recovery process of wounds by controlling cell behavior (such as adhesion, movement, development, and division) and cellular processes in the injured skin to facilitate the regeneration and repair of tissues [201–203]. Furthermore, the dynamic ECM structure serves as a storage and distribution system for various cytokines and growth factors, which are responsible for providing cellular signals. The ECM makeup requires modifications throughout the different stages of healing of wounds, enabling effective interaction between growth factors and cells and enhance the [204] availability of bioactive substances, hence expediting the recovery process [205]. During the hemostasis phase, a temporary ECM composed of fibronectin and fibrin is created to facilitate the attachment and movement of cells. In addition, the ECM has a role in creating a three-dimensional skeletal system that aids in the removal of germs during the inflammatory phase. Subsequently, a temporary ECM is created in the process of blood clotting and is then replaced by the development of granulating tissue (fibronectin) to facilitate the accumulation of collagen. Collagen I is ultimately substituted by collagen III throughout the process of remodeling in order to offer mechanical reinforcement. During the healing procedure, ECM elements interact with cell precursors and stem cells in the skin via integrin and nonintegrin receptors [205]. The connections between the ECM and integrins on stem cells are essential for regulating important biological processes during wound recovery. Integrins play a crucial role in the fundamental processes of healing wounds, including movement, growth, division, and apoptosis, by facilitating outside-in signaling. Consequently, when the elements and rigidity of the ECM become imbalanced, the normal interactions among cells, ECM, and regenerative tissues are interrupted, eventually causing pathogenic diseases. Hence, enhancing our comprehension of the ECM constituents and their functions in the recovery procedure might enhance the efficacy of wound care approaches.

While the ECM-based strategy for stimulating wound recovery is not novel, the mechanisms by which the various ECM parts initiate the recovery phase remain largely unexplored. Over the last several decades, many treatments that use ECM have been used to heal long-term injuries. An effective approach involves using synthetic ECM scaffolds or bioengineered tissue to enhance the process of wound recovery. The ECM-based substrates facilitate the attraction of cells, transport growth factors to the site of injury, and offer mechanical reinforcement. There are many categories of ECM-based structures, including cellular-derived ECM [201], artificial biomaterials-based scaffolds/hydrogels [204, 206], and decellularized ECM (dECM) scaffolds [207, 208]. Out of all the options that utilize the ECM, the dECM-based strategy has shown the greatest potential so far as an option for tackling the challenges related to organ/tissue transplantation, such as a paucity of donors and the danger of disease transmission. ECMs formed from tissues or cells lack cellular elements and operate as storage spaces for interactions between cells and the surrounding matrix, as well as unique bioactive substances. This provides many benefits for regenerating tissues. The conserved nonimmunogenic ECM replicates the natural 3D structure and immune-free conditions necessary for the healing and regrowth of skin tissues [198]. The process of skin restoration is facilitated by the use of dECM scaffolds, which maintain the porous bilayer structure of the skin as well as its bioactive molecules, flexibility, and adhesiveness to the skin and blood vessels [209, 210]. Furthermore, the byproducts of the breakdown of dECM do not cause any harmful effects during the process of skin rejuvenation. Indeed, they actively encourage the creation of tubular structures that aid in the rejuvenation of skin [211]. A study found that dECM-based scaffolds promoted the attachment of keratinocytes by preserving the lower membrane. Furthermore, it has been observed that dECM, while maintaining the integrity of the dermal framework (namely the papillary and reticular dermis), stimulates the proliferation of blood vessels [212, 213]. In a separate investigation, Choi *et al.* provided evidence of a decrease in wound size and a decrease in scarring as a result of the discharge of basic fibroblast developmental factors from the dECM [214]. In addition, Milan *et al.* showed improved recovery of skin wounds by observing decreased deposit of collagen, faster regrowth of the outer layer of skin, and a quicker pace of closing the injuries. This was achieved by employing a scaffold made from a decellularized dermal matrix (DDM) that was introduced with human umbilical cord perivascular cells (HUCPVCs) [215]. Groeber *et al.* conducted research where they created a bilayered skin substitute for grafting skin. They used decellularized pig jejunum and reseeded it with skin fibroblasts, human keratinocytes from the epidermis, and human dermal microvascular endothelial cells [216]. Due to the capacity of dECMs to maintain the natural angiogenic factor and blood vessels, dECM substrates have been used for the management of diabetic pressure ulcers (DPUs). This is particularly beneficial in cases when a high level of blood vessel formation is needed to facilitate recovery [217]. A separate study has shown that rapid wound repair in DPUs may be achieved by reducing the duration of inflammatory processes, controlling the development of tissue granulation, promoting pro-angiogenesis activity, and facilitating epithelial renewal via the use of dECM substrates [215]. The regulatory functions facilitated by dECM substrates might potentially decrease scar development. Over the last several years, several acellular products, like AlloDerm® regenerating tissue matrix and Oasis® wound matrix, have been introduced for therapeutic use and have demonstrated encouraging outcomes [218]. Nevertheless, to serve as substitutes for skin, ECM-based acellular products require meticulous refinement of production and characterization techniques. Decellularized human placenta-derived ECM substrates effectively influenced the recovery process and promoted the development of hair follicles in full thickness wounds [219]. The primary criterion for achieving native skin rejuvenation is the restoration of hair follicles and sweat glands, a feature that was absent in previous clinical products. In summary, the dECM substrates and other ECM mimics have significant promise for treating long-term injuries. They may greatly enhance wound recovery by adding growth regulators and antibiotics.

9.5.5 Platelet-Rich Plasma (PRP)-Based Strategies

Endogenous therapeutic procedures centered on PRP have recently attracted a lot of interest in regenerative medicine due to their capability to promote and speed up tissue repair, especially wound recovery [42, 220]. Greater levels of growth elements, which are necessary for healing injuries, are represented by PRP, an autologous cell substance with more platelets than blood itself [221]. Due to platelets' hemostatic activity and their promotion of skin cell multiplication and tissue development, PRP is essential in the repairing of wounds [222]. The growth factors included in PRP involve VEGF, IGF-1, KGF, FGF, TGF-β1, PDGF, and others. These elements encourage cellular movement, development, and division, which are necessary for the recovery process to begin. The intricate procedures of wound recovery and tissue rejuvenation rely on growth regulators, which regulate cellular metabolism as signaling molecules [223]. Through the wound-healing stages, nearly all kinds of skin cells are engaged in creating growth regulators, and various cells create several kinds of growth elements. For optimal recovery, each growth regulator regulates a number of cell processes, including the movement and development of cells, remodeling of the ECM, and vasculature. One growth factor is insufficient for the management of long-term injuries that are difficult to cure, particularly diabetic foot ulcers, which highlights the critical role of growth factors in wound recovery [224]. The use of PRP techniques in repairing wounds has many benefits, such as a simple approach, safety, and low cost [225]. An easy centrifugation method is used to prepare PRP once it is taken from patients' blood. Hence, the amount of growth regulators supplied by PRP may be controlled by adjusting the centrifugation conditions and the stimulation technique. PRP techniques have enduring results compared to traditional treatments and are considered safer because of their autologous nature, meaning they are derived from the patient's own body. Furthermore, it has antimicrobial properties along with the capability to reduce the financial burden associated with conventional therapy protocols. Indeed, PRP therapeutic-based therapy does not replace traditional treatment but rather functions as an alternate treatment.

The PRP approach has been increasingly used in repairing wounds in the past few years, as shown by a growing number of clinical and preclinical studies. The advantageous impact of PRP-based therapy on wound recovery has been widely documented in dogs [226], horses [227], humans [228, 229] and other animals [230, 231]. The first known medical application of PRP was the treatment for prolonged leg ulcers utilizing platelet proteins incorporated into collagen. This resulted in the stimulation of the formation of vascularized connective tissue in the incisions [232]. Subsequently, several types of PRP, including gel, solution, or injection, are currently used to treat various ulcers with diverse causes. The therapeutic use of PRP has been documented in several human and animal studies for numerous types of wounds, such as acute dermatological injuries, long-term skin ulcers, blisters, and cosmetic and plastic surgery. In addition, PRP therapy did not cause any negative effects such as infection or hypersensitive responses in clinical investigations [233]. PRP therapy is primarily used in human clinical studies to address the challenge of restoring long-term injuries that are difficult to recover. These injuries have impaired recovery due to factors such as low levels of growth elements, ongoing inflammatory processes, an imbalance of pro-inflammatory and anti-inflammatory cytokines, and a high level of reactive oxygen species [234]. PRP, due to its abundance of various growth regulators, has a robust ability to regenerate tissue, thus reducing the duration required for healing injuries [225]. A study found that the use of PRP-based dressing therapy resulted in rapid recovery of ulcerated skin. Additionally, it stimulated the growth and movement of fibroblasts and mesenchymal stem cells, leading to faster development of new blood vessels in clinical patients [235]. A further clinical trial involving 150 patients demonstrated that the use of PRP applied topically effectively led to the prompt and full healing of diabetic foot ulcers, accompanied by the development of normal granulation tissue [236]. Another investigation found that using a mixture of PRP gel applied to the skin and PRP injections under the skin in nonhealing wounds resulted in a substantial reduction in injury size, pain, and inflammatory conditions. This treatment was also found to be safe for all patients involved, with few reported adverse reactions [228]. Furthermore, PRP therapy has effectively shown accelerated recovery in several acute as well as chronic injuries of diverse origins. Nevertheless, in order to create a conclusive medicinal product based on PRP, more standardization and characterization of various PRP production procedures will be necessary [237]. Furthermore, PRP supplemental therapy has been used in conjunction with both cell therapy and skin grafts in several clinical studies, without specifically isolating the actual benefits of PRP administration alone. However, recent clinical investigations have shown that it may promote the recovery of several kinds of long-lasting wounds, indicating its promise as a therapeutic treatment. Figure 9.5 shows a schematic diagram of PRP for wound treatment.

9.5.6 Cold Atmospheric Plasma Therapy-Based Strategies

Medical engineers have used plasma-based therapeutic techniques for inflammation and tumor treatment over the last 20 years. There are two types of plasma, thermal and nonthermal (cold), which together make up the fourth category of matter. All of the elements in thermal plasma, including the electrons and heavier ones, are in thermal equilibrium, in contrast to the elements in cold or nonthermal plasma. Due to its therapeutic features, plasma-based treatment for recovering injuries has recently garnered a lot of interest. Because of its suitable reduced temperature (<40°C) for medical purposes, cold plasmas—also called nonthermal plasma, gas plasma, or physical plasma—are often used in medical studies. Balanced nuclei and charged electrons, and reactive oxygen and nitrogen species, make up cold plasma, an ionized gas that is close to normal temperature [238]. The plasma cocktail also includes ultraviolet (UV) light and an electric field, which, when combined with various ions and reactive molecules, regulate cellular reactions critical to regenerating tissue. Electrons and photons with enough energy hit neutral atoms and molecules in the gas phase, causing plasma to be generated. A dielectric barrier discharge (DBD) and an

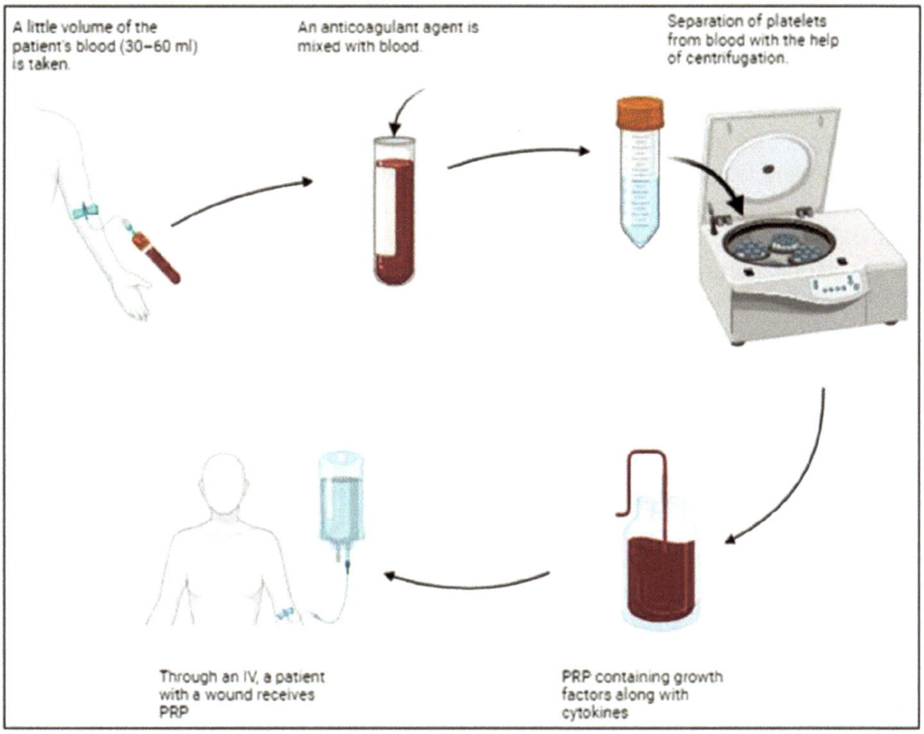

FIGURE 9.5 An illustration demonstrating the use of PRP in wound healing.

atmospheric pressure plasma jet (APPJ) of a non-DBD type are the two main ways low-temperature plasma may be released [239]. When it comes to treating persistent injuries and promoting recovery from injury, cold atmospheric plasma (CAP) is a novel therapy since it uses multifunctional modes of action [240]. Because it generates anti-infective reactive oxygen and nitrogen species, cold plasma may lessen the bacterial burden and stimulate the hemostatic phase of wound rejuvenation. Aside from its shown efficacy in healing injuries, cold plasma therapy also kills microbes by producing reactive nitrogen species (RNS) and ROS. For reducing bacterial burden, promoting ECM remodeling, and blood vessel development, nitric oxide (NO) is the most prevalent reactive nitrogen species [241]. Collaboratively, ROS and nitrogen oxides govern injury size and rejuvenation by mediating the increased production of growth regulators in wounded areas [242]. Cold plasma employs both biological and physical procedures to facilitate wound repair as shown in many experimental research and case reports. The results involve the decrease in the physical component of plasma-based therapy, including the production of free radicals and reactive species to promote recovery from wounds. Meanwhile, the natural procedure utilizes several cellular mechanisms that cause harm to bacterial DNA and cellular membranes. Cold plasma has many benefits for repairing wounds, such as ease of administration, noninvasiveness, injury acidification, antibacterial properties, promotion of cell growth, improved blood flow, and eventually reducing treatment expenses [243–245]. The process of healing wounds is enhanced by tissue acidification, which is caused by a reduction in pH in the aqueous media resulting from cold helium plasma therapy [244]. There exist three cold plasma generators that have been approved and authorized for biological applications. The first plasma device, known as kINPen® MED, was granted certification as a type IIa medical equipment in 2013 by INP Greifswald/neoplas tools GmbH in Greifswald, Germany. This certification followed a thorough assessment including preclinical, scientific, and biochemical investigations. kINPen MED is an APPJ instrument that is the size of a pen. It was originally created for use in laboratories but has now been adapted for commercial applications as well. The kINPen MED is designed for the management of many types of injuries, including severe wounds, infected wounds, pathogen-induced skin disorders, and persistent injuries [246]. Furthermore, the kINPen MED gadget offers the most extensive and well-developed potential for treating diabetic foot ulcers, which are persistent lesions. Subsequently, CINOGY Technologies GmbH, a company located in Germany, created a plasma device called PlasmaDerm® VU-2010, which utilizes a DBD source. The equipment has been validated by MEDCERT under ISO 13485. The medical device SteriPlas was created by Adtec Ltd., London, to reduce bacterial load and cure both acute and long-term injuries [247]. Cold atmospheric plasma treatment's impact on recovery from wounds has been evaluated in numerous clinical research and case reports. The cold plasma therapy in wound therapy has many positive effects, including the decrease in bacterial load, expedited recovery, and accelerated movement of cells. The capacity to produce reactive species like ROS and RNS to stimulate cell renewal offers further advantages. Multiple experimental and medical studies have been carried out in people to treat persistent injuries without causing significant adverse impacts on healthy tissue [248, 249]. The use of CAP therapy helped convert persistent injuries into nonstagnating lesions by regulating inflammatory processes, reactions among tissue-reactive species, and the activation of multiple growth variables [250, 251]. Fathollah *et al.* provided evidence of expedited wound recovery in diabetic wounds by enhanced secretion of transforming growth factor (TGF)-β1 [252]. After cold plasma therapy, the researchers observed an increase in cell proliferation, the development of new blood vessels (neovascularization), and the production of the outermost layer of the skin (epidermis). However, the use of cold plasma therapy to avoid inflammatory skin disorders and improve wound recovery has great potential for future applications. Plasma therapy has contributed to our understanding of improved therapeutic choices for wound recovery in this particular situation.

9.5.7 MicroRNA (miR)-Based Strategies for Wound Healing

The prevention and medical care of long-lasting injuries are often impacted by several complex factors. In order to address these issues, a range of therapy methods have been used to promote the recovery of persistent wounds. Nevertheless, the existing therapeutic methods for long-term injuries are usually insufficient. Recently, there has been significant interest in using a therapeutic strategy based on microRNA (miR) for long-term wound recovery. This technique involves either replacing or inhibiting miR as a curative method, and it is considered a novel treatment option [253]. MicroRNAs are tiny units of RNA that do not participate in coding and are around 18–25 nucleotides long. They have a role in several pathological and biological processes such as division, metabolism, and growth, including the repair of wounds, by regulating the production of miRs [254]. Prior investigations have shown that miRs play a crucial role in cutaneous metabolism by controlling the ability to regenerate and develop cutaneous skin stem cells [255]. Several most abundant miRs in the outer layer of the epidermis include miR-203, miR-214, miR-16, miR-152, miR-191, miR-143, miR-205, miR-27b, miR-34a, miR-30b, miR-21, miR-27a, miR-125a, miR-125b, miR-126, and the miR-200 family (-200a, -200b, -200c, -141, -429) [256]. Prior studies have shown that there is an increase and decrease in the levels of miR throughout the stages of repairing wounds that occur at the same time. Furthermore, in the context of persistent injury recovery, there is a dysregulation in the levels of miRs that are normally produced through the regular procedure of injury recovery. Previous research found that diabetic wounds showed an increase in the expression of 18 miRs and a decrease in the expression of 65 miRs, in contrast to nondiabetic rats in the placebo wound restoration group. This suggests that miRs might be a potential target for therapeutic interventions in long-term injury recovery [257]. MicroRNA-based treatment offers a groundbreaking and sophisticated treatment for persistent injuries due to the capacity of miRs to control a cluster of genes with a single miR. MicroRNA-based treatment for injuries utilizes therapeutic miRs to increase the expression of beneficial miRs and decrease the expression of harmful miRs using various techniques. Beneficial miRs may be increased by using miR mimics, which are artificial two-stranded oligonucleotides. Conversely, the suppression of harmful miRs may be achieved by the use of competitive antagonists called antagomirs, which are complementary oligonucleotides. The efficacy of delivering mimics or antagomirs to the target miR relies on the strength of their adherence and their ability to withstand degradation in various tissues [258]. Furthermore, the discovery and investigation of suitable miRs that play a role in various stages of wound healing might possibly offer a powerful kind of genetic treatment for long-term wound recovery. This treatment would work by controlling several signaling cascades. Nevertheless, to surmount obstacles inside cells and outside of cells, it is crucial to develop a delivery method that is well-designed to guarantee the effectiveness of miR as a gene therapy. MiRs exhibit an important function during the course of wound recovery, including the inflammatory phase and proliferative phase. They are involved in processes such as blood vessel development, rejuvenation, and the creation of tissue granulation, all of which contribute to the acceleration of injury recovery [259, 260]. Moreover, the significance of miRs in controlling the different stages of wound restoration was validated via the observation of a prolonged healing process when a crucial regulator responsible for miR maturation (enzyme Dicer) was lost [261]. Recently, a number of miRs have been identified as regulators of wound recovery by recognizing and attaching to certain genes, therefore promoting the healing process in persistent injuries. miR-132 facilitated the progression of the inflammatory process to the shift from the proliferating phase [262]. MiR-31 facilitates wound recovery by promoting the division and movement of keratinocytes [263]. A previous study found that miR-27b expedited the process of wound recovery in diabetic lesions [264]. Another investigation found that miR-21 has a role in regulating wound size and the accumulation of collagen, hence improving the course of wound healing [265]. Later on, exosomes generated from umbilical cord blood and containing miR-21-3p were used to enhance the activity of fibroblasts and stimulate blood vessel development, leading to improved wound recovery [266]. In a separate study, miR-129 and miR-335 were shown to enhance recovery in diabetic injuries by suppressing the level of Spl-mediated MMP-9 levels [267]. In addition, miR-497 expedited the recovery of diabetic wounds by reducing the levels of pro-inflammatory cytokines (IL-1β, IL-6, and TNF-α) and promoting wound closure [268]. Another research found that blocking miR-133b led to faster wound recovery in individuals with slower healing of wounds due to diabetes. This was achieved by restoring the levels of epidermal growth factor receptor (EGFR) and promoting vasculature [269]. A recent study found that miRs carried by exosomes play a crucial role in controlling vasculature in the repair of diabetic wounds. They do this by controlling the production of VEGF and the downstream signaling cascade [253]. While miR-based treatment has shown encouraging outcomes in in vivo studies for enhancing long-term wound reconstruction, it is still in its early stages for wound recovery applications. Modulating complicated and complex processes that include a network of interacting components with a single miR presents a significant challenge. However, in order to expand this technique as successful treatments for many kinds of long-term injuries, it is necessary to have a deeper understanding and more investigation of the roles and processes of miRs.

9.6 CHALLENGES AND FUTURE PERSPECTIVES

The increased prevalence of chronic wounds has emerged as one of the main causes of death, especially in conditions with underlying diseases. It is inevitable that novel and cutting-edge approaches to wound healing will be developed. Nanomaterial-based therapies, stem cell therapy, and 3D bioprinting approaches are the innovative and cutting-edge techniques that might eliminate many obstacles in wound healing treatment. Notwithstanding all the advancements made in the use of such approaches for wound healing, clinical studies for these technologies are still in their infancy, and there are still numerous constraints to overcome before they can be widely implemented or adopted. The current challenges in adopting such technologies lie in terms of toxicity, physicochemical properties, biological properties, reproducibility, and stability with respect to nanotherapy [270]. However, it's more likely that the combination of different nanomaterials will be the future of this discipline. The incorporation of metals into nonmetals and different nontoxic and suitable polymers may provide efficient wound dressing production techniques [271]. Biocomposites, although being in their early phases of research, may provide new products over the next ten years for efficient wound healing. In general, with the establishment of strong evidence of efficacy by rigorous research utilizing clinical isolates, defined pathogen indicator strains, simulated wound fluid media due to bacterial growth conditions, and mixed pathogen communities, animal wound models will be needed [272]. Stem cell therapy faces challenges like genetic stability, cell reproducibility, in vivo cell survival, 2D cell culture methods that are very expensive, immunogenicity, and safety [273]. Other challenges in this field comprise efficient cell isolation techniques, cell culture, and expansion methodologies. In terms of intricate signaling pathways, epigenetic modulations, and environmental factors, stem cell therapy is still not fully comprehended. For enhanced safety and efficacy, reduced risk of adverse effects, and the identification of new targets for drug design, determination of suitable stem cells and their delivery in an effective manner, clinical and experimental use of stem cells in wound healing further research and development is required in this field. Ex vivo methodologies need to be developed to extract a large number of stem cells for clinical use [274]. 3D bioprinting offers numerous advantages such as high resolution, flexibility, high throughput output, and reproducibility but various drawbacks pose challenges in this technique. First, the significant technical issues and errors with bioink in 3D bioprinting need to be resolved. Since human skin has distinct physical and mechanical properties, the selection of biological materials to match the skin tissue for 3D bioprinting is very critical and challenging. Also, the 3D printing hydrogels have poor mechanical strength, which makes them vulnerable to compression, and their structural integrity gets disrupted. In order to maintain stability and cell viability, it is important to improve or develop new gelation techniques that enable these hydrogels to resist changes in structure without compromising cell viability [275]. 3D bioprinting opens various avenues for research by developing ideal bioink material compositions that may be optimized by introducing various nanomaterials for improving tissue regrowth and identifying materials that are compatible with 3D printing systems [276]. Attention should also move to the study of bioengineered skin substitutes for healing purposes that reduce the risk of allogenic skin grafting failures. Further study in developing biomaterials and 3D printing fields along with understanding the tissue regeneration mechanism will help in the production of more sophisticated skin substitutes [277]. Treatment of chronic wounds requires new approaches in the future, including nanomaterial-based wound dressings, expanding knowledge about 3D bioprinting, and stem cell therapy. The current state of affairs is constantly updated by several initiatives undertaken in various domains such as multidisciplinary science, bioengineering, material science, and gene therapy.

9.7 CONCLUSION

Wound healing is a three-staged, complex, and intricate process involving inflammation, tissue remodeling, and tissue regeneration. Its management is becoming more important due to the increased suffering and deaths caused by chronic wounds globally. In this review, the conventional methods of wound healing and recent approaches were showcased. Although conventional treatments like gene therapy, cell therapy, skin grafting, and wound dressing are widely used for chronic wound treatment, they are accompanied with numerous disadvantages, the most significant being the slow rate of action. Thus, there emerges a need for innovative and novel methods to speed up wound healing. Newly developed cutting-edge technologies like nanotherapy, stem cell therapy, and 3D bioprinting offer innovative and novel approaches and methodologies for the improved treatment of chronic wounds. Combining both conventional and newly developed techniques offers a very promising outlook for the future. The specificity of bioengineered nanomaterials in terms of their physicochemical and biological properties has been significantly investigated for improved wound healing. The combination of natural polymers, and biodegradable and biocompatible nanomaterials are the preferred candidates. 3D bioprinting offers a valuable tool for wound healing, possessing high throughput output, printing repeatability, high resolution, and operational flexibility, but printing highly complex and unique skin structures remains challenging in this field. Development in printing bioinks appropriate for different types of cells is one of the main areas in the advancement of 3D bioprinting. Stem cell therapy is currently an active area of research in wound healing because multipotent cells not only possess substantial proliferative ability but also release cytokines and growth factors crucial for wound healing. Research advancements will result in enhanced wound healing therapies and improved life-threatening circumstances, with the aim of restoring chronic wounds to normalcy, which will have a significantly positive impact on real-world outcomes. In biomedical research, novel therapeutic approaches, tissue repair and regeneration, and improved disease pathways are advancing at a rapid pace due to new cutting-edge technology.

REFERENCES

1. Horch, R. E.; Kopp, J.; Kneser, U.; Beier, J.; Bach, A. D., Tissue engineering of cultured skin substitutes. *Journal of Cellular and Molecular Medicine* **2005,** *9* (3), 592–608.
2. Vig, K.; Chaudhari, A.; Tripathi, S.; Dixit, S.; Sahu, R.; Pillai, S.; Dennis, V. A.; Singh, S. R., Advances in skin regeneration using tissue engineering. *International Journal of Molecular Sciences* **2017,** *18* (4), 789.
3. Zahedi, P.; Rezaeian, I.; Ranaei-Siadat, S. O.; Jafari, S. H.; Supaphol, P., A review on wound dressings with an emphasis on electrospun nanofibrous polymeric bandages. *Polymers for Advanced Technologies* **2010,** *21* (2), 77–95.
4. Weller, C. D.; Team, V.; Sussman, G., First-line interactive wound dressing update: a comprehensive review of the evidence. *Frontiers in Pharmacology* **2020,** *11*, 495485.
5. Aragona, M.; Dekoninck, S.; Rulands, S.; Lenglez, S.; Mascré, G.; Simons, B. D.; Blanpain, C., Defining stem cell dynamics and migration during wound healing in mouse skin epidermis. *Nature Communications* **2017,** *8* (1), 14684.
6. Tottoli, E. M.; Dorati, R.; Genta, I.; Chiesa, E.; Pisani, S.; Conti, B., Skin wound healing process and new emerging technologies for skin wound care and regeneration. *Pharmaceutics* **2020,** *12* (8), 735.
7. Nussbaum, S. R.; Carter, M. J.; Fife, C. E.; DaVanzo, J.; Haught, R.; Nusgart, M.; Cartwright, D., An economic evaluation of the impact, cost, and medicare policy implications of chronic nonhealing wounds. *Value in Health* **2018,** *21* (1), 27–32.
8. Armstrong, D. G.; Swerdlow, M. A.; Armstrong, A. A.; Conte, M. S.; Padula, W. V.; Bus, S. A., Five year mortality and direct costs of care for people with diabetic foot complications are comparable to cancer. *Journal of Foot and Ankle Research* **2020,** *13*, 1–4.
9. Jeffcoate, W. J.; Vileikyte, L.; Boyko, E. J.; Armstrong, D. G.; Boulton, A. J., Current challenges and opportunities in the prevention and management of diabetic foot ulcers. *Diabetes Care* **2018,** *41* (4), 645–652.
10. Huang, Y.-Z.; Gou, M.; Da, L.-C.; Zhang, W.-Q.; Xie, H.-Q., Mesenchymal stem cells for chronic wound healing: current status of preclinical and clinical studies. *Tissue Engineering Part B: Reviews* **2020,** *26* (6), 555–570.
11. Monavarian, M.; Kader, S.; Moeinzadeh, S.; Jabbari, E., Regenerative scar-free skin wound healing. *Tissue Engineering Part B: Reviews* **2019,** *25* (4), 294–311.
12. Han, G.; Ceilley, R., Chronic wound healing: a review of current management and treatments. *Advances in Therapy* **2017,** *34*, 599–610.
13. Okur, M. E.; Karantas, I. D.; Şenyiğit, Z.; Okur, N. Ü.; Siafaka, P. I., Recent trends on wound management: new therapeutic choices based on polymeric carriers. *Asian Journal of Pharmaceutical Sciences* **2020,** *15* (6), 661–684.
14. Wang, E. T.; Sandberg, R.; Luo, S.; Khrebtukova, I.; Zhang, L.; Mayr, C.; Kingsmore, S. F.; Schroth, G. P.; Burge, C. B., Alternative isoform regulation in human tissue transcriptomes. *Nature* **2008,** *456* (7221), 470–476.
15. Shanmugasundaram, T.; Radhakrishnan, M.; Gopikrishnan, V.; Kadirvelu, K.; Balagurunathan, R., In vitro antimicrobial and in vivo wound healing effect of actinobacterially synthesised nanoparticles of silver, gold and their alloy. *RSC Advances* **2017,** *7* (81), 51729–51743.

16. Stoica, A. E.; Chircov, C.; Grumezescu, A. M., Nanomaterials for wound dressings: an up-to-date overview. *Molecules* **2020,** *25* (11), 2699.
17. Kalantari, K.; Mostafavi, E.; Afifi, A. M.; Izadiyan, Z.; Jahangirian, H.; Rafiee-Moghaddam, R.; Webster, T. J., Wound dressings functionalized with silver nanoparticles: promises and pitfalls. *Nanoscale* **2020,** *12* (4), 2268–2291.
18. Allen, T. M.; Cullis, P. R., Liposomal drug delivery systems: from concept to clinical applications. *Advanced Drug Delivery Reviews* **2013,** *65* (1), 36–48.
19. Bespalova, Y.; Kwon, D.; Vasanthan, N., Surface modification and antimicrobial properties of cellulose nanocrystals. *Journal of Applied Polymer Science* **2017,** *134* (18).
20. Gehring, J.; Schleheck, D.; Trepka, B.; Polarz, S., Mesoporous organosilica nanoparticles containing superacid and click functionalities leading to cooperativity in biocidal coatings. *ACS Applied Materials & Interfaces* **2015,** *7* (1), 1021–1029.
21. Seisenbaeva, G. A.; Fromell, K.; Vinogradov, V. V.; Terekhov, A. N.; Pakhomov, A. V.; Nilsson, B.; Ekdahl, K. N.; Vinogradov, V. V.; Kessler, V. G., Dispersion of TiO2 nanoparticles improves burn wound healing and tissue regeneration through specific interaction with blood serum proteins. *Scientific Reports* **2017,** *7* (1), 15448.
22. Jahromi, M. A. M.; Zangabad, P. S.; Basri, S. M. M.; Zangabad, K. S.; Ghamarypour, A.; Aref, A. R.; Karimi, M.; Hamblin, M. R., Nanomedicine and advanced technologies for burns: preventing infection and facilitating wound healing. *Advanced Drug Delivery Reviews* **2018,** *123*, 33–64.
23. Nam, G.; Rangasamy, S.; Purushothaman, B.; Song, J. M., The application of bactericidal silver nanoparticles in wound treatment. *Nanomaterials and Nanotechnology* **2015,** *5*, 23.
24. Lee, N.-Y.; Ko, W.-C.; Hsueh, P.-R., Nanoparticles in the treatment of infections caused by multidrug-resistant organisms. *Frontiers in Pharmacology* **2019,** *10*, 452171.
25. Huh, A. J.; Kwon, Y. J., "Nanoantibiotics": a new paradigm for treating infectious diseases using nanomaterials in the antibiotics resistant era. *Journal of Controlled Release* **2011,** *156* (2), 128–145.
26. Niño-Martínez, N.; Salas Orozco, M. F.; Martínez-Castañón, G.-A.; Torres Méndez, F.; Ruiz, F., Molecular mechanisms of bacterial resistance to metal and metal oxide nanoparticles. *International Journal of Molecular Sciences* **2019,** *20* (11), 2808.
27. Wang, W.; Lu, K.-J.; Yu, C.-H.; Huang, Q.-L.; Du, Y.-Z., Nano-drug delivery systems in wound treatment and skin regeneration. *Journal of Nanobiotechnology* **2019,** *17* (1), 82.
28. Sánchez-López, E.; Gomes, D.; Esteruelas, G.; Bonilla, L.; Lopez-Machado, A. L.; Galindo, R.; Cano, A.; Espina, M.; Ettcheto, M.; Camins, A., Metal-based nanoparticles as antimicrobial agents: an overview. *Nanomaterials* **2020,** *10* (2), 292.
29. Arora, N.; Thangavelu, K.; Karanikolos, G. N., Bimetallic nanoparticles for antimicrobial applications. *Frontiers in Chemistry* **2020,** *8*, 412.
30. Shkodenko, L.; Kassirov, I.; Koshel, E., Metal oxide nanoparticles against bacterial biofilms: perspectives and limitations. *Microorganisms* **2020,** *8* (10), 1545.
31. McDonnell, G.; Russell, A. D., Antiseptics and disinfectants: activity, action, and resistance. *Clinical Microbiology Reviews* **1999,** *12* (1), 147–179.
32. Rodrigues, M.; Kosaric, N.; Bonham, C. A.; Gurtner, G. C., Wound healing: a cellular perspective. *Physiological Reviews* **2019,** *99* (1), 665–706.
33. Natarajan, V. T.; Ganju, P.; Ramkumar, A.; Grover, R.; Gokhale, R. S., Multifaceted pathways protect human skin from UV radiation. *Nature Chemical Biology* **2014,** *10* (7), 542–551.
34. Menon, G. K.; Dryer, L.; Kalafsky, R., Approaches to the development of cosmetic products to counter the effects of skin aging. In Nava Dayan (Ed.), *Skin Aging Handbook*, Elsevier: 2009; pp 265–290.
35. Pasparakis, M.; Haase, I.; Nestle, F. O., Mechanisms regulating skin immunity and inflammation. *Nature Reviews Immunology* **2014,** *14* (5), 289–301.
36. Shirakata, Y., Regulation of epidermal keratinocytes by growth factors. *Journal of Dermatological Science* **2010,** *59* (2), 73–80.
37. Ter Horst, B.; Chouhan, G.; Moiemen, N. S.; Grover, L. M., Advances in keratinocyte delivery in burn wound care. *Advanced Drug Delivery Reviews* **2018,** *123*, 18–32.
38. Candi, E.; Schmidt, R.; Melino, G., The cornified envelope: a model of cell death in the skin. *Nature Reviews Molecular Cell Biology* **2005,** *6* (4), 328–340.
39. Usui, M. L.; Mansbridge, J. N.; Carter, W. G.; Fujita, M.; Olerud, J. E., Keratinocyte migration, proliferation, and differentiation in chronic ulcers from patients with diabetes and normal wounds. *Journal of Histochemistry & Cytochemistry* **2008,** *56* (7), 687–696.
40. Turner, C. T.; Hiroyasu, S.; Granville, D. J., Granzyme B as a therapeutic target for wound healing. *Expert Opinion on Therapeutic Targets* **2019,** *23* (9), 745–754.

41. Stecco, C.; Pirri, C.; Fede, C.; Fan, C.; Giordani, F.; Stecco, L.; Foti, C.; De Caro, R., Dermatome and fasciatome. *Clinical Anatomy* **2019,** *32* (7), 896–902.
42. Kolimi, P.; Narala, S.; Nyavanandi, D.; Youssef, A. A. A.; Dudhipala, N., Innovative treatment strategies to accelerate wound healing: trajectory and recent advancements. *Cells* **2022,** *11* (15), 2439.
43. Stan, D.; Tanase, C.; Avram, M.; Apetrei, R.; Mincu, N. B.; Mateescu, A. L.; Stan, D., Wound healing applications of creams and "smart" hydrogels. *Experimental Dermatology* **2021,** *30* (9), 1218–1232.
44. Hsu, H.; Siwiec, R., *Knee Osteoarthritis.* StatPearls Publishing; 2022.
45. Wong, R.; Geyer, S.; Weninger, W.; Guimberteau, J. C.; Wong, J. K., The dynamic anatomy and patterning of skin. *Experimental Dermatology* **2016,** *25* (2), 92–98.
46. Brunke, M.; Broadhurst, M.; Oliver, K.; Levine, D., Manual therapy in small animal rehabilitation. *Advances in Small Animal Care* **2021,** *2*, 19–30.
47. Kim, J.; Simon, R., Calculated decisions: wound closure classification. *Pediatric Emergency Medicine Practice* **2018,** *14* (Suppl 10), 1–3.
48. Kuhlmann, M.; Wigger-Alberti, W.; Mackensen, Y.; Ebbinghaus, M.; Williams, R.; Krause-Kyora, F.; Wolber, R., Wound healing characteristics of a novel wound healing ointment in an abrasive wound model: a randomised, intra-individual clinical investigation. *Wound Medicine* **2019,** *24* (1), 24–32.
49. Tort, S.; Demiröz, F. T.; Cevher, Ş. C.; Sarıbaş, S.; Özoğul, C.; Acartürk, F., The effect of a new wound dressing on wound healing: biochemical and histopathological evaluation. *Burns* **2020,** *46* (1), 143–155.
50. Das, A.; Abas, M.; Biswas, N.; Banerjee, P.; Ghosh, N.; Rawat, A.; Khanna, S.; Roy, S.; Sen, C. K., A modified collagen dressing induces transition of inflammatory to reparative phenotype of wound macrophages. *Scientific Reports* **2019,** *9* (1), 14293.
51. Liao, X.; Liang, J.-X.; Li, S.-H.; Huang, S.; Yan, J.-X.; Xiao, L.-L.; Song, J.-X.; Liu, H.-W., Allogeneic platelet-rich plasma therapy as an effective and safe adjuvant method for chronic wounds. *Journal of Surgical Research* **2020,** *246*, 284–291.
52. Cañedo-Dorantes, L.; Cañedo-Ayala, M., Skin acute wound healing: a comprehensive review. *International Journal of Inflammation* **2019.**
53. Desmet, C. M.; Préat, V.; Gallez, B., Nanomedicines and gene therapy for the delivery of growth factors to improve perfusion and oxygenation in wound healing. *Advanced Drug Delivery Reviews* **2018,** *129*, 262–284.
54. Eming, S. A.; Martin, P.; Tomic-Canic, M., Wound repair and regeneration: mechanisms, signaling, and translation. *Science Translational Medicine* **2014,** *6* (265), 265sr6–265sr6.
55. Chhabra, S.; Chhabra, N.; Kaur, A.; Gupta, N., Wound healing concepts in clinical practice of OMFS. *Journal of Maxillofacial and Oral Surgery* **2017,** *16*, 403–423.
56. Diller, R. B.; Tabor, A. J., The role of the extracellular matrix (ECM) in wound healing: a review. *Biomimetics* **2022,** *7* (3), 87.
57. El-Ashram, S.; El-Samad, L. M.; Basha, A. A.; El Wakil, A., Naturally-derived targeted therapy for wound healing: beyond classical strategies. *Pharmacological Research* **2021,** *170*, 105749.
58. Kanji, S.; Das, H., Advances of stem cell therapeutics in cutaneous wound healing and regeneration. *Mediators of Inflammation* **2017,** *2017.*
59. Kawasumi, A.; Sagawa, N.; Hayashi, S.; Yokoyama, H.; Tamura, K., Wound healing in mammals and amphibians: toward limb regeneration in mammals. *New Perspectives in Regeneration* **2013**, 33–49.
60. Wu, L.; Chen, X.; Zhao, J.; Martin, B.; Zepp, J. A.; Ko, J. S.; Gu, C.; Cai, G.; Ouyang, W.; Sen, G., A novel IL-17 signaling pathway controlling keratinocyte proliferation and tumorigenesis via the TRAF4–ERK5 axis. *Journal of Experimental Medicine* **2015,** *212* (10), 1571–1587.
61. Zhao, R.; Liang, H.; Clarke, E.; Jackson, C.; Xue, M., Inflammation in chronic wounds. *International Journal of Molecular Sciences* **2016,** *17* (12), 2085.
62. Barman, P. K.; Koh, T. J., Macrophage dysregulation and impaired skin wound healing in diabetes. *Frontiers in Cell and Developmental Biology* **2020,** *8*, 528.
63. Krzyszczyk, P.; Schloss, R.; Palmer, A.; Berthiaume, F., The role of macrophages in acute and chronic wound healing and interventions to promote pro-wound healing phenotypes. *Frontiers in Physiology* **2018,** *9*, 419.
64. Barchitta, M.; Maugeri, A.; Favara, G.; Magnano San Lio, R.; Evola, G.; Agodi, A.; Basile, G., Nutrition and wound healing: an overview focusing on the beneficial effects of curcumin. *International Journal of Molecular Sciences* **2019,** *20* (5), 1119.
65. Den Dekker, A.; Davis, F. M.; Kunkel, S. L.; Gallagher, K. A., Targeting epigenetic mechanisms in diabetic wound healing. *Translational Research* **2019,** *204*, 39–50.
66. Raja, S. K.; Garcia, M. S.; Isseroff, R. R., Wound re-epithelialization: modulating keratinocyte migration in wound healing. *Front Biosci* **2007,** *12* (3), 2849–2868.

67. Garraud, O.; Hozzein, W. N.; Badr, G., Wound healing: time to look for intelligent,'natural' immunological approaches? *BMC Immunology* **2017,** *18*, 1–8.
68. Darby, I. A.; Laverdet, B.; Bonté, F.; Desmoulière, A., Fibroblasts and myofibroblasts in wound healing. *Clinical, Cosmetic and Investigational Dermatology* **2014**, 301–311.
69. Larouche, J.; Sheoran, S.; Maruyama, K.; Martino, M. M., Immune regulation of skin wound healing: mechanisms and novel therapeutic targets. *Advances in Wound Care* **2018,** *7* (7), 209–231.
70. Zhang, K.; Lui, V. C.; Chen, Y.; Lok, C. N.; Wong, K. K., Delayed application of silver nanoparticles reveals the role of early inflammation in burn wound healing. *Scientific Reports* **2020,** *10* (1), 6338.
71. Aydemir, I.; Öztürk, Ş.; Sönmez, P. K.; Tuğlu, M. İ., Mesenchymal stem cells in skin wound healing. *Anatomy* **2016,** *10* (3), 228–234.
72. Leavitt, T.; Hu, M. S.; Marshall, C. D.; Barnes, L. A.; Longaker, M. T.; Lorenz, H. P., Stem cells and chronic wound healing: state of the art. *Chronic Wound Care Management and Research* **2016**, 7–27.
73. Negut, I.; Grumezescu, V.; Grumezescu, A. M., Treatment strategies for infected wounds. *Molecules* **2018,** *23* (9), 2392.
74. Lindholm, C.; Searle, R., Wound management for the 21st century: combining effectiveness and efficiency. *International Wound Journal* **2016,** *13*, 5–15.
75. Dowsett, C.; Newton, H., Wound bed preparation: TIME in practice. *Wounds Uk* **2005,** *1* (3), 58.
76. Dowsett, C.; Gronemann, M.; Harding, K., Taking wound assessment beyond the edge. *Wounds International* **2015,** *6* (1), 19–23.
77. Gushiken, L. F. S.; Beserra, F. P.; Bastos, J. K.; Jackson, C. J.; Pellizzon, C. H., Cutaneous wound healing: an update from physiopathology to current therapies. *Life* **2021,** *11* (7), 665.
78. Bellingeri, A.; Falciani, F.; Traspedini, P.; Moscatelli, A.; Russo, A.; Tino, G.; Chiari, P.; Peghetti, A., Effect of a wound cleansing solution on wound bed preparation and inflammation in chronic wounds: a single-blind RCT. *Journal of Wound Care* **2016,** *25* (3), 160–168.
79. Lumbers, M., Wound debridement: choices and practice. *British Journal of Nursing* **2018,** *27* (15), S16–S20.
80. David, J. A.; Chiu, E. S., Surgical debridement. *Interventional Treatment of Wounds: A Modern Approach for Better Outcomes* **2018**, 3–15.
81. Ferrario, C.; Rusconi, F.; Pulaj, A.; Macchi, R.; Landini, P.; Paroni, M.; Colombo, G.; Martinello, T.; Melotti, L.; Gomiero, C., From food waste to innovative biomaterial: sea urchin-derived collagen for applications in skin regenerative medicine. *Marine Drugs* **2020,** *18* (8), 414.
82. Varkey, M.; Ding, J.; Tredget, E. E., Advances in skin substitutes—potential of tissue engineered skin for facilitating anti-fibrotic healing. *Journal of Functional Biomaterials* **2015,** *6* (3), 547–563.
83. Powers, J. G.; Higham, C.; Broussard, K.; Phillips, T. J., Wound healing and treating wounds: chronic wound care and management. *Journal of the American Academy of Dermatology* **2016,** *74* (4), 607–625.
84. Xu, K.; Chai, B.; Zhang, K.; Xiong, J.; Zhu, Y.; Xu, J.; An, N.; Xia, W.; Ji, H.; Wu, Y., Topical application of fibroblast growth factor 10-PLGA microsphere accelerates wound healing via inhibition of ER stress. *Oxidative Medicine and Cellular Longevity* **2020,** *2020.*
85. Dhivya, S.; Padma, V. V.; Santhini, E., Wound dressings–a review. *BioMedicine* **2015,** *5* (4), 22.
86. Pal, K.; Sarkar, P.; Thakur, G.; Kim, D., Essential oil-loaded biopolymeric films for wound healing applications. In Kunal Pal, Sarika Verma, Pallab Datta, Ananya Barui, S. A. R. Hashmi, & Avanish Kumar Srivastava (Eds.), *Advances in Biomedical Polymers and Composites*. Elsevier; 2023, pp. 541–562.
87. Michelin, R. M.; Ahdoot, E.; Zakhary, B. L.; McDowell, M.; French, M., Choosing the optimal wound dressing for bathing after total knee arthroplasty. *The Journal of Arthroplasty* **2021,** *36* (3), 970–977.
88. Cheng, L.; Cai, Z.; Ye, T.; Yu, X.; Chen, Z.; Yan, Y.; Qi, J.; Wang, L.; Liu, Z.; Cui, W., Injectable polypeptide-protein hydrogels for promoting infected wound healing. *Advanced Functional Materials* **2020,** *30* (25), 2001196.
89. Zhang, M.; Huang, Y.; Pan, W.; Tong, X.; Zeng, Q.; Su, T.; Qi, X.; Shen, J., Polydopamine-incorporated dextran hydrogel drug carrier with tailorable structure for wound healing. *Carbohydrate Polymers* **2021,** *253*, 117213.
90. He, J.; Shi, M.; Liang, Y.; Guo, B., Conductive adhesive self-healing nanocomposite hydrogel wound dressing for photothermal therapy of infected full-thickness skin wounds. *Chemical Engineering Journal* **2020,** *394*, 124888.
91. Goodarzi, P.; Falahzadeh, K.; Nematizadeh, M.; Farazandeh, P.; Payab, M.; Larijani, B.; Tayanloo Beik, A.; Arjmand, B., Tissue engineered skin substitutes. *Cell Biology and Translational Medicine, Volume 3: Stem Cells, Bio-materials and Tissue Engineering* **2018**, 143–188.
92. Still, J.; Glat, P.; Silverstein, P.; Griswold, J.; Mozingo, D., The use of a collagen sponge/living cell composite material to treat donor sites in burn patients. *Burns* **2003,** *29* (8), 837–841.

93. Zeng, R.; Lin, C.; Lin, Z.; Chen, H.; Lu, W.; Lin, C.; Li, H., Approaches to cutaneous wound healing: basics and future directions. *Cell and Tissue Research* **2018,** *374*, 217–232.
94. Li, H.; Li, B.; Ma, J.; Ye, J.; Guo, P.; Li, L., Fate of antibiotic-resistant bacteria and antibiotic resistance genes in the electrokinetic treatment of antibiotic-polluted soil. *Chemical Engineering Journal* **2018,** *337*, 584–594.
95. Gao, W.; Chen, Y.; Zhang, Y.; Zhang, Q.; Zhang, L., Nanoparticle-based local antimicrobial drug delivery. *Advanced Drug Delivery Reviews* **2018,** *127*, 46–57.
96. Rajendran, N. K.; Kumar, S. S. D.; Houreld, N. N.; Abrahamse, H., A review on nanoparticle based treatment for wound healing. *Journal of Drug Delivery Science and Technology* **2018,** *44*, 421–430.
97. Rahman, M.; Kamal, M. A. *Cancer Nanotherapeutics: Targeted Medicine, Therapeutic Vaccination and Challenges with Cancer Nanomedicines.* Seminars in Cancer Biology; 2021, pp 1–4.
98. Huang, R.; Hu, J.; Qian, W.; Chen, L.; Zhang, D., Recent advances in nanotherapeutics for the treatment of burn wounds. *Burns & Trauma* **2021,** *9*, tkab026.
99. Malekzad, H.; Mirshekari, H.; Sahandi Zangabad, P.; Moosavi Basri, S.; Baniasadi, F.; Sharifi Aghdam, M.; Karimi, M.; Hamblin, M. R., Plant protein-based hydrophobic fine and ultrafine carrier particles in drug delivery systems. *Critical Reviews in Biotechnology* **2018,** *38* (1), 47–67.
100. Debone, H. S.; Lopes, P. S.; Severino, P.; Yoshida, C. M. P.; Souto, E. B.; da Silva, C. F., Chitosan/Copaiba oleoresin films for would dressing application. *International Journal of Pharmaceutics* **2019,** *555*, 146–152.
101. Hayat, M.; Nawaz, A.; Chinnam, S.; Muzammal, M.; Latif, M. S.; Yasin, M.; Ashique, S.; Zengin, G.; Farid, A., Formulation development and optimization of herbo synthetic gel: in vitro biological evaluation and in vivo wound healing studies. *Process Biochemistry* **2023** Jul 1**,** 130, 116–126.
102. Safdar, M. H.; Hussain, Z.; Abourehab, M. A.; Hasan, H.; Afzal, S.; Thu, H. E., New developments and clinical transition of hyaluronic acid-based nanotherapeutics for treatment of cancer: reversing multidrug resistance, tumour-specific targetability and improved anticancer efficacy. *Artificial Cells, Nanomedicine, and Biotechnology* **2018,** *46* (8), 1967–1980.
103. Rahim, M. A.; Jan, N.; Khan, S.; Shah, H.; Madni, A.; Khan, A.; Jabar, A.; Khan, S.; Elhissi, A.; Hussain, Z., Recent advancements in stimuli responsive drug delivery platforms for active and passive cancer targeting. *Cancers* **2021,** *13* (4), 670.
104. Hussain, Z.; Thu, H. E.; Rawas-Qalaji, M.; Naseem, M.; Khan, S.; Sohail, M., Recent developments and advanced strategies for promoting burn wound healing. *Journal of Drug Delivery Science and Technology* **2022,** *68*, 103092.
105. Zhao, Y.-Z.; Lu, C.-T.; Zhang, Y.; Xiao, J.; Zhao, Y.-P.; Tian, J.-L.; Xu, Y.-Y.; Feng, Z.-G.; Xu, C.-Y., Selection of high efficient transdermal lipid vesicle for curcumin skin delivery. *International Journal of Pharmaceutics* **2013,** *454* (1), 302–309.
106. Shaikh, S.; Nazam, N.; Rizvi, S. M. D.; Ahmad, K.; Baig, M. H.; Lee, E. J.; Choi, I., Mechanistic insights into the antimicrobial actions of metallic nanoparticles and their implications for multidrug resistance. *International Journal of Molecular Sciences* **2019,** *20* (10), 2468.
107. Sharma, M., Transdermal and intravenous nano drug delivery systems: present and future. In Shyam S. Mohapatra, Shivendu Ranjan, Nandita Dasgupta, Raghvendra Kumar Mishra, & Sabu Thomas (Eds.), *Applications of Targeted Nano Drugs and Delivery Systems.* Elsevier; 2019, pp. 499–550.
108. Gabrielyan, L.; Hovhannisyan, A.; Gevorgyan, V.; Ananyan, M.; Trchounian, A., Antibacterial effects of iron oxide (Fe3O4) nanoparticles: distinguishing concentration-dependent effects with different bacterial cells growth and membrane-associated mechanisms. *Applied Microbiology and Biotechnology* **2019,** *103* (6), 2773–2782.
109. Das, S.; Baker, A. B., Biomaterials and nanotherapeutics for enhancing skin wound healing. *Frontiers in Bioengineering and Biotechnology* **2016,** *4*, 82.
110. Cao, L.; Shao, G.; Ren, F.; Yang, M.; Nie, Y.; Peng, Q.; Zhang, P., Cerium oxide nanoparticle-loaded polyvinyl alcohol nanogels delivery for wound healing care systems on surgery. *Drug Delivery* **2021,** *28* (1), 390.
111. Abazari, M.; Ghaffari, A.; Rashidzadeh, H.; Momeni badeleh, S.; Maleki, Y., Current status and future outlook of nano-based systems for burn wound management. *Journal of Biomedical Materials Research Part B: Applied Biomaterials* **2020,** *108* (5), 1934–1952.
112. Gainza, G.; Villullas, S.; Pedraz, J. L.; Hernandez, R. M.; Igartua, M., Advances in drug delivery systems (DDSs) to release growth factors for wound healing and skin regeneration. *Nanomedicine: Nanotechnology, Biology and Medicine* **2015,** *11* (6), 1551–1573.
113. Bruna, T.; Maldonado-Bravo, F.; Jara, P.; Caro, N., Silver nanoparticles and their antibacterial applications. *International Journal of Molecular Sciences* **2021,** *22* (13), 7202.

114. Riazunnisa, K., Antimicrobial activity of Biosynthesized silver nanoparticles of Bauhinia racemosa leaf extracts. *Research Journal of Pharmacy and Technology* **2023,** *16* (2), 745–749.
115. Hammami, I.; Alabdallah, N. M., Gold nanoparticles: synthesis properties and applications. *Journal of King Saud University-Science* **2021,** *33* (7), 101560.
116. Hashem, A. H.; Shehabeldine, A. M.; Ali, O. M.; Salem, S. S., Synthesis of chitosan-based gold nanoparticles: antimicrobial and wound-healing activities. *Polymers* **2022,** *14* (11), 2293.
117. Salvo, J.; Sandoval, C., Role of copper nanoparticles in wound healing for chronic wounds: literature review. *Burns & Trauma* **2022,** *10*, tkab047.
118. Yimeng, S.; Huilun, X.; Ziming, L.; Kejun, L.; Chaima, M.; Xiangyu, Z.; Yinchun, H.; Yan, W.; Di, H., Copper-based nanoparticles as antibacterial agents. *European Journal of Inorganic Chemistry* **2023,** *26* (4), e202200614.
119. Manzano, M.; Vallet-Regí, M., Mesoporous silica nanoparticles for drug delivery. *Advanced Functional Materials* **2020,** *30* (2), 1902634.
120. Selvarajan, V.; Obuobi, S.; Ee, P. L. R., Silica nanoparticles—a versatile tool for the treatment of bacterial infections. *Frontiers in Chemistry* **2020,** *8*, 602.
121. Souza, R. C. d.; Haberbeck, L. U.; Riella, H. G.; Ribeiro, D. H.; Carciofi, B. A., Antibacterial activity of zinc oxide nanoparticles synthesized by solochemical process. *Brazilian Journal of Chemical Engineering* **2019,** *36*, 885–893.
122. Husain, F. M.; Qais, F. A.; Ahmad, I.; Hakeem, M. J.; Baig, M. H.; Masood Khan, J.; Al-Shabib, N. A., Biosynthesized zinc oxide nanoparticles disrupt established biofilms of pathogenic bacteria. *Applied Sciences* **2022,** *12* (2), 710.
123. Jassal, P. S.; Kaur, D.; Prasad, R.; Singh, J., Green synthesis of titanium dioxide nanoparticles: development and applications. *Journal of Agriculture and Food Research* **2022,** *10*, 100361.
124. Waghmode, M. S.; Gunjal, A. B.; Mulla, J. A.; Patil, N. N.; Nawani, N. N., Studies on the titanium dioxide nanoparticles: biosynthesis, applications and remediation. *SN Applied Sciences* **2019,** *1* (4), 310.
125. Zia ul Quasim, S.; Ali, I.; Irfan, S.; Naveed, A., Advances in drug delivery system: carbon nanotubes. *Research Journal of Pharmacy and Technology* **2013,** *6* (2), 125–129.
126. Bakthavachalam, A.; Remya, P.; Damodharan, N., Review on solid lipid nanoparticles. *Research Journal of Pharmacy and Technology* **2020,** *13* (9), 4430–4434.
127. Prabhakar, C.; Krishna, K. B., A review on polymeric nanoparticles. *Research Journal of Pharmacy and Technology* **2011,** *4* (4), 496–498.
128. Kell, A. J.; Stewart, G.; Ryan, S.; Peytavi, R.; Boissinot, M.; Huletsky, A.; Bergeron, M. G.; Simard, B., Vancomycin-modified nanoparticles for efficient targeting and preconcentration of Gram-positive and Gram-negative bacteria. *Acs Nano* **2008,** *2* (9), 1777–1788.
129. Gong, C.; Wu, Q.; Wang, Y.; Zhang, D.; Luo, F.; Zhao, X.; Wei, Y.; Qian, Z., A biodegradable hydrogel system containing curcumin encapsulated in micelles for cutaneous wound healing. *Biomaterials* **2013,** *34* (27), 6377–6387.
130. Wathoni, N.; Motoyama, K.; Higashi, T.; Okajima, M.; Kaneko, T.; Arima, H., Enhancement of curcumin wound healing ability by complexation with 2-hydroxypropyl-γ-cyclodextrin in sacran hydrogel film. *International Journal of Biological Macromolecules* **2017,** *98*, 268–276.
131. Dickinson, L. E.; Gerecht, S., Engineered biopolymeric scaffolds for chronic wound healing. *Frontiers in Physiology* **2016,** *7*, 213166.
132. Duscher, D.; Barrera, J.; Wong, V. W.; Maan, Z. N.; Whittam, A. J.; Januszyk, M.; Gurtner, G. C., Stem cells in wound healing: the future of regenerative medicine? A mini-review. *Gerontology* **2016,** *62* (2), 216–225.
133. Chen, M.; Przyborowski, M.; Berthiaume, F., Stem cells for skin tissue engineering and wound healing. *Critical Reviews™ in Biomedical Engineering* **2009,** *37* (4–5).
134. Kølle, S.-F. T.; Fischer-Nielsen, A.; Mathiasen, A. B.; Elberg, J. J.; Oliveri, R. S.; Glovinski, P. V.; Kastrup, J.; Kirchhoff, M.; Rasmussen, B. S.; Talman, M.-L. M., Enrichment of autologous fat grafts with ex-vivo expanded adipose tissue-derived stem cells for graft survival: a randomised placebo-controlled trial. *The Lancet* **2013,** *382* (9898), 1113–1120.
135. Tsai, H.-W.; Wang, P.-H.; Tsui, K.-H., Mesenchymal stem cell in wound healing and regeneration. *LWW* 2018**,** *81*, 223–224.
136. Hu, M. S.; Borrelli, M. R.; Lorenz, H. P.; Longaker, M. T.; Wan, D. C., Mesenchymal stromal cells and cutaneous wound healing: a comprehensive review of the background, role, and therapeutic potential. *Stem Cells International* **2018,** *2018*.
137. Dash, B. C.; Xu, Z.; Lin, L.; Koo, A.; Ndon, S.; Berthiaume, F.; Dardik, A.; Hsia, H., Stem cells and engineered scaffolds for regenerative wound healing. *Bioengineering* **2018,** *5* (1), 23.

138. Yu, J.; Wang, M.-Y.; Tai, H.-C.; Cheng, N.-C., Cell sheet composed of adipose-derived stem cells demonstrates enhanced skin wound healing with reduced scar formation. *Acta Biomaterialia* **2018,** *77,* 191–200.
139. Mazini, L.; Rochette, L.; Admou, B.; Amal, S.; Malka, G., Hopes and limits of adipose-derived stem cells (ADSCs) and mesenchymal stem cells (MSCs) in wound healing. *International Journal of Molecular Sciences* **2020,** *21* (4), 1306.
140. Rasulov, M.; Vasil'Chenkov, A.; Onishchenko, N.; Krasheninnikov, M.; Kravchenko, V.; Gorshenin, T.; Pidtsan, R.; Potapov, I., First experience in the use of bone marrow mesenchymal stem cells for the treatment of a patient with deep skin burns. *Bulletin of Experimental Biology and Medicine* **2005,** *139,* 141–144.
141. Xu, Y.; Huang, S.; Fu, X., Autologous transplantation of bone marrow–derived mesenchymal stem cells: a promising therapeutic strategy for prevention of skin-graft contraction. *Clinical and Experimental Dermatology* **2012,** *37* (5), 497–500.
142. Vojtaššák, J.; Danišovič, L.; Kubeš, M.; Bakoš, D.; Jarabek, L.; Uličná, M.; Blaško, M., Autologous biograft and mesenchymal stem cells in treatment of the diabetic foot. *Neuroendocrinology Letters* **2006,** *27* (Supplement 2), 134–137.
143. An, Y.; Liu, W.; Xue, P.; Ma, Y.; Zhang, L.; Zhu, B.; Qi, M.; Li, L.; Zhang, Y.; Wang, Q., Autophagy promotes MSC-mediated vascularization in cutaneous wound healing via regulation of VEGF secretion. *Cell Death & Disease* **2018,** *9* (2), 58.
144. Lee, S.; Choi, E.; Cha, M.-J.; Hwang, K.-C., Cell adhesion and long-term survival of transplanted mesenchymal stem cells: a prerequisite for cell therapy. *Oxidative Medicine and Cellular Longevity* **2015,** *2015.*
145. Hamdan, S.; Pastar, I.; Drakulich, S.; Dikici, E.; Tomic-Canic, M.; Deo, S.; Daunert, S., Nanotechnology-driven therapeutic interventions in wound healing: potential uses and applications. *ACS Central Science* **2017,** *3* (3), 163–175.
146. Navone, S. E.; Pascucci, L.; Dossena, M.; Ferri, A.; Invernici, G.; Acerbi, F.; Cristini, S.; Bedini, G.; Tosetti, V.; Ceserani, V., Decellularized silk fibroin scaffold primed with adipose mesenchymal stromal cells improves wound healing in diabetic mice. *Stem Cell Research & Therapy* **2014,** *5,* 1–15.
147. Tuca, A.-C.; Ertl, J.; Hingerl, K.; Pichlsberger, M.; Fuchs, J.; Wurzer, P.; Pfeiffer, D.; Bubalo, V.; Parvizi, D.; Kamolz, L.-P., Comparison of matrigel and matriderm as a carrier for human amnion-derived mesenchymal stem cells in wound healing. *Placenta* **2016,** *48,* 99–103.
148. Goodarzi, P.; Larijani, B.; Alavi-Moghadam, S.; Tayanloo-Beik, A.; Mohamadi-Jahani, F.; Ranjbaran, N.; Payab, M.; Falahzadeh, K.; Mousavi, M.; Arjmand, B., Mesenchymal stem cells-derived exosomes for wound regeneration. *Cell Biology and Translational Medicine, Volume 4: Stem Cells and Cell Based Strategies in Regeneration* **2018**, 119–131.
149. Wang, M.; Xu, X.; Lei, X.; Tan, J.; Xie, H., Mesenchymal stem cell-based therapy for burn wound healing. *Burns & Trauma* **2021,** *9,* tkab002.
150. Gao, F.; Li, W.; Deng, J.; Kan, J.; Guo, T.; Wang, B.; Hao, S., Recombinant human hair keratin nanoparticles accelerate dermal wound healing. *ACS Applied Materials & Interfaces* **2019,** *11* (20), 18681–18690.
151. Guasch, G., The epithelial stem cell niche in skin. In Ajaykumar Vishwakarma & Jeffrey M. Karp (Eds.), *Biology and Engineering of Stem Cell Niches.* Elsevier; 2017, pp. 127–143.
152. Heidari, F.; Yari, A.; Rasoolijazi, H.; Soleimani, M.; Dehpoor, A.; Sajedi, N.; Veijouye, S. J.; Nobakht, M., Bulge hair follicle stem cells accelerate cutaneous wound healing in rats. *Wounds: A Compendium of Clinical Research and Practice* **2016,** *28* (4), 132–141.
153. Wang, L.; Su, Y.; Huang, C.; Yin, Y.; Chu, A.; Knupp, A.; Tang, Y., NANOG and LIN28 dramatically improve human cell reprogramming by modulating LIN41 and canonical WNT activities. *Biology Open* **2019,** *8* (12), bio047225.
154. Yamanaka, S., Strategies and new developments in the generation of patient-specific pluripotent stem cells. *Cell Stem Cell* **2007,** *1* (1), 39–49.
155. Yang, R.; Zheng, Y.; Burrows, M.; Liu, S.; Wei, Z.; Nace, A.; Guo, W.; Kumar, S.; Cotsarelis, G.; Xu, X., Generation of folliculogenic human epithelial stem cells from induced pluripotent stem cells. *Nature Communications* **2014,** *5* (1), 3071.
156. Itoh, M.; Umegaki-Arao, N.; Guo, Z.; Liu, L.; Higgins, C. A.; Christiano, A. M., Generation of 3D skin equivalents fully reconstituted from human induced pluripotent stem cells (iPSCs). *PloS One* **2013,** *8* (10), e77673.

157. Umegaki-Arao, N.; Pasmooij, A. M.; Itoh, M.; Cerise, J. E.; Guo, Z.; Levy, B.; Gostyński, A.; Rothman, L. R.; Jonkman, M. F.; Christiano, A. M., Induced pluripotent stem cells from human revertant keratinocytes for the treatment of epidermolysis bullosa. *Science Translational Medicine* **2014,** *6* (264), 264ra164.
158. Yan, Y.; Wu, R.; Bo, Y.; Zhang, M.; Chen, Y.; Wang, X.; Huang, M.; Liu, B.; Zhang, L., Induced pluripotent stem cells-derived microvesicles accelerate deep second-degree burn wound healing in mice through miR-16-5p-mediated promotion of keratinocytes migration. *Theranostics* **2020,** *10* (22), 9970.
159. Javaid, M.; Haleem, A., 3D bioprinting applications for the printing of skin: a brief study. *Sensors International* **2021,** *2*, 100123.
160. Rosenberg, E. S.; Dufort, E. M.; Udo, T.; Wilberschied, L. A.; Kumar, J.; Tesoriero, J.; Weinberg, P.; Kirkwood, J.; Muse, A.; DeHovitz, J., Association of treatment with hydroxychloroquine or azithromycin with in-hospital mortality in patients with COVID-19 in New York State. *Jama* **2020,** *323* (24), 2493–2502.
161. He, P.; Zhao, J.; Zhang, J.; Li, B.; Gou, Z.; Gou, M.; Li, X., Bioprinting of skin constructs for wound healing. *Burns & Trauma* **2018,** *6*.
162. Ng, W. L.; Wang, S.; Yeong, W. Y.; Naing, M. W., Skin bioprinting: impending reality or fantasy? *Trends in Biotechnology* **2016,** *34* (9), 689–699.
163. Robin, A.; Nandakumar, K.; Sabu, T., Microbial barrier property and blood compatibility studies of electrospun Poly-ε-caprolactone/zinc oxide nanocomposite scaffolds. *Журнал Сибирского федерального университета. Биология* **2017,** *10* (2), 226–236.
164. Murphy, S. V.; Atala, A., 3D bioprinting of tissues and organs. *Nature Biotechnology* **2014,** *32* (8), 773–785.
165. Fisch, P.; Holub, M.; Zenobi-Wong, M., Improved accuracy and precision of bioprinting through progressive cavity pump-controlled extrusion. *Biofabrication* **2020,** *13* (1), 015012.
166. Vijayavenkataraman, S.; Yan, W.-C.; Lu, W. F.; Wang, C.-H.; Fuh, J. Y. H., 3D bioprinting of tissues and organs for regenerative medicine. *Advanced Drug Delivery Reviews* **2018,** *132*, 296–332.
167. Ishack, S.; Lipner, S. R., A review of 3-dimensional skin bioprinting techniques: applications, approaches, and trends. *Dermatologic Surgery* **2020,** *46* (12), 1500–1505.
168. Perez-Valle, A.; Del Amo, C.; Andia, I., Overview of current advances in extrusion bioprinting for skin applications. *International Journal of Molecular Sciences* **2020,** *21* (18), 6679.
169. Matai, I.; Kaur, G.; Seyedsalehi, A.; McClinton, A.; Laurencin, C. T., Progress in 3D bioprinting technology for tissue/organ regenerative engineering. *Biomaterials* **2020,** *226*, 119536.
170. Ning, L.; Chen, X., A brief review of extrusion-based tissue scaffold bio-printing. *Biotechnology Journal* **2017,** *12* (8), 1600671.
171. Tumbleston, J. R.; Shirvanyants, D.; Ermoshkin, N.; Janusziewicz, R.; Johnson, A. R.; Kelly, D.; Chen, K.; Pinschmidt, R.; Rolland, J. P.; Ermoshkin, A., Continuous liquid interface production of 3D objects. *Science* **2015,** *347* (6228), 1349–1352.
172. Skylar-Scott, M. A.; Mueller, J.; Visser, C. W.; Lewis, J. A., Voxelated soft matter via multimaterial multinozzle 3D printing. *Nature* **2019,** *575* (7782), 330–335.
173. Kelly, B. E.; Bhattacharya, I.; Heidari, H.; Shusteff, M.; Spadaccini, C. M.; Taylor, H. K., Volumetric additive manufacturing via tomographic reconstruction. *Science* **2019,** *363* (6431), 1075–1079.
174. Kim, B. S.; Lee, J.-S.; Gao, G.; Cho, D.-W., Direct 3D cell-printing of human skin with functional transwell system. *Biofabrication* **2017,** *9* (2), 025034.
175. Hospodiuk, M.; Dey, M.; Sosnoski, D.; Ozbolat, I. T., The bioink: a comprehensive review on bioprintable materials. *Biotechnology Advances* **2017,** *35* (2), 217–239.
176. Groll, J.; Burdick, J. A.; Cho, D.-W.; Derby, B.; Gelinsky, M.; Heilshorn, S. C.; Juengst, T.; Malda, J.; Mironov, V. A.; Nakayama, K., A definition of bioinks and their distinction from biomaterial inks. *Biofabrication* **2018,** *11* (1), 013001.
177. Donderwinkel, I.; Van Hest, J. C.; Cameron, N. R., Bio-inks for 3D bioprinting: recent advances and future prospects. *Polymer Chemistry* **2017,** *8* (31), 4451–4471.
178. Masri, S.; Fauzi, M. B., Current insight of printability quality improvement strategies in natural-based bioinks for skin regeneration and wound healing. *Polymers* **2021,** *13* (7), 1011.
179. Masri, S.; Zawani, M.; Zulkiflee, I.; Salleh, A.; Fadilah, N. I. M.; Maarof, M.; Wen, A. P. Y.; Duman, F.; Tabata, Y.; Aziz, I. A., Cellular interaction of human skin cells towards natural bioink via 3D-bioprinting technologies for chronic wound: a comprehensive review. *International Journal of Molecular Sciences* **2022,** *23* (1), 476.

180. Hazur, J.; Endrizzi, N.; Schubert, D. W.; Boccaccini, A. R.; Fabry, B., Stress relaxation amplitude of hydrogels determines migration, proliferation, and morphology of cells in 3-D culture. *Biomaterials Science* **2022,** *10* (1), 270–280.
181. Velasco, D.; Quílez, C.; Garcia, M.; del Cañizo, J. F.; Jorcano, J. L., 3D human skin bioprinting: a view from the bio side. *Journal of 3D Printing in Medicine* **2018**, 141–162.
182. Lee, W.; Debasitis, J. C.; Lee, V. K.; Lee, J.-H.; Fischer, K.; Edminster, K.; Park, J.-K.; Yoo, S.-S., Multi-layered culture of human skin fibroblasts and keratinocytes through three-dimensional freeform fabrication. *Biomaterials* **2009,** *30* (8), 1587–1595.
183. Koch, L.; Kuhn, S.; Sorg, H.; Gruene, M.; Schlie, S.; Gaebel, R.; Polchow, B.; Reimers, K.; Stoelting, S.; Ma, N., Laser printing of skin cells and human stem cells. *Tissue Engineering Part C: Methods* **2010,** *16* (5), 847–854.
184. Binder, K. W.; Zhao, W.; Aboushwareb, T.; Dice, D.; Atala, A.; Yoo, J. J., In situ bioprinting of the skin for burns. *Journal of the American College of Surgeons* **2010,** *211* (3), S76.
185. Albanna, M.; Binder, K. W.; Murphy, S. V.; Kim, J.; Qasem, S. A.; Zhao, W.; Tan, J.; El-Amin, I. B.; Dice, D. D.; Marco, J., In situ bioprinting of autologous skin cells accelerates wound healing of extensive excisional full-thickness wounds. *Scientific Reports* **2019,** *9* (1), 1856.
186. Hakimi, N.; Cheng, R.; Leng, L.; Sotoudehfar, M.; Ba, P. Q.; Bakhtyar, N.; Amini-Nik, S.; Jeschke, M. G.; Günther, A., Handheld skin printer: in situ formation of planar biomaterials and tissues. *Lab on a Chip* **2018,** *18* (10), 1440–1451.
187. Admane, P.; Gupta, A. C.; Jois, P.; Roy, S.; Lakshmanan, C. C.; Kalsi, G.; Bandyopadhyay, B.; Ghosh, S., Direct 3D bioprinted full-thickness skin constructs recapitulate regulatory signaling pathways and physiology of human skin. *Bioprinting* **2019,** *15*, e00051.
188. Derr, K.; Zou, J.; Luo, K.; Song, M. J.; Sittampalam, G. S.; Zhou, C.; Michael, S.; Ferrer, M.; Derr, P., Fully three-dimensional bioprinted skin equivalent constructs with validated morphology and barrier function. *Tissue Engineering Part C: Methods* **2019,** *25* (6), 334–343.
189. Liu, X.; Michael, S.; Bharti, K.; Ferrer, M.; Song, M. J., A biofabricated vascularized skin model of atopic dermatitis for preclinical studies. *Biofabrication* **2020,** *12* (3), 035002.
190. Zidarič, T.; Milojević, M.; Gradišnik, L.; Stana Kleinschek, K.; Maver, U.; Maver, T., Polysaccharide-based bioink formulation for 3D bioprinting of an in vitro model of the human dermis. *Nanomaterials* **2020,** *10* (4), 733.
191. Jorgensen, A. M.; Varkey, M.; Gorkun, A.; Clouse, C.; Xu, L.; Chou, Z.; Murphy, S. V.; Molnar, J.; Lee, S. J.; Yoo, J. J., Bioprinted skin recapitulates normal collagen remodeling in full-thickness wounds. *Tissue Engineering Part A* **2020,** *26* (9–10), 512–526.
192. Kim, B. S.; Gao, G.; Kim, J. Y.; Cho, D. W., 3D cell printing of perfusable vascularized human skin equivalent composed of epidermis, dermis, and hypodermis for better structural recapitulation of native skin. *Advanced Healthcare Materials* **2019,** *8* (7), 1801019.
193. Jang, M.; Bae, S.; Jung, Y.; Kim, J.; Kim, J.; Park, S.; Suh, J.; Yi, S.; Ahn, S.; Lim, J., Enhanced wound healing using a 3D printed VEGF-mimicking peptide incorporated hydrogel patch in a pig model. *Biomedical Materials* **2021,** *16* (4), 045013.
194. Chu, B.; He, J.-m.; Wang, Z.; Liu, L.-l.; Li, X.-l.; Wu, C.-X.; Chen, C.-s.; Tu, M., Proangiogenic peptide nanofiber hydrogel/3D printed scaffold for dermal regeneration. *Chemical Engineering Journal* **2021,** *424*, 128146.
195. Guan, G.; Lv, Q.; Liu, S.; Jiang, Z.; Zhou, C.; Liao, W., 3D-bioprinted peptide coupling patches for wound healing. *Materials Today Bio* **2022,** *13*, 100188.
196. Jara, C. P.; Catarino, C. M.; Lei, Y.; Velloso, L. A.; Karande, P.; Velander, W. H.; de Araujo, E. P., Demonstration of re-epithelialization in a bioprinted human skin equivalent wound model. *Bioprinting* **2021,** *24*, e00102.
197. Phua, Q. H.; Han, H. A.; Soh, B.-S., Translational stem cell therapy: vascularized skin grafts in skin repair and regeneration. *Journal of Translational Medicine* **2021,** *19*, 1–11.
198. Zhang, X.; Chen, X.; Hong, H.; Hu, R.; Liu, J.; Liu, C., Decellularized extracellular matrix scaffolds: recent trends and emerging strategies in tissue engineering. *Bioactive Materials* **2022,** *10*, 15–31.
199. Sutcliffe, J. E.; Thrasivoulou, C.; Serena, T. E.; Madden, L.; Richards, T.; Phillips, A. R.; Becker, D. L., Changes in the extracellular matrix surrounding human chronic wounds revealed by 2-photon imaging. *International Wound Journal* **2017,** *14* (6), 1225–1236.
200. Kusindarta, D. L.; Wihadmadyatami, H., The role of extracellular matrix in tissue regeneration. *Tissue Regeneration* **2018,** *65*.

201. Assunção, M.; Dehghan-Baniani, D.; Yiu, C. H. K.; Später, T.; Beyer, S.; Blocki, A., Cell-derived extracellular matrix for tissue engineering and regenerative medicine. *Frontiers in Bioengineering and Biotechnology* **2020,** *8*, 602009.
202. Turner, N. J., *Application of Extracellular Matrix in Regenerative Medicine.* MDPI; 2021, Vol. 11, p. 3262.
203. Masson-Meyers, D. S.; Andrade, T. A.; Caetano, G. F.; Guimaraes, F. R.; Leite, M. N.; Leite, S. N.; Frade, M. A. C., Experimental models and methods for cutaneous wound healing assessment. *International Journal of Experimental Pathology* **2020,** *101* (1–2), 21–37.
204. Djavid, G. E.; Tabaie, S. M.; Tajali, S. B.; Totounchi, M.; Farhoud, A.; Fateh, M.; Ghafghazi, M.; Koosha, M.; Taghizadeh, S., Application of a collagen matrix dressing on a neuropathic diabetic foot ulcer: a randomised control trial. *Journal of Wound Care* **2020,** *29* (Sup3), S13–S18.
205. Rasouli, M.; Rahimi, A.; Soleimani, M., The interplay between extracellular matrix and progenitor/stem cells during wound healing: opportunities and future directions. *Acta Histochemica* **2021,** *123* (7), 151785.
206. Liu, H.; Li, Z.; Zhao, Y.; Feng, Y.; Zvyagin, A. V.; Wang, J.; Yang, X.; Yang, B.; Lin, Q., Novel diabetic foot wound dressing based on multifunctional hydrogels with extensive temperature-tolerant, durable, adhesive, and intrinsic antibacterial properties. *ACS Applied Materials & Interfaces* **2021,** *13* (23), 26770–26781.
207. Cramer, M. C.; Badylak, S. F., Extracellular matrix-based biomaterials and their influence upon cell behavior. *Annals of Biomedical Engineering* **2020,** *48* (7), 2132–2153.
208. Jiang, Y.; Li, R.; Han, C.; Huang, L., Extracellular matrix grafts: from preparation to application. *International Journal of Molecular Medicine* **2021,** *47* (2), 463–474.
209. Zhang, Q.; Johnson, J. A.; Dunne, L. W.; Chen, Y.; Iyyanki, T.; Wu, Y.; Chang, E. I.; Branch-Brooks, C. D.; Robb, G. L.; Butler, C. E., Decellularized skin/adipose tissue flap matrix for engineering vascularized composite soft tissue flaps. *Acta Biomaterialia* **2016,** *35*, 166–184.
210. Milan, P. B.; Pazouki, A.; Joghataei, M. T.; Mozafari, M.; Amini, N.; Kargozar, S.; Amoupour, M.; Latifi, N.; Samadikuchaksaraei, A., Decellularization and preservation of human skin: a platform for tissue engineering and reconstructive surgery. *Methods* **2020,** *171*, 62–67.
211. Zhang, X.; Yang, J.; Li, Y.; Liu, S.; Long, K.; Zhao, Q.; Zhang, Y.; Deng, Z.; Jin, Y., Functional neovascularization in tissue engineering with porcine acellular dermal matrix and human umbilical vein endothelial cells. *Tissue Engineering Part C: Methods* **2011,** *17* (4), 423–433.
212. Takami, Y.; Yamaguchi, R.; Ono, S.; Hyakusoku, H., Clinical application and histological properties of autologous tissue-engineered skin equivalents using an acellular dermal matrix. *Journal of Nippon Medical School* **2014,** *81* (6), 356–363.
213. Sotnichenko, A.; Gilevich, I.; Melkonian, K.; Yutskevich, Y.; Karakulev, A.; Bogdanov, S.; Bykov, I.; Redko, A.; Porhanov, V.; Alekseenko, S., Techniques for obtaining dermal extracellular matrix scaffold. *Russian Journal of Transplantology and Artificial Organs* **2020,** *21* (4), 81–87.
214. Choi, J. S.; Kim, J. D.; Yoon, H. S.; Cho, Y. W., Full-thickness skin wound healing using human placenta-derived extracellular matrix containing bioactive molecules. *Tissue Engineering Part A* **2013,** *19* (3–4), 329–339.
215. Milan, P. B.; Lotfibakhshaiesh, N.; Joghataie, M.; Ai, J.; Pazouki, A.; Kaplan, D.; Kargozar, S.; Amini, N.; Hamblin, M.; Mozafari, M., Accelerated wound healing in a diabetic rat model using decellularized dermal matrix and human umbilical cord perivascular cells. *Acta Biomaterialia* **2016,** *45*, 234–246.
216. Groeber, F.; Engelhardt, L.; Lange, J.; Kurdyn, S.; Schmid, F. F.; Rücker, C.; Mielke, S.; Walles, H.; Hansmann, J., A first vascularized skin equivalent as an alternative to animal experimentation. *ALTEX-Alternatives to Animal Experimentation* **2016,** *33* (4).
217. Hoganson, D. M.; O'Doherty, E. M.; Owens, G. E.; Harilal, D. O.; Goldman, S. M.; Bowley, C. M.; Neville, C. M.; Kronengold, R. T.; Vacanti, J. P., The retention of extracellular matrix proteins and angiogenic and mitogenic cytokines in a decellularized porcine dermis. *Biomaterials* **2010,** *31* (26), 6730–6737.
218. Parmaksiz, M.; Dogan, A.; Odabas, S.; Elçin, A. E.; Elçin, Y. M., Clinical applications of decellularized extracellular matrices for tissue engineering and regenerative medicine. *Biomedical Materials* **2016,** *11* (2), 022003.
219. Su, L.; Zheng, J.; Wang, Y.; Zhang, W.; Hu, D., Emerging progress on the mechanism and technology in wound repair. *Biomedicine & Pharmacotherapy* **2019,** *117*, 109191.

220. Helmy, Y.; Farouk, N.; Ali Dahy, A.; Abu-Elsoud, A.; Fouad Khattab, R.; Elshahat Mohammed, S.; Abdullbary Gad, L.; Altramsy, A.; Hussein, E.; Farahat, A., Objective assessment of Platelet-Rich Plasma (PRP) potentiality in the treatment of chronic leg ulcer: RCT on 80 patients with Venous ulcer. *Journal of Cosmetic Dermatology* **2021,** *20* (10), 3257–3263.
221. Ahmad, Z.; Howard, D.; Brooks, R. A.; Wardale, J.; Henson, F.; Getgood, A.; Rushton, N., The role of platelet rich plasma in musculoskeletal science. *JRSM Short Reports* **2012,** *3* (6), 1–9.
222. Jinming, W.; Caiyue, L.; Baojin, W.; Antang, L.; Yingfan, Z.; Hui, W.; Lie, Z.; Hua, J., Effects of platelet-rich plasma on tissue expansion in rabbits. *Aesthetic Plastic Surgery* **2017,** *41*, 454–460.
223. Chicharro-Alcántara, D.; Rubio-Zaragoza, M.; Damiá-Giménez, E.; Carrillo-Poveda, J. M.; Cuervo-Serrato, B.; Peláez-Gorrea, P.; Sopena-Juncosa, J. J., Platelet rich plasma: new insights for cutaneous wound healing management. *Journal of Functional Biomaterials* **2018,** *9* (1), 10.
224. Gonchar, I.; Lipunov, A.; Afanasov, I.; Larina, V.; Faller, A.; Kibardin, A., Platelet rich plasma and growth factors cocktails for diabetic foot ulcers treatment: state of art developments and future prospects. *Diabetes & Metabolic Syndrome: Clinical Research & Reviews* **2018,** *12* (2), 189–194.
225. Griffeth, R. J.; García-Párraga, D.; Mellado-López, M.; Crespo-Picazo, J. L.; Soriano-Navarro, M.; Martinez-Romero, A.; Moreno-Manzano, V., Platelet-rich plasma and adipose-derived mesenchymal stem cells for regenerative medicine-associated treatments in bottlenose dolphins (Tursiops truncatus). *PLoS One* **2014,** *9* (9), e108439.
226. Jee, C.-H.; Eom, N.-Y.; Jang, H.-M.; Jung, H.-W.; Choi, E.-S.; Won, J.-H.; Hong, I.-H.; Kang, B.-T.; Jeong, D. W.; Jung, D.-I., Effect of autologous platelet-rich plasma application on cutaneous wound healing in dogs. *Journal of Veterinary Science* **2016,** *17* (1), 79.
227. Carter, C. A.; Jolly, D. G.; Worden Sr, C. E.; Hendren, D. G.; Kane, C. J., Platelet-rich plasma gel promotes differentiation and regeneration during equine wound healing. *Experimental and Molecular Pathology* **2003,** *74* (3), 244–255.
228. Suthar, M.; Gupta, S.; Bukhari, S.; Ponemone, V., Treatment of chronic non-healing ulcers using autologous platelet rich plasma: a case series. *Journal of Biomedical Science* **2017,** *24*, 1–10.
229. Dionyssiou, D.; Demiri, E.; Foroglou, P.; Cheva, A.; Saratzis, N.; Aivazidis, C.; Karkavelas, G., The effectiveness of intralesional injection of platelet-rich plasma in accelerating the healing of chronic ulcers: an experimental and clinical study. *International Wound Journal* **2013,** *10* (4), 397–406.
230. Lee, H. W.; Reddy, M. S.; Geurs, N.; Palcanis, K. G.; Lemons, J. E.; Rahemtulla, F. G.; Ho, K. J.; Chen, D. T.; Davis, C. R.; Feldman, D. S., Efficacy of platelet-rich plasma on wound healing in rabbits. *Journal of Periodontology* **2008,** *79* (4), 691–696.
231. Xu, P.; Wu, Y.; Zhou, L.; Yang, Z.; Zhang, X.; Hu, X.; Yang, J.; Wang, M.; Wang, B.; Luo, G., Platelet-rich plasma accelerates skin wound healing by promoting re-epithelialization. *Burns & Trauma* **2020,** *8*, tkaa028.
232. Knighton, D. R.; Ciresi, K.; Fiegel, V.; Schumerth, S.; Butler, E.; Cerra, F., Stimulation of repair in chronic, nonhealing, cutaneous ulcers using platelet-derived wound healing formula. *Surgery, Gynecology & Obstetrics* **1990,** *170* (1), 56–60.
233. Martinez-Zapata, M. J.; Martí-Carvajal, A. J.; Solà, I.; Expósito, J. A.; Bolíbar, I.; Rodríguez, L.; Garcia, J.; Zaror, C., Autologous platelet-rich plasma for treating chronic wounds. *Cochrane Database of Systematic Reviews* **2016,** (5).
234. Lacci, K. M.; Dardik, A., Platelet-rich plasma: support for its use in wound healing. *The Yale Journal of Biology and Medicine* **2010,** *83* (1), 1.
235. Roubelakis, M. G.; Trohatou, O.; Roubelakis, A.; Mili, E.; Kalaitzopoulos, I.; Papazoglou, G.; Pappa, K. I.; Anagnou, N. P., Platelet-rich plasma (PRP) promotes fetal mesenchymal stem/stromal cell migration and wound healing process. *Stem Cell Reviews and Reports* **2014,** *10*, 417–428.
236. Babaei, V.; Afradi, H.; Gohardani, H.; Nasseri, F.; Azarafza, M.; Teimourian, S., Management of chronic diabetic foot ulcers using platelet-rich plasma. *Journal of Wound Care* **2017,** *26* (12), 784–787.
237. Popescu, M. N.; Iliescu, M. G.; Beiu, C.; Popa, L. G.; Mihai, M. M.; Berteanu, M.; Ionescu, A. M., Autologous platelet-rich plasma efficacy in the field of regenerative medicine: product and quality control. *BioMed Research International* **2021,** *2021*, 1–6.
238. Emmert, S.; Pantermehl, S.; Foth, A.; Waletzko-Hellwig, J.; Hellwig, G.; Bader, R.; Illner, S.; Grabow, N.; Bekeschus, S.; Weltmann, K.-D., Combining biocompatible and biodegradable scaffolds and cold atmospheric plasma for chronic wound regeneration. *International Journal of Molecular Sciences* **2021,** *22* (17), 9199.
239. Lu, X.; Laroussi, M.; Puech, V., On atmospheric-pressure non-equilibrium plasma jets and plasma bullets. *Plasma Sources Science and Technology* **2012,** *21* (3), 034005.

240. Braný, D.; Dvorská, D.; Halašová, E.; Škovierová, H., Cold atmospheric plasma: a powerful tool for modern medicine. *International Journal of Molecular Sciences* **2020,** *21* (8), 2932.
241. Cheng, K.-Y.; Lin, Z.-H.; Cheng, Y.-P.; Chiu, H.-Y.; Yeh, N.-L.; Wu, T.-K.; Wu, J.-S., Wound healing in streptozotocin-induced diabetic rats using atmospheric-pressure argon plasma jet. *Scientific Reports* **2018,** *8* (1), 12214.
242. Nakajima, Y.; Mukai, K.; Rahayu, H. S. E.; Nur, M.; Ishijima, T.; Enomoto, H.; Uesugi, Y.; Sugama, J.; Nakatani, T., Cold plasma on full-thickness cutaneous wound accelerates healing through promoting inflammation, re-epithelialization and wound contraction. *Clinical Plasma Medicine* **2014,** *2* (1), 28–35.
243. Darmawati, S.; Rohmani, A.; Nurani, L. H.; Prastiyanto, M. E.; Dewi, S. S.; Salsabila, N.; Wahyuningtyas, E. S.; Murdiya, F.; Sikumbang, I. M.; Rohmah, R. N., When plasma jet is effective for chronic wound bacteria inactivation, is it also effective for wound healing? *Clinical Plasma Medicine* **2019,** *14*, 100085.
244. Martinez, L.; Dhruv, A.; Lin, L.; Balaras, E.; Keidar, M., Interaction between a helium atmospheric plasma jet and targets and dynamics of the interface. *Plasma Sources Science and Technology* **2019,** *28* (11), 115002.
245. Hiller, J.; Stratmann, B.; Timm, J.; Costea, T. C.; Tschoepe, D., Enhanced growth factor expression in chronic diabetic wounds treated by cold atmospheric plasma. *Diabetic Medicine* **2022,** *39* (6), e14787.
246. Tiede, R.; Hirschberg, J.; Daeschlein, G.; von Woedtke, T.; Vioel, W.; Emmert, S., Plasma applications: a dermatological view. *Contributions to Plasma Physics* **2014,** *54* (2), 118–130.
247. Bernhardt, T.; Semmler, M. L.; Schäfer, M.; Bekeschus, S.; Emmert, S.; Boeckmann, L., Plasma medicine: applications of cold atmospheric pressure plasma in dermatology. *Oxidative Medicine and Cellular Longevity* **2019,** *2019*.
248. Isbary, G.; Stolz, W.; Shimizu, T.; Monetti, R.; Bunk, W.; Schmidt, H.-U.; Morfill, G. E.; Klämpfl, T.; Steffes, B.; Thomas, H., Cold atmospheric argon plasma treatment may accelerate wound healing in chronic wounds: results of an open retrospective randomized controlled study in vivo. *Clinical Plasma Medicine* **2013,** *1* (2), 25–30.
249. Chuangsuwanich, A.; Assadamongkol, T.; Boonyawan, D., The healing effect of low-temperature atmospheric-pressure plasma in pressure ulcer: a randomized controlled trial. *The International Journal of Lower Extremity Wounds* **2016,** *15* (4), 313–319.
250. Arndt, S.; Unger, P.; Wacker, E.; Shimizu, T.; Heinlin, J.; Li, Y.-F.; Thomas, H. M.; Morfill, G. E.; Zimmermann, J. L.; Bosserhoff, A.-K., Cold atmospheric plasma (CAP) changes gene expression of key molecules of the wound healing machinery and improves wound healing in vitro and in vivo. *PLoS One* **2013,** *8* (11), e79325.
251. Kisch, T.; Helmke, A.; Schleusser, S.; Song, J.; Liodaki, E.; Stang, F. H.; Mailaender, P.; Kraemer, R., Improvement of cutaneous microcirculation by cold atmospheric plasma (CAP): results of a controlled, prospective cohort study. *Microvascular Research* **2016,** *104*, 55–62.
252. Fathollah, S.; Mirpour, S.; Mansouri, P.; Dehpour, A. R.; Ghoranneviss, M.; Rahimi, N.; Safaie Naraghi, Z.; Chalangari, R.; Chalangari, K. M., Investigation on the effects of the atmospheric pressure plasma on wound healing in diabetic rats. *Scientific Reports* **2016,** *6* (1), 19144.
253. Lou, R.; Chen, J.; Zhou, F.; Wang, C.; Leung, C.-H.; Lin, L., Exosome-cargoed microRNAs: potential therapeutic molecules for diabetic wound healing. *Drug Discovery Today* **2022,** *27* (10), 103323.
254. Ashoori, M. R.; Rahmati-Yamchi, M.; Ostadrahimi, A.; Aval, S. F.; Zarghami, N., MicroRNAs and adipocytokines: promising biomarkers for pharmacological targets in diabetes mellitus and its complications. *Biomedicine & Pharmacotherapy* **2017,** *93*, 1326–1336.
255. Blanpain, C.; Fuchs, E., Epidermal homeostasis: a balancing act of stem cells in the skin. *Nature Reviews Molecular Cell Biology* **2009,** *10* (3), 207–217.
256. Banerjee, J.; Chan, Y. C.; Sen, C. K., MicroRNAs in skin and wound healing. *Physiological Genomics* **2011,** *43* (10), 543–556.
257. Liu, Y.; Ding, M.; Liu, D.; Liu, Y.; Mao, Y.; Peng, Y., MicroRNA profiling in cutaneous wounds of diabetic rats. *Genetics and Molecular Research* **2015,** *14* (3), 9614–9625.
258. Soliman, A. M.; Das, S.; Abd Ghafar, N.; Teoh, S. L., Role of microRNA in proliferation phase of wound healing. *Frontiers in Genetics* **2018,** *9*, 335936.
259. Mulholland, E. J.; Dunne, N.; McCarthy, H. O., MicroRNA as therapeutic targets for chronic wound healing. *Molecular Therapy-Nucleic Acids* **2017,** *8*, 46–55.
260. Jiang, Y.; Xu, X.; Xiao, L.; Wang, L.; Qiang, S., The role of microRNA in the inflammatory response of wound healing. *Frontiers in Immunology* **2022,** *13*, 852419.
261. Ghatak, S.; Chan, Y. C.; Khanna, S.; Banerjee, J.; Weist, J.; Roy, S.; Sen, C. K., Barrier function of the repaired skin is disrupted following arrest of dicer in keratinocytes. *Molecular Therapy* **2015,** *23* (7), 1201–1210.

262. Li, X.; Li, D.; Wang, A.; Chu, T.; Lohcharoenkal, W.; Zheng, X.; Grünler, J.; Narayanan, S.; Eliasson, S.; Herter, E. K., MicroRNA-132 with therapeutic potential in chronic wounds. *Journal of Investigative Dermatology* **2017,** *137* (12), 2630–2638.
263. Li, D.; Li, X.; Wang, A.; Meisgen, F.; Pivarcsi, A.; Sonkoly, E.; Ståhle, M.; Landén, N. X., MicroRNA-31 promotes skin wound healing by enhancing keratinocyte proliferation and migration. *Journal of Investigative Dermatology* **2015,** *135* (6), 1676–1685.
264. Wang, J.-M.; Tao, J.; Chen, D.-D.; Cai, J.-J.; Irani, K.; Wang, Q.; Yuan, H.; Chen, A. F., MicroRNA miR-27b rescues bone marrow–derived angiogenic cell function and accelerates wound healing in type 2 diabetes mellitus. *Arteriosclerosis, Thrombosis, and Vascular Biology* **2014,** *34* (1), 99–109.
265. Wang, T.; Feng, Y.; Sun, H.; Zhang, L.; Hao, L.; Shi, C.; Wang, J.; Li, R.; Ran, X.; Su, Y., miR-21 regulates skin wound healing by targeting multiple aspects of the healing process. *The American Journal of Pathology* **2012,** *181* (6), 1911–1920.
266. Hu, Y.; Rao, S.-S.; Wang, Z.-X.; Cao, J.; Tan, Y.-J.; Luo, J.; Li, H.-M.; Zhang, W.-S.; Chen, C.-Y.; Xie, H., Exosomes from human umbilical cord blood accelerate cutaneous wound healing through miR-21-3p-mediated promotion of angiogenesis and fibroblast function. *Theranostics* **2018,** *8* (1), 169.
267. Wang, W.; Yang, C.; Wang, X. Y.; Zhou, L. Y.; Lao, G. J.; Liu, D.; Wang, C.; Hu, M. D.; Zeng, T. T.; Yan, L., MicroRNA-129 and-335 promote diabetic wound healing by inhibiting Spl-mediated MMP-9 expression. *Diabetes* **2018,** *67* (8), 1627–1638.
268. Ban, E.; Jeong, S.; Park, M.; Kwon, H.; Park, J.; Song, E. J.; Kim, A., Accelerated wound healing in diabetic mice by miRNA-497 and its anti-inflammatory activity. *Biomedicine & Pharmacotherapy* **2020,** *121*, 109613.
269. Zhong, H.; Qian, J.; Xiao, Z.; Chen, Y.; He, X.; Sun, C.; Zhao, Z., MicroRNA-133b inhibition restores EGFR expression and accelerates diabetes-impaired wound healing. *Oxidative Medicine and Cellular Longevity* **2021,** *2021.*
270. Pang, Q.; Jiang, Z.; Wu, K.; Hou, R.; Zhu, Y. J. A., Nanomaterials-based wound dressing for advanced management of infected wound. *Antibiotics* **2023,** *12* (2), 351.
271. Das, S.; Thakur, P. B.; Harri, A.; Bagade, M.; Patra, C. R., Nanomedicine for Wound Healing: challenges and Future Perspectives. In Pradipta Ranjan Rauta, Yugal Kishore Mohanta & Debasis Nayak (Eds.), *Nanotechnology in Biology and Medicine.* CRC Press; 2019, pp. 115–131.
272. Kadam, S.; Madhusoodhanan, V.; Dhekane, R.; Bhide, D.; Ugale, R.; Tikhole, U.; Kaushik, K. S. J. B., Milieu matters: an in vitro wound milieu to recapitulate key features of, and probe new insights into, mixed-species bacterial biofilms. *Biofilm* **2021,** *3*, 100047.
273. Gupta, D. R.; Singh, S. J. I. J. O. M. S., Stem cells: current applications and future prospects. *Indian Journal of Medical Sciences* **2024,** *76*, 1–5.
274. Kosaric, N.; Kiwanuka, H.; Gurtner, G. C. J. E. O. O. B. T., Stem cell therapies for wound healing. *Expert Opinion on Biological Therapy* **2019,** *19* (6), 575–585.
275. Weng, T.; Zhang, W.; Xia, Y.; Wu, P.; Yang, M.; Jin, R.; Xia, S.; Wang, J.; You, C.; Han, C. J. J. O. T. E., 3D bioprinting for skin tissue engineering: current status and perspectives. *Journal of Tissue Engineering* **2021,** *12*, 20417314211028574.
276. Tottoli, E. M.; Dorati, R.; Genta, I.; Chiesa, E.; Pisani, S.; Conti, B. J. P., Skin wound healing process and new emerging technologies for skin wound care and regeneration. *Pharmaceutics* **2020,** *12* (8), 735.
277. Tavakoli, S.; Klar, A. S. J. A. S., Bioengineered skin substitutes: advances and future trends. *Applied Sciences* **2021,** *11* (4), 1493.

10 Nanobiomaterials
Role in Diabetic Wound Healing and Skin Tissue Engineering

Izzat Ullah[1], Ihsan Ullah[1], Amin Ullah[2], Faheem Ahmed Khan[2], Iftikhar Ahmad[3], and Arshad Farid[3]*

[1] Abasyn University, Peshawar, Pakistan
[2] Iqra National University, Peshawar, Pakistan
[3] Gomal University, D. I. Khan, Pakistan
*Corresponding Author

10.1 INTRODUCTION

10.1.1 Wound Healing

Different skills of nanobiotechnology, resulting from industry, biology, and chemistry, are now being applied in biological practice. Here, we analyze the application, scheme, and scientific trials for nanotechnology-based treatments for long-lasting wound healing, indicating the importance of the scientific perspective of banobiotechnology in such treatments. By summarizing former nanobiotechnology research, we lay the basis for future wound care via a nanotech-based, extensive smart system [1]. Wound healing is a challenge addressed by surgeons in their daily clinical work and is also a debated point of research in the arenas of trauma, orthopedics, and general surgery. Numerous novel therapeutic developments have been seen in medicine and scientific disciplines since the 1990s, and nanomedicine is one of the newer and stirring fields that has been successfully used in theranostics [2].

A methodical assessment of the use of bionanomaterials in the recovery of wounds is outlined. First, the biomedicines currently available for wound healing are presented. Next, the features of wound healing are summarized. Then, the mode of action of nanomaterials on wound healing and their benefits are discussed. Further, the importance of bionanomaterials in wound healing in medical practice is studied. Moreover, an argument on the promising ramifications and toxicological effects of nanomaterials is also involved [3].

10.1.2 Overview of Diabetic Wound Healing

The number of patients with diabetes is frequently increasing around the globe. A great proportion of these individuals are likely to get diabetic ulcers. The financial, communal, and community health problems of these ulcers are massive. It is a public issue that is tackled daily by numerous healthcare employees. Numerous pioneers contributed to the noble standard of care we currently have when treating diabetes and its complications. The present therapy for diabetic foot ulcers includes a multiskilled team approach, addresses the fundamental disease, and treats the wounds using debridement, diverse wound bandages, and distributes pressure off the wound [4].

DOI: 10.1201/9781003504948-10

In diabetic patients, wound recovery is very slow or the wound may not heal at all, resulting in an unwanted condition that is one of the most common problems. Diabetic wounds, if they become long-lasting and infected, may possibly lead to lower extremities amputation, sepsis, and even death. Therefore, timely diagnosis, careful inspection, relieving pressure, and controlling the complications of diabetic lesions are significant elements in the successful management of the condition. Over the decades, numerous promising wound bandages integrating antimicrobial molecules, growth factors, and wound healing agents have been developed to treat diabetic wounds. However, the choice of bandage is mostly based on the skills of the physician [5].

10.1.3 Brief Overview of Nanotechnology in Medicine

Nanotechnology is a tremendously influential developing technology that is predicted to have a considerable influence on therapeutic technology today and in the future. The possible influence of innovative biomedicine applications on infection analysis, treatment, and inhibition is predicted to reform health maintenance in a central way. Moreover, medicinal choices can progressively be tailored to every patient. This report represents the state-of-the-art in the area of advanced nanotechnology tactics for therapeutic technology [6].

Nanotechnology now has, or is nearly ready to, transfigure diverse phases of therapeutic and scientific exploration, with an emphasis on the analysis and observation of illness and the delivery of medicine to cure infection. During this phase, the field of science that has benefited most from developments in nanotechnology is scientific diagnostic imaging, and nanoparticle (NP) contrast agents have previously been used regularly or are presently under scientific trials for consideration. Drug delivery methods are not far behind, though, and in addition to numerous Food and Drug Administration-approved nanomedicines, several formulations that decrease off-site toxicity or increase the buildup of a medicine at an exact target or both are being explored. Now, the effort to advance community health and well-being management has realized that with the help of nanotechnology, analysis and therapy are not essentially distinct measures [7].

10.1.4 Role of Nanobiomaterials

From the material's perspective, biological tissue can be measured as a complex consisting of nanobiomaterials, for instance, nanoapatite grains and nanomembranes. Nanobiomedicine has long been an effort of active research owing to its unique assets that can be used for biomedical benefits such as tissue engineering, bioimaging, gene and drug delivery, medical transplants, and analytic methods like protein and DNA microarrays. The fast development of nanotechnology has allowed investigators to produce NPs, nanocoatings, and nanocomposites for biomedical purposes and has suggested that a nanofunctionalized surface has beneficial biotic properties. Hence, scientific applications of bionanomaterials can be enhanced by creating a nanostructured surface [8].

In the current period, the production of NPs and the examination of their properties have involved the consideration of entire divisions of science, for example, physicists, biologists, engineers, and even medical physicians. Interest in NPs arises from the fact that their mechanical, chemical, optical, magnetic, and magneto-optical properties are totally diverse from their majority features, and the prearranged changes are based on the physical and chemical features of the NPs. Now there are many aspects where NPs are of systematic and technical importance, especially for the therapeutic community, where the physical and biological domains come together and lead to significant concern for the production of harmless nanomaterials [9]. Table 10.1 presents a summary of the different roles shown by nanobiomaterials in biomedicine.

TABLE 10.1
The Role of Nanobiomaterials in Advancing Biomedical Science

Role	Description
Drug delivery	Nanobiomaterials enable precise targeting and controlled release of therapeutic agents to specific tissues.
Tissue engineering	Scaffold-based nanostructures support cell adhesion, proliferation, and differentiation for tissue regeneration.
Imaging agents	Nanoparticles with distinctive optical, magnetic, or radioactive properties enhance contrast for investigative imaging.
Biosensors	Nanostructured materials provide high sensitivity and selectivity for detecting biomolecules in biological samples.
Theranostics	Integration of diagnostics and therapeutics within a single nanobiomaterial platform for personalized medicine.
Targeted therapy	Functionalized nanoparticles deliver drugs selectively to diseased cells, minimizing off-target effects.
Gene therapy	Nanocarriers facilitate efficient and safe delivery of genetic material for gene editing and manipulation.

10.1.5 Importance of Nanobiomaterials in Diabetic Wound Healing and Skin Tissue Engineering

The occurrence of chronic wounds has amplified quickly, causing a significant increase in healthcare expenses and patient illness. These long-lasting wounds vary in their appearance and etiology. Frequently, there is great protease activity, extensive contamination and biofilm development, hypoxia in the tissue, recurrent wounds due to neuropathy, or cellular failure resulting in gangrene related to prolonged injuries. Chronic lesions most frequently occur as arterial leg sores or foot abscesses and are found at high rates in patients with diabetes and obesity. The occurrence of type 2 diabetes has also increased significantly in recent years, which exacerbates the tricky issue of persistent wound growth [10].

Tissue engineering is a multidisciplinary area of nanobiomedicine in which nanomaterials and therapeutic disciplines understand diseased tissue, and the ideologies used to achieve this understanding are employed in refining or sustaining the role of tissue in the development of biotic alternates. To accomplish these objectives, an artificial extracellular matrix and a foundation of cells are obligatory. Though skin tissue constitutes a minor quantity of body weight, skin injuries such as burns and diabetic illnesses can cause serious health problems. Tissue-engineered replacements play a very important role in the management of long-lasting skin wounds. Skin tissue replacement methods need to be relaxed for applying to the wound site. Skin tissue replacements must be readily adherent, have good features and mechanical characteristics, and be nonantigenic [11]. Table 10.2 illustrates the important role of nanobiomaterials in addressing the challenges of diabetic wound healing and advancing skin tissue engineering.

10.2 DIABETES AND WOUND HEALING

Long-lasting diabetic injuries, the utmost public problem of diabetes, have turned into a significant reason for nontraumatic amputations, harming not just the health and standard of living of the patient but also enforcing a financial load on the medical infrastructure and the patients themselves. Commonly, wound repair includes four separate, corresponding phases: hemostasis, soreness, cell differentiation, and matrix restoration. Though the curing procedure for diabetic wounds varies from that of common injuries, diabetic wounds tend to stay stuck in the inflammatory stage due to chronic hyperglycemia, which significantly raises the risk of medical infection. Therefore, the

TABLE 10.2
Harnessing Nanobiomaterials for Diabetic Wound Healing and Skin Tissue Engineering

Importance	Description
Enhanced healing	Nanobiomaterials facilitate faster wound closure by promoting cell migration, proliferation, and angiogenesis.
Antibacterial properties	Nanoparticles with inherent or incorporated antibacterial agents prevent infection and promote aseptic wound healing.
Controlled drug delivery	Nanocarriers enable sustained release of growth factors or antimicrobial agents, optimizing therapeutic efficacy.
Mimicking ECM structure	Nanostructured scaffolds mimic the extracellular matrix (ECM), providing mechanical support and guiding tissue regeneration.
Immunomodulation	Nanomaterials modulate immune responses, reducing inflammation and promoting a regenerative microenvironment in diabetic wounds.
Bioactive dressings	Functionalized nanomaterials offer bioactive properties, such as antioxidative or anti-inflammatory effects, enhancing wound healing.
Personalized medicine	Tailored nanobiomaterial formulations address patient-specific wound characteristics, optimizing treatment outcomes.

synthesis of nanomedicine with antibiofilm, anti-inflammatory, and wound healing purposes is very significant for the healing of diabetic infected wounds [12].

10.2.1 Impact of Diabetes on Wound Healing Process

Long-lasting, nonhealing diabetic wounds are one of the main problems of both forms of diabetes, which are serious concerns for well-being and can potentially influence one's standard of living Overall, diabetic wounds are characterized by low chemokine production, a rare inflammatory reaction, the absence of angiogenesis and epithelialization, and the malfunctioning of fibroblasts. Accumulative systematic evidence from accessible investigational studies on animal and cell models strongly relates damaged wound healing in diabetes with nonfunctional fibroblast differentiation to myofibroblasts, scattered myofibroblast action, and scarce extracellular matrix production [13].

10.3 NANOBIOMATERIALS

The "nanotechnology era" has begun with the invention of biomaterials. Bionanomaterials shaped by nanotechnology are among the fields at the intersection of physical science and biology that are quickly emerging. Such biomaterials have an amazing influence on biomedicine. The term "nanotechnology-derived biomaterials" precisely refers to those bionanomaterials. These structures determine dramatically novel and improved features, with their sizes at the nanoscale. Bionanomaterials shaped by nanotechnology have been broadly applied in a wide range of nanomedical and biotic practices, from therapeutic imaging to drug delivery to artificial implants [14]. The developing models of nanomaterials (such as NPs, nanoemulsions, nanofibrous scaffolds, and nanocomposites) and bionanomaterials (like peptides and oligosaccharides) have huge importance in the field of wound care. Multipurpose, distinctive nanowound care preparations have gained major attention by improving wound healing practices [15].

10.3.1 Characteristics of Nanobiomaterials

Nanobiomaterials have an important role in targeting therapies. Throughout the years, numerous bionanomaterials have received extensive attention in the cure and analysis of infections. The contact of nanobiomaterials with natural systems happens at diverse length-scales, specifically at the nanoscale. The design of nanomaterials and structures with nanometric accuracy has confirmed their massive potential in regulatory cellular processes like adhesion, proliferation, and differentiation. Conversely, nanosized carriers have shown improved efficacy and precision in delivering healing or analytic payloads to exact living environments, thus demonstrating prevailing tools in both the analysis and cure of pathological states [16].

Nanobiomaterials can be amorphous or crystalline, so their electronic features can be easily revised. The perfect features of nanobiomaterials are equal strength, a vigorous surface, and distinct energy levels. The appropriateness of nanoparticles for drug administration and targeting escalates with reducing size. Due to their small size and flexibility, nanoparticles have a higher cellular absorption rate than microparticles, which allows them to target a diversity of intracellular and cellular targets. Since the targeted cells must adopt the nanoparticles, the type of particle is also significant for biodistribution. Moreover, they are superior candidates for endosomal uptake, as immune system cells may identify rod-shaped cationic particles as rod-shaped bacteria [17].

10.3.2 Types of Nanobiomaterials Used in Diabetic Wound Healing and Skin Tissue Engineering

Chronic wounds have been a worldwide health concern over the past few years, requiring vital therapeutic and research care. The factors that delay wound healing progression include obesity, stress, microbial infection, poor oxygenation, diabetes, and graft problems. Biomaterials are being developed and fabricated to accelerate the healing of long-lasting wounds, including hydrogels, nanofibrous materials, composites, and trilayered scaffolds. Some new innovations in biomaterial development for healing both long-lasting and severe wounds are widely discussed here. In addition, numerous features of nanobiomaterials for wound healing applications and how they influence their performance are studied [18].

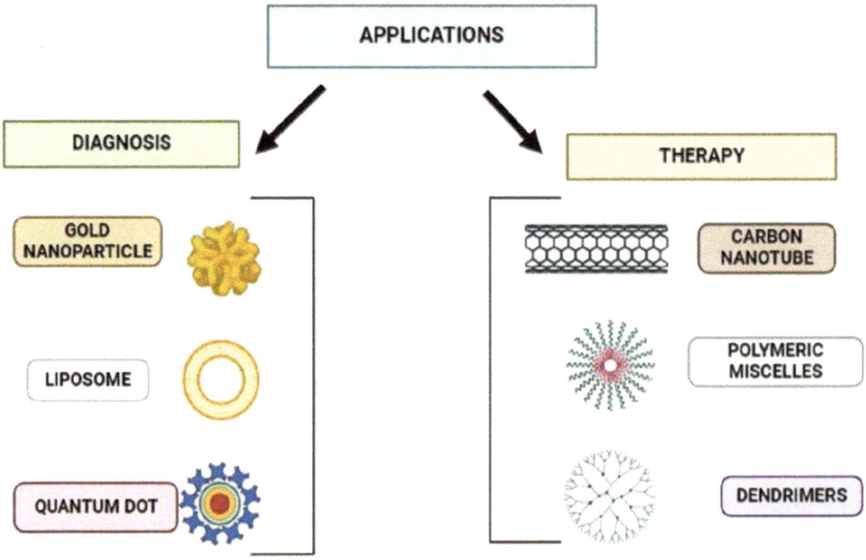

FIGURE 10.1 Application of nanomaterials in wound diagnosis and therapy.

10.3.2.1 Nanoparticles

Nanotechnology is a fast-growing scientific field that links the disciplines of physical science and engineering. NPs, typically ranging in dimension from 1 to 100 nm, have characteristics distinctive from their bulk equivalents. They have exclusive physical, chemical, optical, and biotic properties that can be manipulated to achieve desired outcomes. In previous times, elements such as Ag, Au, Cu, and titanium have been utilized to treat a variety of human conditions. In recent times, scientists have established awareness and consciousness of NPs and how they might be utilized for drug transfer, investigation and imaging, biosensors, and cosmetic purposes. Numerous nanobiomaterials for therapeutic purposes have been effectively examined during the last several years. These have comprised liposomes, dendrimers, quantum dots, fullerenes, iron and titanium oxide, and Au and Ag NPs [19].

Nanotechnology is a developing and significant field for therapeutic purposes and provides a prodigious advancement in therapy for diverse types of diseases, including wound healing. NPs permit the exterior delivery of materials that are frequently generated at the site of the wound. With a maximum half-life of seconds, nitric oxide (NO) is one such essential endogenous chemical. In delayed, chronic wounds such as diabetic lesions, the capability to produce NO is reduced, so that the supply of exogenous NO at the site of the wound is a very favorable treatment, although its everyday application is restricted due to the absence of potent delivery molecules [20].

Benefits including modern delivery of drugs, cell specificity, and the viable and controlled release of encapsulated pharmaceuticals until the wound heals are deliberate features of nanotechnology-based wound healing techniques. NPs have a significant role in topical distribution during wound healing, promoting improved biotic target interactions and increased penetration at the wound sites. Furthermore, by modifying the NP delivery, encapsulated medications might be transported in a sustained mode, and their administration rate could be appropriately adjusted. Consequently, the use of nanomedicine in wound healing therapy to deliver biomolecules opens up a fantastic opportunity to address the issues associated with diabetic wound healing [21].

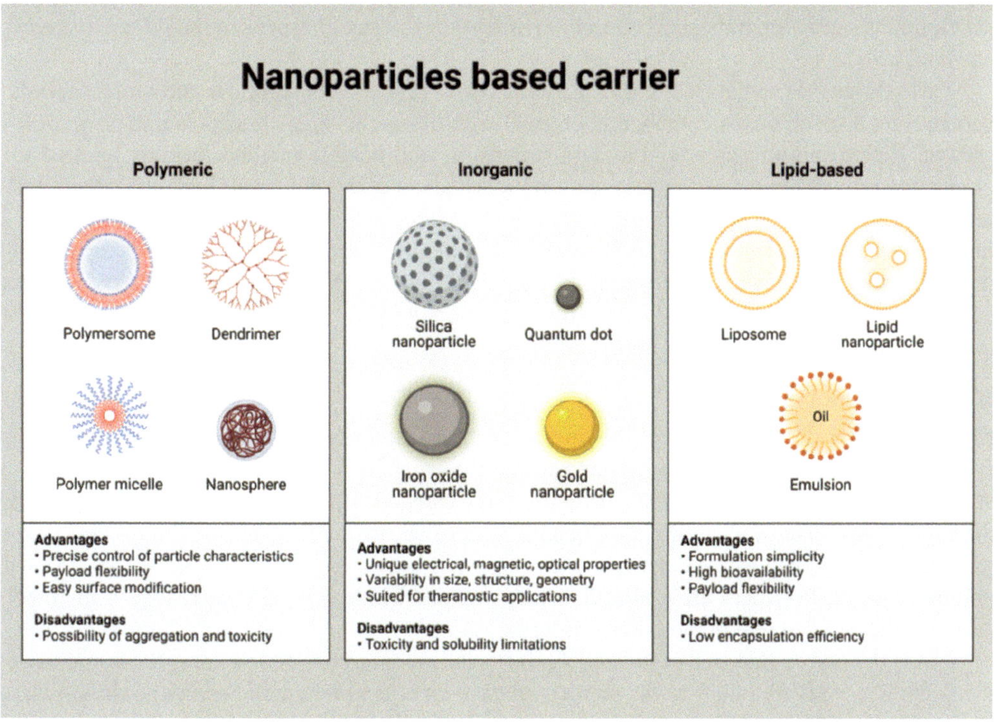

FIGURE 10.2 Nanoparticle-based carriers.

10.3.2.2 Nanofibers

Because of their close structural similarities to the natural extracellular matrix (ECM) environment, nanofibers have drawn more interest. Due to their high surface area-to-volume ratio, tunable mechanical properties, enlarged porosity, and capacity to encapsulate NPs and biological molecules for controlled release, nanofibers have been shown to promote wound healing by encouraging cells to actively interact with the matrix during functionalization and renovation [22].

Despite the fact that there are many therapies available, knowledge of complementary diabetes-related wound healing strategies as therapeutic options has greatly advanced in recent years. Although traditional bandages are less expensive, their effectiveness is insufficient to promote the general practice of diabetes wound healing. Nanocomposites and nanofibers, which mix traditional healing agents with cutting-edge materials, are one of the current trends in enhanced wound dressings, made possible by advancements in tissue engineering. Large porosity, exceptional moisture absorption, an efficient oxygen exchange rate, and antibacterial properties are more desirable in electrospun nanofibers. Starting with a careful selection of polymers and loading them with potent medicinal materials like peptides, API, stem cells, etc. to make them latent dose forms in the therapy of diabetic wounds can boost the benefit of these nanofibers [23].

10.3.2.3 Nanostructured Scaffolds

The study of drug delivery and tissue engineering has made significant progress in recent times due to its high potential to improve human health. The development of nanotechnology increases the chances of classifying, controlling, and assembling the orderly adjustment of nanomaterials at the nanoscale level. Nanobiomaterials with nanoscale applications have been administered to tissue engineering as synthetic matrices and controlled-release drug reservoirs. Both the medicine delivery systems and the scaffolds utilized in tissue engineering must be biodegradable from a material perspective. By producing nanobiomaterials with controlled properties, the biological abilities of encapsulated drugs and cells may be potentially improved. The cutting-edge developments in using nanostructured constituents for tissue engineering and drug delivery are outlined [24].

Scaffolds have a crucial role in this tactic in supporting cells to compose and direct their development into a specific tissue; thus, it is of great significance to shape scaffolds that encourage cell growth. Electrospinning is a simple, less expensive, and flexible method recently applied to the production of tissue engineering nanofeatured scaffolds. It offers numerous benefits over traditional scaffolding approaches by emulating a natural extracellular matrix. Nanobiomaterials with a size range of 100 nanometers, leading to 3D scaffolds, can mimic normal tissue conditions by increasing tissue development, diversity, propagation, cell signaling, etc. Moreover, 3D scaffolds comprising nanofibers, nanotubes, and NPs made from polymers such as PLA, PLGA, and PVA have been found to be potent in influencing stem cell fate [25].

10.3.2.4 Nanocomposites

Nanocomposites are nanomaterials having at least one dimension, which is in nanometers. These are categorized according to their structure. For the creation of nanocomposites, a variety of NPs and matrix ingredients are utilized. There are three categories of these nanomaterials according to the matrix used. They are ceramic matrix nanocomposites, metal matrix nanocomposites, and polymer matrix nanocomposites. Polymer nanocomposites are nanomaterials composed of NPs with a polymer matrix. Polymers are a mixture of monomers. These are the composites of microscopic particles that are discrete in the polymer matrix [26].

Skin is the body's first line of defense against external microbes and sustains the homeostasis of the body. Every severe injury to the skin could have an effect on human health and the standard of living. The aim of tissue engineering is to improve the level of impaired tissue regeneration. The usage of nanocomposite scaffolds to advance angiogenesis in the field of skin tissue engineering has considerably improved over the past few years. Biologic nanocomposites are a class of scaffolds with suitable biomedical applications (drug delivery and cardiac cartilage as well as skin tissue engineering), facilitating one to attain the required level of skin regeneration and diabetic wound healing [27]. Table 10.3 presents a brief overview of different types of nanobiomaterials, their key features, and common applications.

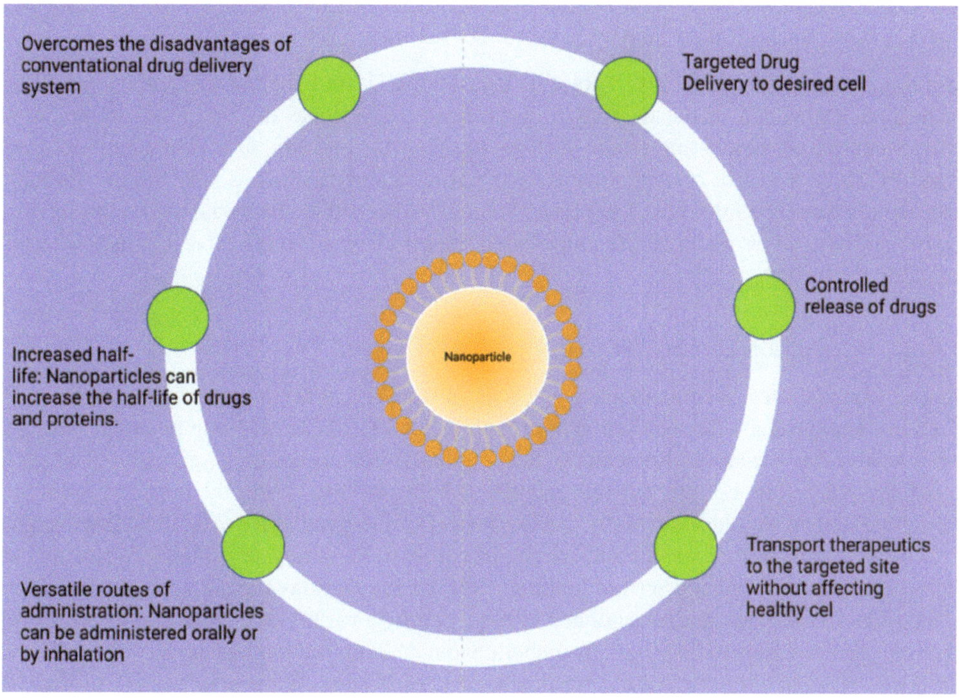

FIGURE 10.3 Nanoparticles in drug delivery.

TABLE 10.3
Advancements in Nanobiomaterials: Innovations and Applications

Nanobiomaterial	Properties	Applications
Quantum dots	Small size, tunable optical properties, high surface area	Bioimaging, drug delivery, sensing
Nanoparticles	Versatile surface chemistry, high surface-to-volume ratio	Targeted drug delivery, tissue engineering, biosensing
Carbon nanotubes	High aspect ratio, mechanical strength, electrical conductivity	Drug delivery, tissue scaffolds, biosensors
Liposomes	Biocompatible, self-assembling, versatile cargo encapsulation	Drug delivery, gene therapy, cosmetics
Nanofibers	High surface area, mimic extracellular matrix, tunable mechanical properties	Wound healing, tissue engineering, drug delivery
Nanocomposites	Enhanced mechanical properties, multifunctionality	Bone regeneration, dental materials, wound dressings
Dendrimers	Highly branched structure, precise molecular weight control, multivalency	Drug delivery, imaging agents, gene therapy
Nanogels	High water content, tunable swelling properties, drug loading capacity	Drug delivery, tissue engineering, wound healing

10.4 MECHANISMS OF ACTION

Nanobiomaterials have shown effective outcomes in wound healing due to their distinctive properties, such as their high surface area-to-volume ratio and tunable surface chemistry. Nanobiomaterials can enhance wound healing by stimulating cell migration, proliferation, and differentiation, as well as moderating the inflammatory response and promoting angiogenesis [28] .

10.4.1 Nanobiomaterials Enhance Wound Healing

Wound surfaces are the site of several biofilm formations. Despite the fact that some unique biofilms can become embedded in the deep skin tissue layers, traditional detection and treatment issues occur. Metal-based nanobiomaterials, such as Ag, Cu, Au, and Mg, are widely employed against wound biofilms. Because of their tiny size, metallic nanoparticles are able to pass through and penetrate microbial membranes. They have the power to eliminate bacterial respiratory chains and enzyme activity. Silver nanoparticles have been shown to be more effective against bacterial biofilms [29].

10.4.2 Interaction of Nanobiomaterials with Diabetic Wounds

Diabetic wounds often need an extensive healing period due to persistent inflammation, microbial infections, and the degraded expression of growth factors. Furthermore, long-lasting hyperglycemia makes new vessel formation problematic, thus regulating the access of oxygen and nutrients to the wound site. Therefore, certain substances need to be transported from the outside to promote the healing of diabetic wounds. Currently, a wide variety of inorganic materials, including oxygen and metal ions, as well as biological components, like growth factors, genes, proteins, etc., are widely employed to support diabetic wound healing. Cellular interactions during normal wound healing are facilitated, coordinated, and regulated in large part by a variety of growth factors. However, the imbalance of several growth factors in diabetic wounds damages angiogenesis, disrupts the extracellular matrix, and ultimately delays the healing process. Using endogenous healing growth factors

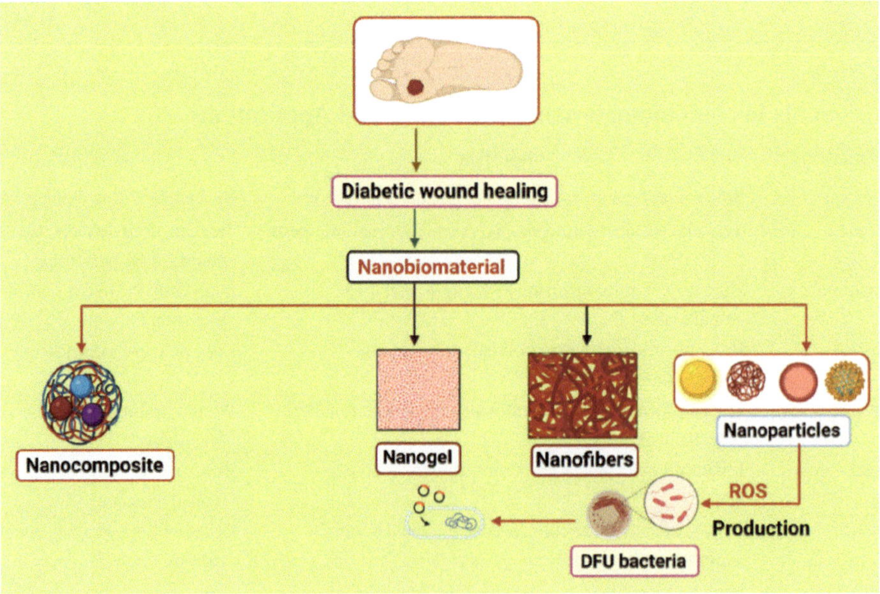

FIGURE 10.4 Mechanisms of action of nanobiomaterials.

locally is one way to restore diabetic wound cell signaling. Therefore, in addition to maintaining growth factor activity, a delivery mechanism is required to enable the continuous and regulated release of growth factors to the target. Many technologies, such as hydrogels and NPs, are currently being employed to provide growth factors to diabetic wounds [30].

10.4.3 Molecular Pathways Involved in Nanobiomaterial-Assisted Healing

All nanobiomaterials cause reactions in cells and tissues when they become embedded in vivo. These reactions are distinct from foreign body reactions (FBR), inflammatory and wound-healing processes, and fibrous encapsulation of the entrenched components. The immune system's primary response to tissue damage and the implantation of artificial or biological nanobiomaterials is displayed by macrophages in particular. For the best possible oral wound healing and tissue regeneration, inflammation control is essential. Many nanobiomaterials have been applied to enhance the healing outcomes; yet, an immune-inflammatory response can be confirmed by the biomaterial embedding. The outcome of soft and hard tissue repair is largely dependent on the interaction between the cells and the biomaterial surface. Frequent inflammation impairs the wound healing response, although the early inflammatory reaction following biomaterial embedding aids in tissue repair and regeneration. Cells interact with nanobiomaterials via extracellular matrix proteins, which cause protein adsorption and, in turn, promote the connection, migration, and multiplication of various immune-inflammatory cells. Cell development, differentiation, and the creation of the extracellular matrix are all dependent on protein adsorption on the outer layers of the biomaterial. The response to inflammation, known as FBR, is thought to be primarily mediated by macrophages. Interleukin-1β (IL-1β) and tumor necrosis factor-α (TNF-α) are two examples of pro-inflammatory cytokines that can be secreted when their surfaces come into contact with them [31].

Table 10.4 lists the mode of action of some common nanobiomaterials.

10.5 APPLICATIONS IN DIABETIC WOUND HEALING

Recent research has demonstrated that nanostructures can significantly improve diabetes patients' ability to recover from their wounds. Between 2011 and 2021, there was a significant advancement in the use of nanoparticles in the recovery process of diabetic wounds. It is critically necessary to develop new therapies, such as those based on nanotechnology, in order to enhance patients' standard of life. Numerous investigations have been conducted subsequently to confirm that

TABLE 10.4
Mechanisms of Action of Nanobiomaterials

Nanobiomaterial	Mechanism of Action
Liposomes	Targeted drug delivery, controlled release
Nanoparticles	Enhanced cellular uptake, sustained release
Nanofibers	Mimicking extracellular matrix, tissue regeneration
Quantum dots	Imaging, diagnostic applications
Nanotubes	Drug delivery, gene therapy
Nanocapsules	Encapsulation of drugs, controlled release
Dendrimers	Drug delivery, molecular imaging
Nanogels	Hydrogel properties for drug delivery, wound healing
Nanorods	Photothermal therapy, imaging
Nanosponges	Toxin neutralization, drug delivery
Nanowires	Sensing applications, tissue engineering

nanostructures with various biological properties can achieve promising outcomes in the healing of diabetic wounds. Certain dressings for wounds made of nanomaterials, like nanofibers, mimic the extracellular matrix and offer a moist environment that promotes the healing of wounds and facilitates the proliferation and migration of cells by encapsulating medications, proteins, and cytokines. Certain nanostructures have strong antimicrobial and angiogenesis properties, like metal NPs [32].

10.5.1 Role of Nanobiomaterials in Promoting Wound Closure

Due to their unique applicability, nanobiomaterials are crucial in advancing wound healing approaches. These nanoengineered substances have several advantages, such as better tissue regeneration, regulated medication administration, and increased biological compatibility. Through precise manipulation of their physical and chemical attributes, including dimensions, form, and surface features, nanomaterials are able to replicate the extracellular matrix and engage in molecular interactions with cells, thereby facilitating cell adhesion, growth, and migration. Additionally, the large surface area-to-volume ratio makes it easier for bioactive compounds like growth factors and medicines to be loaded and released over time, creating the ideal environment for healing wounds. By reducing scarring, promoting faster healing, and improving patient outcomes, the use of nanobiomaterials has the potential to completely transform wound treatment. Administration of Ag NPs has been demonstrated to alter the production of cytokines that are anti-inflammatory, leading to reports of fast wound healing with minimal damage. They may also encourage the growth of keratinocytes, which is necessary for the re-epithelialization of the epiderm [33].

10.5.2 Prevention of Infection and Inflammation

Skin wounds are among the most prevalent kinds of tissue damage and can result from burns, surgeries, persistent diseases, malignancies, or trauma. Wounds often become persistent in unfavorable situations. The main goals of treating persistent wounds are to speed up the repair of wounds and enhance the healing duration. Yet, as the field of nanotechnology and our knowledge of wound closure processes have advanced, a variety of nanobiotechnology-based wound treatment systems, such as cell-carrying systems, drug and gene delivery platforms, and antibacterial systems, have been invented and demonstrated their potential uses. However, in spite of these developments, wound dressings are still primarily antiquated and devoid of features that enable flexible wound reactions and wound management. As a result, researchers have looked into smart hydrogels and dressing systems made of nanoscale biological materials that can react to stimuli or track the condition of persistent injuries [34].

For a while, autogenic and allogenic grafts were among the many transplants used in skin tissue engineering. The search for reliable and biodegradable scaffolds, such as hydrogels, has taken center stage due to the growing drawbacks of these materials. The donor site communicates these limits, which include excruciating pain, repeated ailments, and a slow recovery that eventually leaves scars. The search for effective and biodegradable scaffolds, including hydrogels, has taken center stage in skin tissue engineering for a number of years. These limitations—which are indicated by the donor site—include excruciating pain, frequent illness, and an extended healing process that eventually leaves scars. Skin regeneration is primarily limited by the absence of autogenic skin, as well as the rejection problems that allografts or xenografts face. Consequently, the application of bioengineered synthetic skin in wound recovery has increased significantly, particularly in cases of burn damage [35].

10.5.3 Enhancement of Angiogenesis and Tissue Regeneration

Vascular development, angiogenesis, and arteriogenesis are processes that result in the development and modification of blood vessels in vivo. The process by which blood vessels develop throughout embryogenesis involves several steps. The initial stage involves the development of angioblasts, which are a subpopulation of primordial mesodermal cells that are dedicated to the lineage of endothelial cells. The next stage involves the angioblasts at or close to the place of their origin, forming primitive blood vessels. Vasculogenesis is the term for this two-step process that leads to circulation. Adult vasculogenesis occurs when EPCs settle in an injured region and develop into endothelial cells to construct rudimentary blood vessel networks. This process occurs during tissue regeneration or tumor formation. However, angiogenesis—mostly resulting from existing blood vessels—is the adult phase of blood vessel production in response to tissue demand. The development of angiogenesis involves an increase in blood flow, which permits the passage of blood proteins. Additionally, the disintegration of the basement membrane with the modified ECM releases growth factors controlled by the ECM and allows for connections between pericytes and endothelial cells. Endothelial cells infiltrate adjacent tissue and form cellular tubes with lumens that function. Angiogenesis represents the most effective and expedient means for a damaged site to generate new blood vessels in tissue healing processes, such as regeneration of bone, which is necessary to maintain the function of cells by providing nourishment and oxygen [36].

In recent times, an exciting approach for enhancing the recovery of diabetic wounds has appeared: stimulating angiogenesis via various techniques, most notably biomaterial-based therapies. The increasing attention given to nanobiomaterial-driven approaches highlights their promising potential as agents for achieving meaningful improvements in wound healing. Biomaterials offer a very adaptable and adjustable framework to tackle the complex issues raised by diabetic wounds. They can include growth factors and biological chemicals that are essential for stimulating angiogenesis, or they can be biologically developed to mimic the organic extracellular matrix. By doing this, nanomaterials promote improved blood flow, which is necessary to supply oxygen as well as nutrients to the injury site and promote efficient tissue repair [37].

10.6 SKIN TISSUE ENGINEERING

The body's first line of defense includes the skin, which is the largest organ in the human body. It maintains bodily fluids, controls body temperature, and senses external stimulation, in addition to keeping invaders out. While external injuries and disease infections are most likely to occur on the skin, human life and health are in danger as soon as there is a noticeable injury to the skin. Skin tissue damage puts the patient's life in danger and adds to the financial and social burden of their care by causing a disturbance of body homeostasis. Currently, wound healing—especially for chronic wounds—remains a crucial clinical concern. Chronic wounds that take a long time to heal frequently afflict patients, the medical system, and the social economy as a whole. As of right now, there's no practical way to treat persistent wounds in a clinic. Angiogenesis is a critical phase in the recovery of wounds. Nanobiomaterials have been thoroughly examined in angiogenesis because they possess features not found in ordinary materials. Persistent wounds triggered by aging, inadequate healing of acute injuries, diabetic foot ulcers (DFUs), and pressure ulcers are the most common types. The International Diabetes Federation has released data indicating that 425 million individuals globally had diabetes in 2017. This figure will increase to 629 million by 2045. DFUs, a leading source of complications from diabetes, patient amputations, and recurrent hospital stays, affects 25% of patients with diabetes. Additionally, the morbidity and fatality rates for DFUs are very high [38].

10.6.1 Overview of Skin Tissue Engineering

The Washington National Science Foundation bioengineering panel meeting first formally introduced the word "tissue engineering" in 1987. In the year 1993, Langer and Vacanti methodically expanded on its meaning. In order to create tissue analogs for skin repairs and enhancements, skin tissue engineering combines the ideas of biology and engineering. The fields of medicine and dermatology have greatly benefited from the advancement of skin tissue engineering. There are two basic approaches to vascularized tissue engineering. The initial technique is the use of endothelial cells with their capacity to generate new blood vessels; the other is the use of artificial tubular scaffolds or vessel systems originating from biology. The constraints of common artificial vessels are anticipated to be overcome by nanobiomaterial vessels, which closely mimic the structure of natural vessels.

The structure of the nanobiomaterial vessels closely resembles that of biological vessels, and they are likely capable of overcoming the limitations of traditional synthetic vessels. For instance, pro-angiogenic peptides can be released locally and continuously by using NPs as carriers for drugs and targeting damaged tissue. In the meantime, nanoparticles can significantly improve target cells' replication of angiogenic genes as nonviral carriers with comparatively few safety issues. NPs have the potential to improve cell survival, continuation, and angiogenic factor production when combined with cell treatment. The use of nanoparticles in tissue engineering has created several opportunities recently, particularly in the area of blood vessel development, which presents a major obstacle to the field of regenerative medicine. In this study, the link between angiogenesis and the healing of wounds is outlined, the molecular mechanisms of angiogenesis are summarized, and the types, benefits, and limitations of associated nanomaterials are highlighted [39].

10.6.2 Integration of Nanobiomaterials in Tissue Engineering Approaches

Nanobiomaterials are so small—less than 100 nm—that they may communicate with and absorb into cells, triggering a biological reaction. It is possible for biomaterials to activate a cell's receptors and direct it to take on specific actions. Carbon nanotubes are considered to be the most widely employed nanomaterials in tissue engineering. They are composed of coiled graphene sheets with a cylindrical shape and sp2-hybridized carbon atoms that are confined in a hexagonal pattern. One or several concentric rolled graphene layers can be used to create carbon nanotubes (CNTs). They go by the moniker of multiwalled CNTs (MWCNTs) in the final scenario. Their typical dimensions are in the micrometer range for length and up to 100 nanometers for width. Their contact with cells is facilitated by their significant aspect ratio. The surface of carbon nanotubes can be modified to further improve biological compatibility and to take advantage of their exceptional electrical qualities in neuronal tissue engineering to hasten the repair of injured neurons [40]. Nanostructures can be engineered to serve as carriers for bioactive molecules, growth factors, or drugs. By controlling the release kinetics, nanobiomaterials enable sustained and localized delivery of therapeutic agents, promoting tissue regeneration while minimizing systemic side effects. Nanostructured biomaterials can influence cellular behaviors such as adhesion, migration, and differentiation. Surface topography, stiffness, and chemical composition of nanobiomaterials can be tailored to guide specific cell responses, facilitating tissue regeneration. Some nanostructures possess inherent imaging properties (e.g., quantum dots, gold nanoparticles) that enable noninvasive monitoring of tissue regeneration processes. These nanomaterials can be used for real-time tracking of scaffold degradation, cell migration, and tissue formation. Incorporation of bioactive nanoparticles, such as nanoscale hydroxyapatite or bioactive glasses, can promote osteogenesis or dentinogenesis in bone and dental tissue engineering applications, respectively [41].

10.6.3 Advantages of Nanobiomaterials in Skin Regeneration

Nanobiomaterials offer several advantages in the field of skin regeneration due to their unique properties at the nanoscale. Nanobiomaterials can penetrate the skin barrier more effectively compared to conventional materials, allowing them to deliver therapeutic agents, growth factors, or stem cells directly to the target site for regeneration. They enable the controlled release of bioactive molecules, such as growth factors or drugs, over extended periods. This controlled release kinetics can mimic the natural healing process of the skin and improve regeneration outcomes. In order to improve biological compatibility and lower the possibility of immunological rejection or adverse reactions, nanostructured materials can be made that are comparable to the extracellular matrix of the skin [41]. In order to facilitate particular biological responses, such as the adhesion of cells, proliferation, or differentiation, which are essential for skin regeneration, nanobiomaterials can be readily functionalized with bioactive chemicals, such as peptides. Theranostic techniques for customized healthcare can be made possible by nanomaterials, which integrate medicinal and diagnostic characteristics onto a single platform. This enables physicians to customize treatment plans based on the reactions of specific patients [42].

10.7 CLINICAL IMPLICATIONS AND FUTURE DIRECTIONS

Research in the field of nanotechnology has grown at an unparalleled rate in recent years. There is growing hope that the application of nanomaterials in dentistry and medicine may result in major advancements in illness avoidance, detection, and treatment. A new area known as nanomedicine is emerging as a result of growing interest in the potential medical uses of nanotechnology. In order to be applied more widely, nanomedicine must overcome its current obstacles, advance our knowledge of the pathophysiologic causes of disease, provide more advanced diagnostic options, and produce more potent treatments and preventative measures. Nanotechnology has the potential to significantly impact dental care, medicine, and human existence in ways that many previous advancements have not. Like any technology, nanotechnology has a great deal of scope for abuse and exploitation on a scale and extent that has never been observed before. However, it could also result in important advantages, including better health, more efficient use of resources from nature, and less contamination of the environment. These are indeed amazing and miraculous days. The current focus of research is on the latest advancements in dental management, specifically with regard to NPs and nanotubes. Materials derived from these developments, including nanocomposites, nanoporous materials, and nanomembranes, will become increasingly important in the fabrication of dental industry substances [43].

10.7.1 Clinical Trials and Studies on Nanobiomaterials in Diabetic Wound Healing

The physiological process is extremely dynamic and complex throughout wound healing. Diabetic wounds are harder to recover from than nondiabetic wounds. In contrast to typical wounds, diabetic wounds exhibit hypoxia, decreased angiogenesis, persistent inflammation, and elevated matrix metalloproteinase production, all of which contribute to the wounds' delayed recovery. Biologic NPs exhibit important promise in the healing of diabetic wounds because they have rich functional group structures that enable them to attach to medicines and growth factors more effectively, whereas metallic NPs' ions have excellent antimicrobial capabilities. Investigators have found NPs with a variety of biological characteristics that can be used in diabetic wound care to address some of the shortcomings of medical care as a result of the growth of society and science [44].

10.7.2 Challenges and Limitations

The use of nanobiomaterials has increased within medicine over the last ten years, and a variety of NP systems have been employed in clinical settings. Polymeric, metallic, and lipid-based NPs have all found a place in medicine, despite their inadequate performance in clinical research. Generally, these NPs offer improved medicinal or pharmacological efficacy in comparison to normal drug therapies. Despite persistent requests from the scientific community, there is right now very little regulatory guidance in this area, regardless of the high level of excitement related to the field of nanotechnology and its potential impact on pharmaceutical companies. This is crucial because it will give manufacturers, legislators, medical professionals, and the general public legal certainty.

The manufacture of nanomaterials and nanomedicines for use in health-related applications is not formally regulated.

They may be categorized as medical devices or medications, and different international regulators may have different opinions on this. Because nanobiomedicine may be classified as a medical device in some countries and as a pharmaceutical in others, different countries may have different regulations that need to be followed. Another challenge that regulators face is nanotoxicology and biological response. Before being approved for clinical usage, the mechanism and action of drug delivery in nanomedicine development require extensive preclinical safety data, including information on side effects. High dosages of drugs in nanoform can have harmful effects on specific cells or organs, which can be fatal for people with diabetes or chronic renal failure, or the developing resistance to antibiotics. Furthermore, because the particles are smaller than their larger counterparts, they could be dangerous to patients. This enables them to get through the blood–brain barrier, which may eventually impair neurological function or, at the very least, result in edema [45].

10.7.3 Future Prospects and Potential Advancements

In the future, it may be possible to create nanobiomaterials that will release growth factors, cytokines, and the extracellular matrix, thereby accelerating the healing of wounds through the promotion of blood vessel development, tissue regeneration, and cell proliferation. It is possible to engineer nanostructures to modify the immune response, supporting an appropriate healing cycle and lowering the level of inflammation. People with compromised immune systems and those with persistent wounds may benefit from this method. By incorporating nanosensors into dressings for wounds, it is possible to track the advancement of wound healing, and identify infections or other problems instantly [46].

10.8 CONCLUSION

In conclusion, nanobiomaterials are being employed as a therapeutic approach for skin tissue engineering and wound healing because they enable the regulated and prolonged release of growth factors. By encapsulating growth factors in nanomaterials, this method encourages tissue regeneration and cell multiplication. The broad range of applications that nanomaterials may have in the treatment of both acute and chronic wounds is encouraging for the direction that wound care is taking. Nanoparticles have demonstrated potential as a means to distribute and release therapeutic drugs over time in wound dressings. Moreover, light can be absorbed by wound healing nanoparticles and transformed into reactive oxygen species, which destroy any bacteria that might be present in the wound. Thus, there is hope for the future in the identification and management of wound infections with nanoparticle-based wound therapy.

REFERENCES

1. Jiang, T., Li, Q., Qiu, J., Chen, J., Du, S., Xu, X., & Chen, T. (2022). Nanobiotechnology: Applications in chronic wound healing. *International Journal of Nanomedicine*, *17*, 3125.
2. Du, J., & Wong, K. K. (2019). Nanomaterials for wound healing: Scope and advances. In Wenguo Cui & Xin Zhao (Eds.), *Theranostic bionanomaterials* (pp. 211–230). Elsevier.
3. Chhabra, P., & Bhati, K. (2020). Bionanomaterials: Advancements in wound healing and tissue regeneration. *Recent Advances in Wound Healing*.
4. Naves, C. C. (2016). The diabetic foot: A historical overview and gaps in current treatment. *Advances in Wound Care*, *5*(5), 191–197.
5. Okur, M. E., Bülbül, E. Ö., Mutlu, G., Eleftherıadou, K., Karantas, I. D., Okur, N. Ü., & Siafaka, P. I. (2022). An updated review for the diabetic wound healing systems. *Current Drug Targets*, *23*(4), 393–419.

6. Roszek, B., De Jong, W. H., & Geertsma, R. E. (2005). *Nanotechnology in medical applications: state-of-the-art in materials and devices*. National Institute for Public Health and the Environment.
7. Mansfeld, F. M., Davis, T. P., & Kavallaris, M. (2017). Nanotechnology in medical research. In Mariusz Skwarczynski & Istvan Toth (Eds.), *Micro and nanotechnology in vaccine development* (pp. 21–45). William Andrew Publishing.
8. Singh, N. E. E. T. U., & Sharma, U. C. (2014). Introduction to nano-biomaterials. *Nanotechnology, 11*, 1–12.
9. Rahman, M., Laurent, S., Tawil, N., Yahia, L. H., Mahmoudi, M., Rahman, M., ... Mahmoudi, M. (2013). The biological significance of "Nano"-interactions. *Protein-Nanoparticle Interactions: The Bio-Nano Interface*, 1–20.
10. Das, S., & Baker, A. B. (2016). Biomaterials and nanotherapeutics for enhancing skin wound healing. *Frontiers in Bioengineering and Biotechnology, 4*, 82.
11. Singh, V. A., Gade, W., Vats, T., Lenardi, C., & Milani, P. (2010). Nanomaterials: New generation therapeutics in wound healing and tissue repair. *Current Nanoscience, 6*(6), 577–586.
12. Xu, Y., Jiang, J., Lv, X., Li, H., Yang, D., Wang, W., & Cai, Y. (2024). Environment-triggered nanoagent with programmed gas release performance for accelerating diabetic infected wound healing. *Chemical Engineering Journal, 479*, 147645.
13. Wan, R., Weissman, J. P., Grundman, K., Lang, L., Grybowski, D. J., & Galiano, R. D. (2021). Diabetic wound healing: The impact of diabetes on myofibroblast activity and its potential therapeutic treatments. *Wound Repair and Regeneration, 29*(4), 573–581.
14. Khalilov, R. (2023). A comprehensive review of advanced nano-biomaterials in regenerative medicine and drug delivery. *Advances in Biology & Earth Sciences, 8*(1).
15. Chakrabarti, S., Chattopadhyay, P., Islam, J., Ray, S., Raju, P. S., & Mazumder, B. (2019). Aspects of nanomaterials in wound healing. *Current Drug Delivery, 16*(1), 26–41.
16. Muneeswaran, T., Maruthupandy, M., Vennila, T., Sundararaju, S., & Quero, F. (2023). Nano-biomaterials for therapeutic and diagnostic applications. In *Applications of multifunctional nanomaterials* (pp. 617–649). Elsevier.
17. Yasir, M., Mishra, R., Tripathi, A. S., Maurya, R. K., Shahi, A., Zaki, M. E., & Masand, V. H. (2024). Theranostics: A multifaceted approach utilizing nano-biomaterials. *Discover Nano, 19*(1), 35.
18. Findik, F. (2021). Nanomaterials and their applications. *Periodicals of Engineering and Natural Sciences, 9*(3), 62–75.
19. Naderi, N., Karponis, D., Mosahebi, A., & Seifalian, A. M. (2018). Nanoparticles in wound healing; from hope to promise, from promise to routine. *Frontiers in Bioscience, 23*, 1038–1059.
20. Rajendran, N. K., Kumar, S. S. D., Houreld, N. N., & Abrahamse, H. (2018). A review on nanoparticle based treatment for wound healing. *Journal of Drug Delivery Science and Technology, 44*, 421–430.
21. Haque, S. T., Saha, S. K., Haque, M. E., & Biswas, N. (2021). Nanotechnology-based therapeutic applications: In vitro and in vivo clinical studies for diabetic wound healing. *Biomaterials Science, 9*(23), 7705–7747.
22. Ezhilarasu, H., Vishalli, D., Dheen, S. T., Bay, B. H., & Srinivasan, D. K. (2020). Nanoparticle-based therapeutic approach for diabetic wound healing. *Nanomaterials, 10*(6), 1234.
23. Pamu, D., Tallapaneni, V., Karri, V. V. S. R., & Singh, S. K. (2022). Biomedical applications of electrospun nanofibers in the management of diabetic wounds. *Drug Delivery and Translational Research*, 1–9.
24. Hayat, M., Nawaz, A., Chinnam, S., Muzammal, M., Latif, M. S., Yasin, M., Ashique, S., Zengin, G., & Farid, A. (2023, July 1). Formulation development and optimization of herbo synthetic gel: In vitro biological evaluation and in vivo wound healing studies. *Process Biochemistry, 130*, 116–126.
25. Adil, A. M. A. (2020). Scaffolds in tissue engineering. *Annals of RSCB, 24*(1), 142–152.
26. Shameem, M. M., Sasikanth, S. M., Annamalai, R., & Raman, R. G. (2021). A brief review on polymer nanocomposites and its applications. *Materials Today: Proceedings, 45*, 2536–2539.
27. Nosrati, H., Aramideh Khouy, R., Nosrati, A., Khodaei, M., Banitalebi-Dehkordi, M., Ashrafi-Dehkordi, K., & Alizadeh, Z. (2021). Nanocomposite scaffolds for accelerating chronic wound healing by enhancing angiogenesis. *Journal of Nanobiotechnology, 19*, 1–21.
28. Nandhini, J., Karthikeyan, E., & Rajeshkumar, S. (2024). Nanomaterials for wound healing: Current status and futuristic frontier. *Biomedical Technology, 6*, 26–45.
29. Jiang, T., Li, Q., Qiu, J., Chen, J., Du, S., Xu, X., ... Chen, T. (2022). Nanobiotechnology: Applications in chronic wound healing. *International Journal of Nanomedicine, 17*, 3125.
30. Bai, Q., Han, K., Dong, K., Zheng, C., Zhang, Y., Long, Q., & Lu, T. (2020). Potential applications of nanomaterials and technology for diabetic wound healing. *International Journal of Nanomedicine*, 9717–9743.

31. Batool, F., Özçelik, H., Stutz, C., Gegout, P. Y., Benkirane-Jessel, N., Petit, C., & Huck, O. (2021). Modulation of immune-inflammatory responses through surface modifications of biomaterials to promote bone healing and regeneration. *Journal of Tissue Engineering*, *12*, 20417314211041428.
32. Zhang, J., Liu, H., Che, T., Zheng, Y., Nan, X., & Wu, Z. (2023). Nanomaterials for diabetic wound healing: Visualization and bibliometric analysis from 2011 to 2021. *Frontiers in Endocrinology*, *14*, 1124027.
33. Kushwaha, A., Goswami, L., & Kim, B. S. (2022). Nanomaterial-based therapy for wound healing. *Nanomaterials*, *12*(4), 618.
34. Rawat, N., Ahmad, N., Raturi, P., Singhvi, N., Sahai, N., & Kothiyal, P. (2023). Nanobiomaterials: Exploring mechanistic roles in combating microbial infections and cancer. *Discover Nano*, *18*(1), 158.
35. El-Nablaway, M., Rashed, F., Taher, E. S., Foda, T., Abdeen, A., Abdo, M., ... Atia, G. A. (2024). Prospectives and challenges of nano-tailored biomaterials-assisted biological molecules delivery for tissue engineering purposes. *Life Sciences*, 122671.
36. Azari, Z., Gorgani, S., Hosseini, S. A., Wang, A. Z., Kim, H. W., & Kargozar, S. (2023). The role of immune cells in therapeutic angiogenesis: Concepts in tissue engineering. *Current Opinion in Biomedical Engineering*, 100470.
37. Wang, X., Li, R., & Zhao, H. (2024). Enhancing angiogenesis: Innovative drug delivery systems to facilitate diabetic wound healing. *Biomedicine & Pharmacotherapy*, *170*, 116035.
38. Nosrati, H., Aramideh Khouy, R., Nosrati, A., Khodaei, M., Banitalebi-Dehkordi, M., Ashrafi-Dehkordi, K., ... Alizadeh, Z. (2021). Nanocomposite scaffolds for accelerating chronic wound healing by enhancing angiogenesis. *Journal of Nanobiotechnology*, *19*, 1–21.
39. Cui, L., Liang, J., Liu, H., Zhang, K., & Li, J. (2020). Nanomaterials for angiogenesis in skin tissue engineering. *Tissue Engineering Part B: Reviews*, *26*(3), 203–216.
40. Di Marzio, N., Eglin, D., Serra, T., & Moroni, L. (2020). Bio-fabrication: Convergence of 3D bioprinting and nano-biomaterials in tissue engineering and regenerative medicine. *Frontiers in Bioengineering and Biotechnology*, *8*, 326.
41. Kumar, P., Saini, M., Dehiya, B. S., Sindhu, A., Kumar, V., Kumar, R., ... Thakur, R. (2020). Comprehensive survey on nanobiomaterials for bone tissue engineering applications. *Nanomaterials*, *10*(10), 2019.
42. Jajal, D., & Shah, S. (2022). Application of nanobiomaterials in soft tissue engineering. In Huinan H. Liu, Tolou Shokuhfar, Sougata Ghos (Eds.), *Nanotechnology in medicine and biology* (pp. 79–108). Elsevier.
43. Jandt, K. D., & Watts, D. C. (2020). Nanotechnology in dentistry: Present and future perspectives on dental nanomaterials. *Dental Materials*, *36*(11), 1365–1378.
44. Qin, W., Wu, Y., Liu, J., Yuan, X., & Gao, J. (2022). A comprehensive review of the application of nanoparticles in diabetic wound healing: Therapeutic potential and future perspectives. *International Journal of Nanomedicine*, *17*, 6007.
45. Foulkes, R., Man, E., Thind, J., Yeung, S., Joy, A., & Hoskins, C. (2020). The regulation of nanomaterials and nanomedicines for clinical application: Current and future perspectives. *Biomaterials Science*, *8*(17), 4653–4664.
46. Nandhini, J., Karthikeyan, E., & Rajeshkumar, S. (2024). Nanomaterials for wound healing: Current status and futuristic frontier. *Biomedical Technology*, *6*, 26–45.

11 Role of Antibacterial Nanomaterials in Wound Healing

Roomah Javed[1], Maria Faraz[1], Arshad Farid[1], Alamgir Khan[1], and Julius Tibyangye[2,3]*

[1] Gomal University, D.I. Khan, Pakistan
[2] Muni University, Arua, Uganda
[3] University of Aberdeen, Aberdeen, UK
[*] Corresponding Author

11.1 INTRODUCTION

Damage or injury to healthy skin is referred to as a wound. It is caused by different traumatic events. There are three main types of wounds: mechanical disturbance ensuing from exterior aspects like surgical incisions, war injuries, and traffic accidents; shock due to exposure to extreme temperatures or harmful chemicals, such as frostbite; and chronic ulcerative wounds and burns (Ma, Cheng et al., 2019; Sorg, Tilkorn et al., 2017). Given the prevalence of trauma in society, effective wound healing remains a critical aspect of medical care. Advances in modern medicine have led to diverse wound healing approaches. Green nanomaterials have exhibited a substantial role in different biomedical applications, as well as wound healing. These biodegradable resources possess inimitable properties that support the process of wound healing. Many green nanomaterials, extracted from plants, exhibit essential antimicrobial properties, helping to prevent wound infections that could impede healing (Cui, Yu et al., 2022). Additionally, green nanomaterials may have anti-inflammatory effects, reducing pain and promoting faster healing at the wound site. When merged with wound dressings or scaffolds, these materials can assist as drug carriers and growth factor regulators (Nasibova, 2023). Nanomaterials are significantly used because they are environmentally friendly (Ahmadov, Silva, Arit, Rahman et al., 2021).

The collection and preparation of nanomaterials are influenced by the specific nature of the lesion, the requirements of patients, and the preferred healing result. Moreover, controlling deliberations are dominant when retaining nanomaterials in medicinal applications. As a result, these substances undergo rigorous testing and evaluation to verify their safety, protection, and expediency in promoting wound healing. Notably, zinc oxide nanoparticles and silver nanoparticles show strong antibacterial properties (Godoy-Gallardo, Eckhard et al., 2021). These nanoparticles directly interact with microbes triggering harm to their cellular structures and inhibiting their propagation. By preventing and treating infections, nanomaterials create an ideal environment for wound healing of wounds (Makvandi, Wang et al., 2020).

DOI: 10.1201/9781003504948-11

11.2 WOUND HEALING

A skin wound disrupts the skin's structure and impairs its normal functions. A variety of factors, including cuts, chemicals, burns, and certain diseases, are responsible for causing such wounds, and may also involve damage to deeper tissues (Nosrati, Heydari et al., 2023; Nosrati, Aramideh Khouy et al., 2021).

Skin wounds are typically categorized as either acute or chronic. Acute wounds heal relatively quickly, usually showing signs of recovery within three months. They are often the result of trauma, surgical procedures, or burns, and their healing follows the standard phases of the wound healing process.

Chronic wounds, though, take longer than 90 days to settle and are often characterized by significant inflammation and scarring (Falanga, Isseroff et al. ,2022). Examples of these include diabetic foot ulcers, pressure ulcers, venous ulcers, and arterial insufficiency ulcers (Zeng, Lin et al., 2018). Sometimes, acute wounds that do not heal properly can also become chronic wounds. It is estimated that about 35% of all wounds fall into the chronic category. Chronic wounds are of particular clinical significance due to their rising prevalence and the increasing awareness of the morbidity and socioeconomic impact associated with them (Gupta, Sagar et al., 2021). Given the vital roles that the skin plays, it is crucial to restore its integrity as swiftly and efficiently as possible.

Wound healing is tightly controlled and involves the regeneration of wounded skin through four overlying steps: hemostasis, inflammation, proliferation, and remodeling (Rodrigues, Kosaric et al., 2019).

a) **Hemostasis**: Hemostasis is the initial stage of the curative process of wound healing, acting as the body's immediate response to injury (Han, 2023). During this initial phase, which lasts only a few minutes, the main goal is to stop bleeding. Platelets are crucial players at this stage. When platelets become activated, they form a fibrin clot. These platelets also release growth factors and cytokines, involved in the regulation of the healing process. The fibrin clot, although temporary, is essential for stopping bleeding and provides a framework for other cells involved in wound healing to move in and do their job (Opneja, Kapoor et al., 2019).

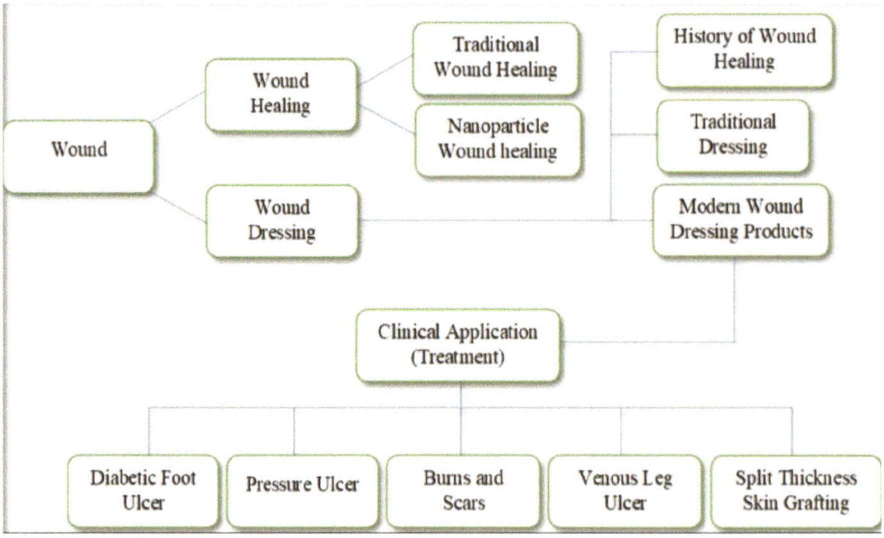

FIGURE 11.1 Wound healing.

b) **Inflammation**: The inflammatory phase, which aims primarily to stop infection, occurs at the same time as the hemostasis phase and can last up to 48 hours after wounding (Strudwick and Cowin, 2018). During this phase, neutrophils, lymphocytes, and monocytes infiltrate and migrate to the wound site (Singh, Young et al., 2017). These cells manufacture reactive oxygen species and proteinases to protect against microbial infections. The proliferative phase of wound healing begins with growth factors and cytokines secreted by these inflammatory cells (Behm, Babilas et al., 2012).
c) **Proliferation**: This phase ensues practically one and a half weeks after an injury and is a multifaceted stage involving cell propagation, maturation, development of granulation tissue, re-epithelialization, and wound reduction (Behm, Babilas et al., 2012). During angiogenesis, new capillaries form from existing blood vessels, creating an intricate network. A major growth factor in angiogenesis, known as vascular endothelial growth factor (VEGF), is secreted in response to hypoxia (Adair and Montani, 2011). Endothelial cell proliferation, migration, and formation of tubular structures are stimulated by VEGF. Matrix metalloproteinases (MMPs), a group of calcium-dependent zinc-containing endopeptidases, play an important role in angiogenesis through extracellular matrix (ECM) remodeling. As maturation progresses, capillary sprouts permeate the fibrin clot, producing a dense vascular network. Fibroblasts and macrophages are involved in the advancement of granulation tissue (Adair and Montani, 2011). Fibroblasts progressively hide ECM apparatuses. The configuration of collagen is also crucial for the retrieval of skin integrity. When enough ECM has been produced, some fibroblasts, influenced by macrophages, differentiate into a myofibroblast phenotype. These contractile cells help in wound contraction and are also involved in collagen deposition and maturation in wound healing. Ultimately, granulation tissue is replaced by normal tissue in the successive healing phase (Rodrigues, Kosaric et al., 2019).
d) **Remodeling**: Remodeling is the last phase that results in the advancement of scar tissue and normal epithelium (Spielman, Griffin et al., 2023). During this phase, apoptosis is caused in many myofibroblasts, macrophages, and endothelial cells. The tensile strength of the tissue is enhanced as type III collagen is substituted by type I collagen, enhancing strength (Spielman, Griffin et al., 2023). However, despite these changes, the repaired tissue will never recuperate the tensile strength of the original, healthy tissue preceding the injury.

11.3 NANOMATERIALS IN WOUND HEALING

Nanotechnology, a vital component of tissue engineering, has seen swift advancements, leading to the creation of nanomaterials containing several applications in biomedicine. These include transportation of drugs, antibacterial and anti-inflammatory applications, gene transfection, and imaging technologies (Min, Lee et al., 2018). Nanomaterials possess significant characteristics such as the ability to stimulate cell differentiation and control cellular immunity (Hayat et al., 2023).

Nanomaterials are changing wound healing by leveraging their unique properties. These tiny materials, with specific surface characteristics and high surface-to-volume ratios, can influence cellular and molecular processes. Researchers are excited about using nanomaterials to create advanced wound healing therapies. By incorporating antibacterial properties, controlled drug release, and signaling molecules, nanomaterials offer new possibilities for tissue rejuvenation, contamination control, and minimizing scars (Nandhini, Karthikeyan et al., 2024; Kushwaha, Goswami et al., 2022; Tocco, Zavan et al., 2012).

Nanomaterials can be engineered to possess antibacterial, anti-inflammatory, and regenerative properties, all of which are crucial for effective wound management. These tiny particles represent a new era in drug delivery for wound healing (Naskar and Kim, 2020). Specifically designed to be nonantigenic, durable under shear stress, and supportive of gas exchange, nanomaterials offer advantages for wound dressings (Naskar and Kim, 2020). Researchers are exploring their potential

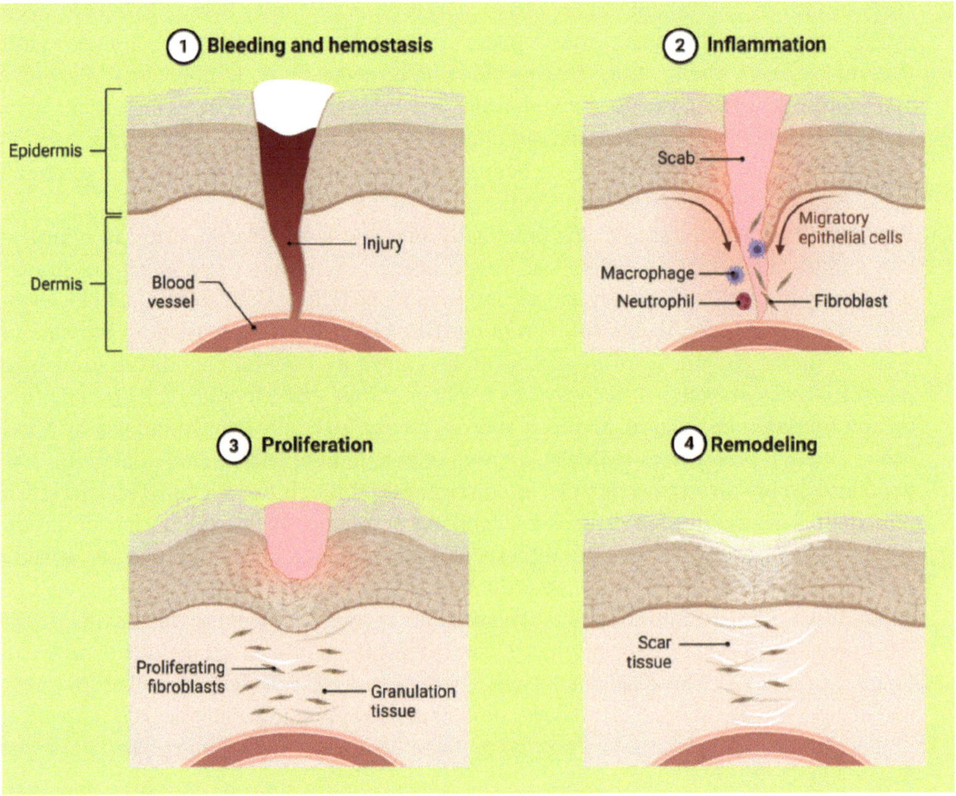

FIGURE 11.2 Stages of wounds healing.

to prevent infections, especially in the context of biofilms, which are notorious for their antibiotic resistance (Wang, Lu et al., 2019). Certain pharmaceutical nanoparticles have shown remarkable efficacy in promoting skin regeneration and preventing infections. Nanomaterials play a role in all four phases of wound healing: hemostasis, inflammation, proliferation, and remodeling. Their application holds promise for more effective and targeted wound therapies across various wound types (Wang, Lu et al., 2019). Two common types of nanomaterials in wound therapy include those with intrinsic wound-healing properties and those used as carriers for therapeutic agents.

Researchers have been studying how tiny materials called nanomaterials can help wounds heal faster. These nanomaterials can be made from metals like silver, gold, and zinc. When these metal nanoparticles are added to wound dressings, they can enhance the process of wound healing and also have antibacterial effects. Nanomaterials can be categorized into two main types: inorganic and organic materials (Chopra, Kumar et al., 2022).

11.4 INORGANIC NANOMATERIALS

Inorganic nanomaterials lack significant carbon content and can consist of either metal or nonmetal elements. These materials are typically stable and inert, making them attractive because of their optical and magnetic features (Dong, Chen et al., 2015). Inorganic nanomaterials fall into three primary categories: metallic, polymeric, and ceramic nanomaterials (Rizwan, Shoukat et al., 2021). Metal and metal oxide nanoparticles are frequently used for the treatment of wounds. Compared to traditional methods, nanoparticles are favorably used for the treatment of several types of lesions or injuries. They inherently have comparable wound-healing and antibacterial properties (Rizwan, Shoukat et al., 2021). Gold, silver, and zinc oxide are included in metallic nanoparticles.

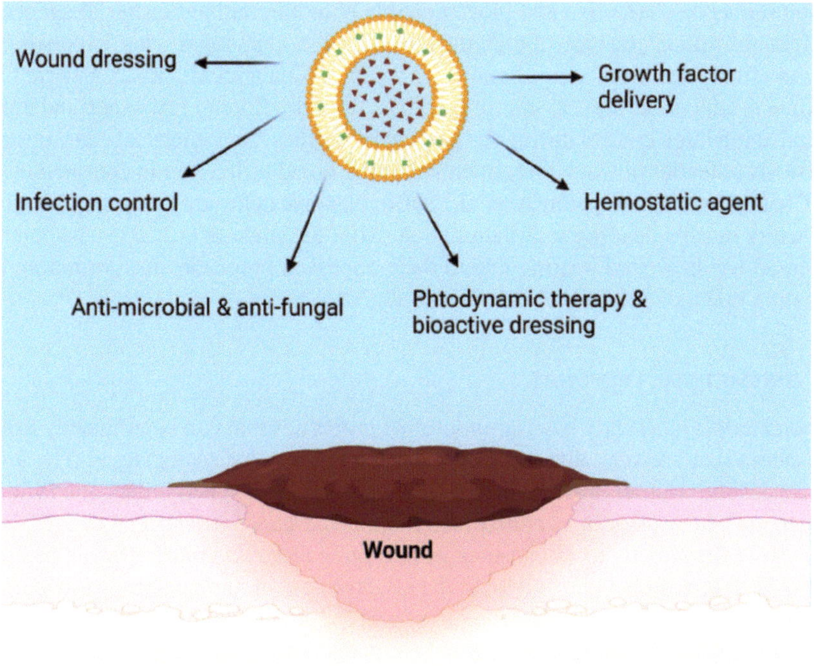

FIGURE 11.3 Applications of nanomaterials in wound healing.

Inorganic nanomaterials, naturally composed of different materials, are used extensively in a variety of biomedicine applications due to their inimitable thermodynamic possessions (Rizwan, Shoukat et al., 2021). Inorganic nanomaterials carry therapeutic agents in a meticulous fashion or exhibit inherent properties, such as antiseptic, antiphlogistic, and oxidation inhibition, serving as a therapeutic agent (Wang, Zhong et al., 2021; Rajendran, Kumar et al., 2018). They facilitate communication and dissemination at the site of the wound, support growth of cells and signaling, and inspire the formation of blood vessels. Among their therapeutic applications, inorganic nanomaterials are particularly studied for their antibacterial effects (Li, Chen et al., 2021). Commonly used metal ions for wound healing include Ag+, Au+, Cu2+, Zn2+, and Co2+ (Qi, Shen et al., 2023). Ag+ is proficient in combating bacteria, including both gram-positive and gram-negative bacteria, without inducing antimicrobial resistance. Presently, the practice of silver ions for antibacterial applications remains an active area of research. The antimicrobial activities of Ag+ nanoparticles are due to their reported toxicity to human and animal cells. Au+ nanoparticles have gained substantial attention in wound healing because they have less toxic effects and are biocompatible (Qi, Shen et al., 2023). Some inorganic nanoparticles can generate reactive oxygen species (ROS) with the help of photodynamic therapy or chemodynamic therapy. Reactive oxygen species are used for infection control during the wound healing process of wounds because of their antimicrobial properties. Photodynamic therapy is a noninvasive approach where ROS are generated by photosensitizers, which react with nearby oxygen species. In general, photosensitizers are mostly made of inorganic semiconductor materials, which exhibit tremendous antibacterial properties through the generation of ROS when exposed to UV light (Ren, Liu et al., 2020).

Photothermal therapy has recently been recognized as an operative substitutive treatment method for infectious bacterial infections. Specific inorganic nanoparticles can function as photothermal particles, producing substantial heat that denatures vital proteins and subsequently eliminates bacteria (Ren, Liu et al., 2020). Photothermal therapy quickly kills bacteria, preventing the emergence of antibiotic-resistant bacteria. MoS^2, CuS, and 2D materials, which are noble metals, are commonly used as cost-effective and biocompatible photothermal antibacterial agents due to their superior light absorption and ease of manufacture. The hybrid nanoparticles demonstrated excellent antibacterial effects against *S. aureus* and *E. coli* (Zhou, Zhang et al., 2022). Hybrid nanoparticles with multiple mechanisms prevent microorganisms from developing resistance and are extensively used for active infection control during the healing of wounds. Nanomaterials face obstacles despite their enormous potential in wound treatment and their broad utilization in commercial and clinical settings (Chakrabarti, Chattopadhyay et al., 2019). Nanoparticles are noxious, potentially causing oxidative stress in cells, leading to inflammation (Min, Suminda et al., 2023). Nanomaterials have seen increased use in wound healing due to their improved properties for promoting tissue repair and combating infection (Kalashnikova, Das et al., 2015).

11.4.1 Antibacterial Properties

Some nanomaterials inherently have antimicrobial properties that can be efficiently used to counter various bacteria that are very difficult to treat with conventional medications. The amalgamation of nanoparticles into wound dressings can aid in infection control and reduce the formation of biofilms, which are often resistant to antibiotics. Engineered nanoparticles can deliver drugs, growth factors, or stem cells directly to wound sites, encouraging faster and more efficient healing (Mallick, Nag et al., 2022). Researchers are studying the effectiveness of nanomaterials, including antibacterial silver nanoparticles, in inhibiting the growth of common wound-infecting bacteria such as *E. coli* and *S. aureus*. Furthermore, research is being conducted on carbon-based nanomaterials like graphene and carbon nanotubes for their potential to deliver antibiotics and other therapeutic agents to enhance the wound healing process. The development of nanomaterial-based wound dressings is an active area of research, with the goal of creating dressings that not only prevent infection but also support the body's natural healing processes (Girija, Balasubramanian et al., 2022). These advancements could lead to more effective wound treatments, improving patient outcomes and potentially reducing the strain on healthcare systems (Girija, Balasubramanian et al., 2022). Antimicrobial therapy efficiently eradicates wound infection by increasing the operative concentration of antibiotics. Hence, in order to reduce drug resistance and harmful effects from antibiotics, it should be ideally used only at the application site rather than systemic and long-term treatments. In wound injuries, the skin no longer serves as a barrier, meaning that any medication that is topically applied, including nanoparticles, could be greatly absorbed. Additionally, specific to topical applications, deliberation toward the half-life of nanoparticles in the wound site is important, which is also limited by the site cleaning services.

11.4.2 Types of Antibacterial Nanomaterials

11.4.2.1 Metal Nanoparticles

Metallic nanoparticles (NPs) play a significant role in numerous scientific disciplines, including medicine, physics, mechanics, and pharmaceutics (Shnoudeh, Hamad et al., 2019). Metallic nanoparticles are being explored for utilization in clinical settings because of their cost-effectiveness, high surface-to-volume ratio, stability, and safety (Uzair, Liaqat et al., 2020). In wound healing, nanoparticles are assessed based on three key factors: their role as delivery agents, their participation in the repair process, and their antimicrobial properties (Rajendran, Kumar et al., 2018). Metal nanoparticles, including gold, silver, copper, and silica, stand out as excellent choices for wound treatment and integration into bandages (Pormohammad, Monych et al., 2021). Metal nanoparticles have the ability to interact with different bacterial biochemical pathways, making them promising agents to prevent antimicrobial resistance in wounds. Integrating metals with other nanomaterials and polymers is an effective approach for designing wound dressings (Pormohammad, Monych et al., 2021).

11.4.2.1.1 Silver Nanoparticles (AgNPs)

Silver nanoparticles (AgNPs) are commonly used in the treatment of wound infections. Historical records indicate that silver has been employed for wound care since 1850 BC in Egypt, and its healing properties were documented in the medical texts of Hippocrates. Recently, growing interest has developed in the utilization of silver nanoparticles for various applications in the medical field because of their antibacterial properties. These NPs are now integrated into wound dressings, infection-preventing artificial implants, and to promote wound healing. Importantly, long-term therapy with AgNP-based dressings has not been associated with any negative effects (Nandhini, Karthikeyan et al., 2024). The small size and large surface area of AgNPs make it able to enter bacterial cells and prevent their growth. When combined with biopolymers, AgNPs are safe and effective for wound dressings. These combinations not only prevent infection but also accelerate wound healing (Liao, Li et al., 2019). Researchers have demonstrated their effectiveness in both laboratory tests (in vitro) and living organisms (in vivo) (Alavi and Varma, 2021; Alavi and infections, 2020). Negative pressure wound therapy promoting blood flow, removes excess fluid and results in faster tissue regeneration (Alavi and infections, 2020). This approach reduces scarring and infection risk, eventually enhancing the quality of life and reducing healthcare costs.

11.4.2.1.2 Gold Nanoparticles (AuNPs)

Gold nanoparticles (AuNPs) play a very important function in medicine. They enhance the healing of wounds, aid regeneration of tissue, and enable precise drug delivery (Ghafari, Asefnejad et al., 2024). By adjusting their surface plasmon resonance, we can boost their antimicrobial properties. These tiny particles prevent bacterial DNA from unwinding, inhibiting replication and transcription. As a result, they curb the development of harmful pathogens like *P. aeruginosa* and *S. aureus*. Additionally, AuNPs disrupt bacterial energy metabolism by altering membrane potential and obstructing ATP synthase activity, effectively eliminating bacteria (Nisar, Ali et al., 2019). Their antioxidant properties accelerate wound closure and prevent harmful reactive oxygen species generated by bacteria (Nisar, Ali et al., 2019).

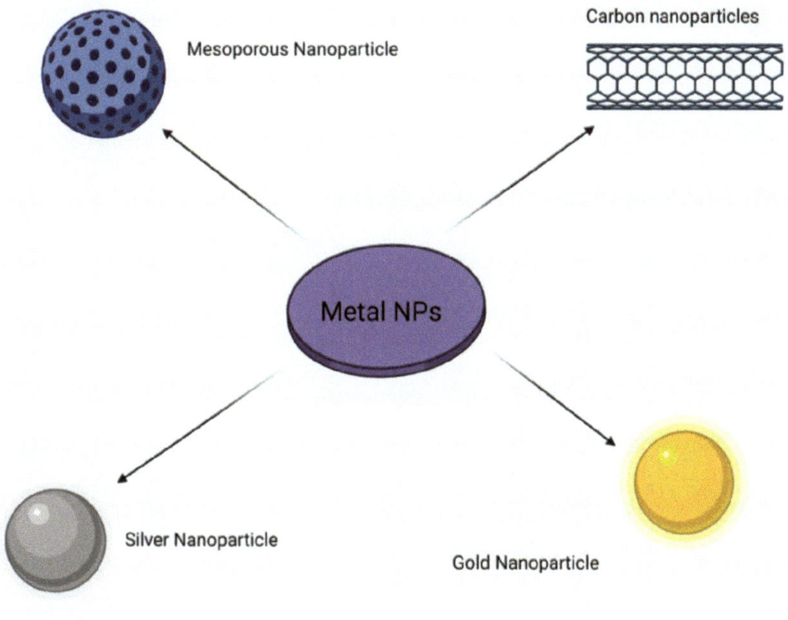

FIGURE 11.4 Types of metal nanoparticles.

The process of integrating AuNPs with materials like collagen, gelatin, and chitosan, followed by cross-linking, enhances the properties of these substances. By functionalizing AuNPs, we improve their compatibility with biological systems and their ability to break down naturally (Sharifiaghdam, Shaabani et al., 2022). When combined with chitosan, AuNPs significantly enhance wound healing. They scavenge free radicals and promote better tissue growth, outperforming chitosan alone in rat surgical wound models. Additionally, blending AuNPs with cryopreserved human fibroblasts accelerates burn wound healing, reduces inflammation duration, and speeds up collagen deposition.

Studies conducted on diabetic mice revealed that combining AuNPs with chitosan significantly enhanced wound healing (Sharifiaghdam, Shaabani et al., 2022). This nanocomposite promoted angiogenesis, accelerated ulcer closure, and exhibited antimicrobial properties, effectively protecting against wound infections and reducing inflammation, surpassing the effects of chitosan alone.

In test center experiments, scientists observed that integrating AuNPs into a chitosan hydrogel significantly boosted its antimicrobial effects, effectively protecting against wound infections and decreasing inflammation. Clinical trials have underscored the efficacy of platelet-derived growth factor (PDGF) hydrogels in picking up the task of enduring ulcers in patients with exterior artery disease, promoting fibroblast movement and collagen production, thus accelerating ulcer healing and reducing inflammation (Cherng, Lin et al., 2022).

11.4.2.1.3 Copper Nanoparticles (CuNPs)

Copper nanoparticles (CuNPs) are known for their effective antimicrobial properties, mainly in the context of healing wounds. These nanoparticles release copper ions (Cu2+), which coagulate protein structures and change enzymatic functions. The effectiveness of CuNPs is associated with their extent and absorption, with higher levels totally preventing bacteria like *E. coli*. CuNPs can adhere to and enter bacterial cell sheaths, with the released Cu2+ first breaking down the cell walls, then the cytoplasm, and eventually instigating bacterial death (Yu, Liu et al., 2023).

CuNPs have been found to be safe for the skin. CuNPs help prevent the early formation of biofilms by *Pseudomonas aeruginosa*, which can cause infections. When applied as a coating on wound dressings, CuNPs rapidly and efficiently destroy *Acinetobacter*, commonly found in skin wounds. CuNPs promote the growth of new blood vessels. CuNPs have recently been established into a multifunctional gel with wound healing properties (Nandhini, Karthikeyan et al., 2024). The rate of wound regeneration increased by 30–40%. In vitro cell line studies established

TABLE 11.1
Applications of Different Types of Nanomaterials

Nanomaterial	Applications
AgNPs	• Advancement of maturation • Microbial disturbance • Decrease irritation by controlling immune response • Excite propagation and movement of cells
AuNPs	• Delivers electrical stimulation to the wound • Reassures tissue regeneration • Nanocarriers • Gene therapy
CuNPs	• Wound closure in wound healing • Keratinocyte proliferation and collagen synthesis
SiNPs	• Recovers topical wound healing • Stability

fibroblast proliferation, signifying the biocompatible nature of the copper nanoparticles (Nandhini, Karthikeyan et al., 2024). CuNPs are known for triggering the wound healing signaling pathway by increasing nitric oxide levels in the wound tissue through electron paramagnetic resonance (Anandhi and Rajeshkumar, 2023).

CuNPs have also been employed in treating enduring lesions. CuNP-infused dressings offered a substantial reduction in wound size and enhanced outcomes in the healing process versus normal dressings. Moreover, CuNPs have been used efficiently to increase the expediency of antibiotic treatments in the healing of different wounds.

11.4.2.1.4 Silica Nanoparticles (SiNPs)

Silica-based materials have been employed particularly in wound care and scaffolds. Silicic acid released from silica nanoparticles (SiNPs) has been found effective in improving the healing process of wounds specifically on the skin surface (Janjua, Cao et al., 2023). This enhancement is accomplished by stimulating the movement and proliferation of skin fibroblast cells in the wounded area.

The actions of colloidal species simultaneous to silicic acid, along with their impact on cells, are influenced by the size and condensation level. When merged into wound dressings, nanocomposites containing SiNPs and collagen recover stability while reducing the toxicity of macrophages (Dam, Celik et al., 2023). The collagen rehydrated frozen tissues, while the globules of silica shortened sustained antibiotics and stimulated proliferation.

Researchers have discovered that combining silica nanoparticles and polyvinylpyrrolidone (PVP) as carriers can speed up wound healing (Wang, Chen et al., 2023; Nandhini, Karthikeyan et al., 2024). Wound healing was significantly enhanced by silica microspheres and collagen. Using silica nanoparticles and PVP as carriers led to even faster wound healing in diabetic patients (Yu, Zhu et al., 2023). PVP acted as a defensive barrier, averting infections and preserving the necessary moist environment for optimal healing. Moreover, the large surface area of the silica particles helps absorb any fluids from the wound, creating a clean and saturated atmosphere that speeds up the healing process (Haririan, Asefnejad et al., 2023).

11.4.2.2 Metal Oxide Nanomaterials

Due to their inimitable chemical and physical properties, nanomaterials have been considered suitable agents for therapeutic applications. Developments in nanotechnology and interdisciplinary fields have made it possible to use nanomaterials for the healing of wounds (Chung et al., 2019). Mostly, nanotechnology has enabled the production of materials of less than 200 nm in size, and therefore can be used for many nanomedical applications such as wound treatment and targeted delivery of medication (Biswas et al., 2014). Due to their small size and large surface area, nanoparticles have the ability to pass the blood–brain barrier and enter the pulmonary system by absorption via endothelial cells. These properties of nanoparticles provide better colloidal stability and better bioavailability (Rizvi and Zaid, 2001). Particularly metal oxide nanoparticles (MONPs) have been suggested as potential therapeutic agents in the medical field of medical because of their high stability, easy manipulation to the required size, no swelling variation, easy functionalization by various molecules, easy inclusion into hydrophilic and hydrophobic systems, and negative surface charge (Sanchez-Moreno et al., 2016). Depending on their size, stability, shape, purity, and surface characteristics, MONPs interact with different in vivo systems. Therefore, it is crucial to assess the morphology of MONPs. There is rapid advancement in the biomedical application of nanoparticles in recent years for the treatment and diagnosis of different diseases. Nanoparticles can be utilized for safe medical delivery, the healing of chronic wounds, and enhanced medical device performance.

There are many advantages of metal oxide nanoparticles in medicines for both diagnostic and therapeutic applications. Through the process of water-dispersible NP engineering, these particles can be employed in a vast array of essential and practical biomedical research projects. NPs are being used in targeted medication delivery, genetic and autoimmune illnesses, cancer, photosensitizers in photodynamic treatment, and imaging of different molecular markers in diagnostics (Sukhanova et al., 2018).

11.4.2.2.1 Zinc Oxide Nanoparticles (ZnONPs)

Since they are nontoxic, compatible with living organisms, and effective against germs, zinc oxide nanoparticles, or ZnONPs, are significantly used in the cosmetics and medicinal sectors (Jones et al., 2008). Studies indicate the dose-dependent inhibitory effect of ZnONPs on *P. aeruginosa* and *S. aureus* biofilm formation (Husain et al., 2022). ZnONPs have been revealed to be useful in the healing process and lowering the risk of infection when added to materials such as chitosan hydrogel (George et al., 2019), collagen dressings (Andonegi et al., 2020), or cellulose sheets (Khatri et al., 2014). For example, antibacterial activity has been shown by ZnO when added to aqueous solutions (Yimeng et al., 2023). Studies revealed that ZnONPs have high lethality against *Mycobacterium* through direct contact by generating reactive oxygen species. Moreover, throughout the healing process, researchers coated surgical implants with ZnONPs for the prevention of bacterial infection. ZnO minimizes the risk of implant-associated infection by effectively inhibiting the formation of biofilm by *S. aureus* and *P. aeruginosa* (Poscher and Salinas, 2020). It has been concluded from this example that ZnONPs are useful for wound healing by inhibiting biofilm formation and colonization of bacteria (LewisOscar et al., 2015). The ions released by ZnO-coated surgical implants inhibit bacterial growth within implants (McCarty and Percival, 2013). ZnONPs inhibit the biofilm formation of *S. aureus* and *P. aeruginosa* on the surface of the implants, therefore reducing the risk of developing dangerous infections linked to implants. Consequently, the patient's wound heals faster and more efficiently, and no bacterial colonization occurs at the site of the wound (Ahmad et al., 2022).

11.4.2.2.2 Titanium Dioxide Nanoparticles (TiO2NPs)

Another great choice for wound healing is the use of titanium dioxide nanoparticles (TiO2NPs) because they have the ability to greatly reduce inflammation and have potent antibacterial qualities. They are also hydrophilic, biocompatible, and have a high swelling index (Sivaranjani and Philominathan, 2016). TiO2NPs boost the production and release of chemokines, promoting wound healing (Batt et al., 2018). Bacterial cellulose–TiO2 nanoconjugates that are biocompatible and antibacterial have potential as wound dressings (Khalid et al., 2017). Staining of mice with burn injuries demonstrated that TiO2 facilitated angiogenesis, fibroblast migration, and re-epithelialization—all of which aided in quicker healing of wounds (David et al., 2022). TiO2NP wound dressings coated in chitosan and pectin exhibit promising wound-healing and antimicrobial properties (Archana et al., 2013). In one study, a diabetic rat with a chronic wound was treated with artificial skin made of nano-TiO2–chitosan–collagen (Peng et al., 2008). Results revealed that the use of nano-TiO2 wound dressing significantly increased the production of collagen, reduced the wound size, and increased re-epithelialization, indicating the potential use of the titanium dioxide nanomaterial for healing chronic wound (Veiseh et al., 2010). In a clinical investigation, pectin and chitosan wound dressings were used to treat patients with infected wounds (Suramwar et al., 2016). The pectin and chitosan wound dressing effectively enhanced the healing process and inhibited bacterial growth at the wound site (Sandoval et al., 2022). In another investigation, a special artificial skin made of nano-TiO2, chitosan, and collagen was employed to treat diabetic rats with chronic wounds. This unique dressing significantly reduced the size of the wounds, increased collagen production, and increased tissue regeneration. As a result of this treatment, the chronic wound healed faster, infection was also reduced, thus improving the overall health of the diabetic rat (Hashem et al., 2022). As a result of this treatment the overall health of the rats improved because the chances of infection were greatly reduced and the wounds healed more quickly.

11.4.2.2.3 Iron Oxide Nanoparticles (IONs)

Another best choice for wound treatment and enhanced delivery of medications is the utilization of iron oxide nanoparticles because of their antimicrobial and enhanced physicochemical features (Veiseh et al., 2010). Small magnetic nanoparticles of iron, known as superparamagnetic iron oxide nanoparticles (SPIONs), have a core of iron oxide or cores of iron oxide combined with manganese, cobalt, or sometimes copper (Wahajuddin and Arora, 2012). SPIONs with gold coating become more noxious and effectively enter the biofilm of bacteria, thus greatly inhibiting bacterial biofilm growth. The antibacterial property of SPIONs is further increased when their core or intermediate gold shells are heated by applying laser and magnetic fields (Dulińska-Litewka et al., 2019). Broad-ranging antibacterial properties and enhanced wound healing properties have been shown by iron oxide-coated carbon nanotubes (Khashan et al., 2017). In rat incisional wounds, the dressing of SPIONs increased the human plasma thrombin half-life, which therefore promoted the healing process and increased the skin tensile strength (Arakha et al., 2015). SPIONs with polyethylene glycol and dextran coating, which are biocompatible polymers, are effective for wound healing (De Souza et al., 2019). Medicinal chemicals can be conjugated with these coated nanoparticles, resulting in an accelerated healing process and enhanced dispersion of medication in the blood. In rat incisional wounds, conjugation of these nanoparticles with thrombin increased the half-life of thrombin in blood plasma. As a result of this conjugation, the overall wound healing process is improved and accelerated by increasing skin tensile strength (Perumal et al., 2014). In another investigation, tumor-bearing mice were treated with anticancer drugs conjugated with nanoparticles (iron oxide) coated with biocompatible polymers, resulting in nanoparticle accumulation at the tumor site due to increased dispersion in the bloodstream. This overall improved the survival rate of the mice by significantly reducing tumor size through targeted delivery of the anticancer drugs by these nanomaterials complex (Kannan et al., 2021). Scientists in another study were able to mimic the extracellular matrix of the skin by developing nanofiber scaffolds. This innovative and novel approach has shown enhanced in vitro and in vivo results, making it a best therapeutic candidate for chronic wound healing and severe burns by promoting cellular migration and proliferation (Sanchez-Moreno et al., 2016). The healing process is further enhanced by adding growth hormones and stem cells to the scaffolds as compared to standard wound healing procedures. The scaffold was evaluated in animal research on mice that had suffered severe burns; the results showed that the scaffold not only encouraged skin regeneration but also reduced the production of scars, improving overall healing outcomes (Li et al., 2017).

11.4.2.2.4 Magnesium Oxide Nanoparticles

To treat epidermal injuries, many types of dressings for wounds have been developed worldwide. Traditionally, wound exudates have been absorbed using materials such as cotton wool, lint, bandages, and gauze to keep the wound dry and avoid bacterial infection (Gosain and DiPietro, 2004). An optimal dressing should have exceptional biocompatibility to promote regeneration of tissue, given the complexity of wound healing (Jiang et al., 2016).

Thus, it is necessary to create new materials that satisfy these requirements. Among them, magnesium oxide (MgO)-based nanomaterials have garnered a lot of scientific interest in the recent ten years. Magnesium oxide is a potential option for various applications in the biomedical field since it is inexpensive, ecologically benign, biodegradable, and physiologically safe (Verma et al., 2022). In research, Verma and colleagues effectively created and assessed films made of carboxymethyl cellulose and polyvinyl alcohol for possible application in skin tissue engineering. They discovered through detailed analysis that the most beneficial composition ratio was 50:50 for carboxymethyl cellulose and polyvinyl alcohol. In both dry and wet settings, this ratio demonstrated tensile strength similar to human skin as well as optimum hydrophilicity (Verma et al., 2022).

The suggested film also had a rough surface that promoted swelling, controlled breakdown, and cell attachment. The polymer film gained antibacterial characteristics with the addition of MgO nanoparticles. Additionally, the inclusion of MgO nanoparticles improved cell survival, especially in the film that included 1.5 weight percent of these particles. This suggests that MgO nanoparticles might be a useful material for future tailored skin tissue transplants.

Insufficient angiogenesis is one of the main challenges of diabetic wound healing. According to recent research, electrospun nanofiber membranes show capability as a treatment for wounds. These membranes must, however, encourage wound angiogenesis in order to treat diabetic wounds efficiently. The current approaches mainly focus on augmenting the angiogenic characteristics of these membranes via pro-angiogenic growth factors. Nevertheless, there are technological difficulties in incorporating growth factors into electrospun nanofibers and maintaining their activity over time.

Liu and colleagues (2022) established membranes of electrospun including gelatin, magnesium oxide nanoparticles, and polycaprolactone in response to these difficulties. Mg2+ ions are released by these membranes to promote angiogenesis even further. The membranes that were created facilitated the development of human umbilical vein endothelial cells (HUVECs) and improved the in vitro generation of vascular endothelial growth factor. According to implantation tests conducted in a rat model, the magnesium oxide-containing membranes promoted the development of strong blood capillaries one week after surgery, which eventually resulted in richer capillary networks within the disintegrating membranes.

Furthermore, by lowering inflammatory responses, encouraging angiogenesis, and improving granulation formation, these membranes significantly sped up the healing of diabetic wounds. Their findings suggested that magnesium-infused electrospun membranes with appropriate engineering might improve angiogenesis and be utilized to treat diabetic wounds.

11.4.2.2.5 Carbon-Based Nanomaterials

All four stages of wound healing can be treated with carbon-based nanomaterials, such as zero-dimensional nanodiamonds, fullerenes, and diamond-like carbon; one-dimensional carbon nanotubes; and two-dimensional graphene. Antibiotics, stem cells, and growth factors can also be carried by these materials. Due to their special physicochemical features and wide surface area, these nanoparticles are utilized in tissue engineering, targeted medication delivery, and skincare. Carbon-based nanomaterials are naturally antibacterial due to their hydrophobic nature and ROS generation.

11.4.2.2.5.1 Graphene Nanoparticles Graphene nanomaterials exhibit excellent antibacterial activity. Strong ROS are produced when they interact with bacteria, leading to oxidative damage (Seifi and Kamali, 2021). The membrane potential of graphene is able to transfer electrons to microbial cells because of the configuration of its carbon atoms (Yousefi et al., 2017). Epoxy-coated graphene oxide nanofilm displays enhanced biocompatibility, kills both *S. aureus* and *E. coli,* and speeds up the healing process of wounds in infected mice (Zheng et al., 2021). Graphene oxide

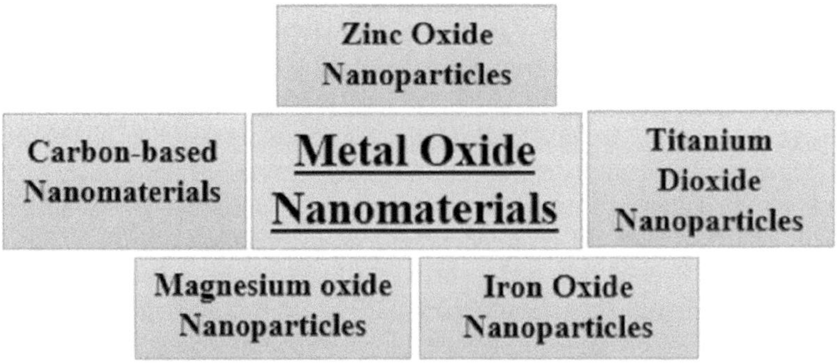

FIGURE 11.5 Types of metal oxide nanomaterials.

with laser wounds heals infectious wounds of bacteria or fungus more quickly (Sundaram and Abrahamse, 2020). Chitosan and polyvinyl alcohol (PVA) nanofibers incorporated with graphene were shown to successfully suppress the development of *Agrobacterium* and *E. coli* cells in Van Beveren rabbits and mice but not in yeast (Lu et al., 2012). A copper-infused chitosan/hyaluronic acid nanocomposite with graphene oxide was shown to be biocompatible in both in vitro and in vivo settings, which enhanced its efficacy in treating difficult bacterial wound infections (Yang et al., 2020). Infection control and cutaneous healing of wounds are improved by a mixture of mesenchymal stem cells and graphene dressing scaffolds through mechanical and chemical stimuli (Lasocka et al., 2019). However, despite the many therapeutic advantages of graphene, structural flaws affect the therapeutic potential of graphene. For example, hexagons may convert to pentagons, and during the manufacturing process, single or multiple vacancies arise (Eckmann et al., 2012).

For instance, when a dressing scaffold of graphene inserted with mesenchymal stem cells as well as antimicrobial agents is applied to a chronic infection at the wound site caused by multidrug-resistant bacteria, it significantly improves infection control through the release of antimicrobial substances that are stimulated by the graphene dressing (Pati et al., 2014). The graphene dressing increases integrity and stability, and the effectiveness of the scaffolds in tissue regeneration and antimicrobial release is greatly affected by structural flaws such as the conversion of pentagons into hexagons during the manufacturing process (Stoica et al., 2020). More investigation and developmental efforts are essential to minimize structural flaws and to enhance the graphene dressing's performance in wound healing of wound (Li et al., 2017).

11.4.2.2.5.2 Carbon Nanotubes Integrity of the bacterial membrane, metabolic activity, and morphology is significantly affected by single-walled and multiwalled carbon nanotubes through heavy metal residues and reactive oxygen species (Kittana et al., 2018). In particular single-walled nanotubes as compared to multiwalled nanotubes have enhanced antibacterial activity because of their tiny size, which results in a large surface area that significantly disrupts the bacterial membrane. According to a recent in vitro study, strong antimicrobial activity against various species of *Acinetobacter baumannii* causing burn-related infections has been demonstrated by carbon nanotube–mercury composite and thus can be used as a therapeutic agent for such burn-related infection treatments. It has been shown that the mixture of carbon nanotubes with titanium dioxide nanoparticles and silver nanoparticles increases the antibacterial effects (Niu et al., 2010). Enhanced antimicrobial activity against *Paracoccus denitrificans* has been shown by hydroxyl-functionalized single-walled nanotubes (Su et al., 2015). Fibroblasts proliferation, and re-epithelialization of wounds were all enhanced by chitosan incorporated with carbon nanotubes (single- and multiwalled) (Kittana et al., 2018). However, fibrosis was increased both in vitro and in vivo. It has been suggested that wound healing was expedited by the composite of electrospun carbon nanotube/PVA/EGF dressing, which also showed a regulated release of EGF that was positively active in stimulating fibroblast proliferation (Liao et al., 2017). By changing adhesion, cell spreading, survival, and migration, exposure to carbon nanotubes alters dermal fibroblasts' capacity to repair wounds and causes genotoxicity and inflammation (Luanpitpong et al., 2014). Carbon nanotubes (multiwalled) damaged the cytoskeleton and thus disrupted actin stress fibers, inhibiting DNA synthesis and adhesion-related gene expression (Zhang et al., 2011). Topical multiwalled carbon nanotubes at low concentrations have also been shown to exacerbate skin allergies and cause keratinocyte cytotoxicity (Palmer et al., 2019).

11.4.2.2.5.3 Fullerenes In order to target wounds that are difficult to treat, fullerenes and their derivatives may reduce inflammation, fight germs, change the way mast cells produce toxic mediators, promote the migration of cells, and speed up the healing process of healing (Zhou et al., 2010). In vitro ROS-independent oxidative stress is initiated by fullerene water solutions that are adsorbed on *Bacillus subtilis* bacterial membranes (Fang et al., 2007). The most researched fullerene, buckminsterfullerene (often referred to as C60), is somewhat water-soluble and very hydrophobic (Lyon et al., 2005). Wound healing was accelerated by the fullerene derivative C70-(ethylenediamine)8, which also dramatically suppressed *E. coli* and regulated immune system and growth factor production. The ethylenediamine moiety's amino groups create electrostatic contacts with the outer bacterial membrane, while the inner membrane's strong hydrophobic interactions with the C70 hydrophobic surface cause cytoplast leakage (Man and Ho, 2012).

11.4.2.2.5.4 Nanodiamonds The nanodiamond's tiny size, biocompatibility, adsorption capabilities, improved physical and adsorption qualities, capacity to disperse water-insoluble pharmaceuticals more easily, and overall durability make it a desirable ingredient for drug nanoformulations (Namdar and Nafisi, 2018). Additionally, while protecting healthy cells and tissues, nanodiamonds quicken the healing process (Ansari et al., 2016). They might develop additional methods for the sustained delivery of drugs and release of growth factors and biomolecules, as well as a thermosensitive, biocompatible, multifunctional hydrogel platform. Without changing the thermosensitive gelation properties of the injectable hydrogels, nanodiamonds enhance their mechanical properties. For example, the release of exogenous human VEGF is sustained by the thermosensitive hydrogel containing nanodiamonds, chitosan, and gelatin (Pacelli et al., 2017). In vitro, doxorubicin concentrations were higher and the cytotoxicity was decreased in a membrane made of carboxylated nanodiamond-cellulose nanocomposite loaded with doxorubicin (Luo et al., 2016). The biocompatible and flexible chitosan/bacterial cellulose composite sheets with nanodiamonds were used as wound dressings (Ostadhossein et al., 2015). Wound healing was also facilitated by the electrospun chitosan-based biopolymers that contained bacterial cellulose and medical-grade nanodiamonds (Fang et al., 2007).

11.4.2.2.5.5 Diamond-Like Carbon Coating Diamond-like carbon film coatings are hydrophobic, which change the cell membrane and kill bacteria. They are chemically stable and have excellent mechanical, biocompatible, and antibacterial efficacious qualities (Jelinek et al., 2013). Furthermore, improving bioactivity and cell adhesion and reducing bacterial attachment and proliferation, diamond-like carbon also regulates inflammation (Lopes et al., 2017). Enhancement of diamond-like carbon film coatings with harmful elements (e.g., Ag or Cu) has greater antimicrobial action against bacterial colonies (e.g., *S. aureus* and *S. epidermidis*), which raises the possibility of potentially fatal bacteremia (AlBuhairn et al., 2008). The application of a synthetic silk tissue bandage coated with diamond-like carbon–AgNPs effectively inhibited nearly all bacterial growth, including methicillin-resistant *S. aureus* (MRSA) (Juknius et al., 2016). Angiogenesis was stimulated because of the quick release of iron ions from the diamond-like carbon impregnated with AgNPs, and therefore, *S. aureus* and *S. epidermidis* were unable to colonize, enabling in vitro long-term tissue regeneration. Because of its astringent and antibacterial properties, the diamond-like carbon–TiO2 composite was biocompatible, promoted cell adhesion, and enhanced wound healing in vivo (Gorzelanny et al., 2016).

11.5 ORGANIC AND HYBRID NANOPARTICLES IN WOUND HEALING

11.5.1 Lipid-Based Nanoparticles

Lipid nanoparticles can be an appropriate delivery route for growth hormones, pharmaceuticals, and small interfering RNA (siRNA) (Choi et al., 2017). For effective transportation of drugs, two types of lipid nanoparticles, known as solid lipid nanoparticles (SLNs) and nanostructured lipid carriers (NLCs), have been used for the wound site (Chaudhari et al., 2023; Hashem et al., 2022). These carriers effectively transport biocompatible components, thus preventing medicines from breaking down and providing sustained and targeted drug release (Dutta et al., 2012). The performance of

SLNs and NLCs in the delivery of medications is exceptionally good compared to conventional delivery vehicles. They are able to enter the biofilm and thus bring the medication close to the bacterial cells because of their small size (Fumakia and Ho, 2016). In the human body, the important first line of defense is the antimicrobial peptide LL37 against various microbial pathogens (Barroso et al., 2020). Peptide LL37 is delivered and applied by SLNs on the wound, and its balanced release over the course of 14 days is monitored. Strong antibacterial properties have been demonstrated by the continuous release of LL37 against *E. coli* and *S. aureus*, which eventually enhance the healing process. Different forms of NLCs or SLNs, such as sponge (Eckmann et al., 2012), hydrogel (Patel et al., 2019), and film (Arantes et al., 2020), have been used for controlled drug delivery to enhance the process of wound healing (Husain et al., 2022). For example, the administration of silver sulfadiazine (SSD)-loaded SLN incorporated in a matrix chitosan hydrogel stimulates the spread and dispersion of SSD inside the biofilm, resulting in improved and extended antibacterial activity.

11.5.2 Role of Nanomaterials in Effective Drug Delivery

Properties like large surface area, small size, and high drug loading capacity make nanomaterials suitable candidates for drug delivery for wound healing (Iqubal et al., 2020). Therapeutic agents are efficiently encapsulated and delivered to the target site by these nanomaterials. Functionalization of nanomaterials with certain ligands or antibodies can improve targeting capabilities and precise delivery to the desired site (He et al., 2021).

Nanomaterials are used for the controlled administration of therapeutic chemicals because of their small size. They make healing faster and more effective, enhance benefits, and decrease adverse effects (X. Wang et al., 2018). Nanomaterials promote cell division and growth, which ultimately enhance tissue regeneration. Nanoparticles are able to mimic the skin extracellular matrix, thereby stimulating blood vessel formation, accelerating the regeneration of skin tissues, and aiding in the migration of cells to adhere to one another (De Souza et al., 2019). Moreover, nanoparticles decrease the complications of the wound healing process and prevent infection as they are functionalized with antibacterial properties (Nqakala et al., 2021). Nanoparticles enhance the safe and stable transportation of growth factors transportation and drugs to the wound site; as a result they totally change the healing process (Gudkov et al., 2021). Growth factors are also added to nanoparticles to ensure sustained and targeted release, thus allowing the targeted and regulated release of drugs to the wound site (Alvarez et al., 2014). Nanoparticles scaffold release chemicals or drugs in response to certain stimuli in the wound environment, which ultimately accelerate the process of healing (Raffi et al., 2010). Advancement in nanotechnology provide the basis for personalized medicine, which will offer unique and specific medication and treatment for each patient (Jones et al., 2008). Nanomaterials are considered potent antimicrobials, therefore they are efficient in reducing the risk of infections. Nanomaterials significantly promote wound healing because of their antibacterial properties, and targeted and sustained drug delivery to the target site, and overall enhance patient outcomes (Singh et al., 2018). Nanomaterials not only provide a scaffold for tissue regeneration and cell growth of the cells but also promote vessel formation and mimic the extracellular matrix, which ultimately accelerates the wound treatment process. Therefore, application of nanomaterials in wound healing provides the potential to revolutionize this field and enhance and improve patient outcomes (De Souza et al., 2019). In one example, silver nanoparticles are loaded with SLN in a chitosan hydrogel for the safe and sustained transfer of SLN to the wound site (Nqakala et al., 2021). This method leads to prolonged antimicrobial activity and enhanced wound healing by offering the safe and regulated release and diffusion of the SLN to the biofilm (Gudkov et al., 2021). Wound healing process is accelerated, and antibacterial activity is increased when organic nanomaterials such as SLNs loaded with SSD are used in conjunction with a chitosan hydrogel matrix (Lorenzo-Anota et al., 2021). Regulated release and distribution of SSD into the biofilm enhance antibacterial actions, thereby improving overall healing outcomes.

11.5.3 Polymeric Nanoparticles

The importance of polymeric nanoparticles in biomedical applications has been demonstrated in bioengineering and biomedicine (Huang and Fu, 2010). The majority of polymeric nanoparticles are manufactured by a combination of polylactic-co-glycolic acid (PLGA), chitosan, alginate, and other polymers (Jacob et al., 2018). Polymeric nanoparticles are very efficient in the transportation of biomolecules such as growth hormones, antibiotics, and DNA (Yun et al., 2004). In the wound microenvironment, the presence of proteolytic enzymes, such as matrix metalloproteinases (MMPs), significantly reduces their half-lives (Chu et al., 2010). Fortunately, the incorporation within polymer nanoparticles allows for the maintenance of these growth factors' bioactivity and release. Many investigations have been conducted on the integration of antimicrobial compounds into polymeric nanoparticles (Dave et al., 2017). For example, PLGA nanoparticles loaded with LL37 exhibit antibacterial action and accelerate wound healing without affecting the cells' metabolism. PLGA-LL37 nanoparticles administered to a mouse model of a full-thickness excisional wound resulted in quicker healing with increased deposition of collagen, formation of tissue granulation, re-epithelialization, and neovascularization (Raghupathi et al., 2011). IL-6 and VEGF levels were raised as a result of both angiogenesis and the control of the inflammatory process (Chereddy et al., 2014).

Several in vitro studies revealed the use of norfloxacin-loaded PLGA nanoparticles as an effective candidate for infectious wound healing. It has been shown that norfloxacin-loaded nanoparticles are able to prevent *S. aureus* and *P. aeruginosa* growth; therefore, they are a new best choice for burn-related infections. These nanoparticles efficiently penetrate through the skin and are able to release the drugs for up to 24 hours (Dave et al., 2017). Studies revealed that these nanoparticle complexes are effective against many drug-resistant bacteria, including *S. aureus* and *P. aeruginosa*, and therefore can be used as alternative treatments for burn-related infections (Nqakala et al., 2021). A significant reduction in bacterial growth and colonization was seen in burn patients who were treated with PLGA nanoparticles loaded with norfloxacin, and as result the wound healing became faster compared to wounds treated with standard antibiotic treatments (He et al., 2021). Less frequent administration of the drugs due to its sustained release results in the reduction of patient discomfort and improved compliance with the treatment.

11.5.4 Nanobiotics

Another effective and game-changing treatment of chronic wounds or infections caused by multidrug-resistant bacteria is antibiotic-loaded nanoparticles, also known as nanobiotics. An in vitro study revealed the use of vancomycin-modified nanoparticles against a vast range of bacteria (Kell et al., 2008). Chronic infections caused by MRSA in different body parts are effectively treated by polyacrylate nanoparticles containing antibiotics, which pose less toxic effects compared to other standard treatments (Cruz et al., 2020). Hydrogel of poly(glutamic acid)/gelatin, which is a complex made of gelatin, epigallocatechin gallate nanoparticles, chitosan, and gentamycin, is responsible for the re-epithelialization of tissue at the site of the wound. It is also observed that this antibiotic nanoparticle complex did not stick to the wound site and thus can easily be removed from the site of the wound (Mostafavi et al., 2021). The potential of gentamicin-containing PLGA nanoparticles in the healing of infectious and chronic wounds is due to the double emulsion process of evaporation in gentamicin-containing PLGA synthesis (Sun et al., 2021). For example, a study revealed the effectiveness of poly(glutamic acid)/gelatin hydrogel in wound healing (He et al., 2021). Synthesis of poly(glutamic acid)/gelatin hydrogel by the double emulsion process of evaporation makes it a great and efficient dressing for chronic wound healing (Kannan et al., 2021). Addition of gentamicin and epigallocatechin gallate nanoparticles to the hydrogel offers great antibacterial activity against many bacteria and, therefore, becomes very useful for the treatment of infectious wounds.

11.5.5 Micelles

Micelles are nanomaterials that are the best choice for carrying both hydrophilic and hydrophobic drugs. This is because in an aqueous solution, micelles have an inside hydrophobic core and an outer hydrophilic shell (Sun et al., 2021). The favorable characteristics of polymeric micelles include better colloidal stability, enhanced cargo capacity, biocompatibility, controlled drug release, and nontoxicity (Alberti et al., 2017). In rat incision wound models, micelles loaded with curcumin demonstrated considerably reduced granulation, low collagen content, and wound maturity scores compared to micellar suspensions in polyethylene glycol-polycaprolactone hydrogel. Chitosan oleate micelles were used to encapsulate silver sulfadiazine in a different case study (Gong et al., 2013). This combination reduces the cytotoxic effects of drugs on human cells while maintaining the antibiotic's antibacterial efficiency, all while enhancing the solubility and bioavailability of antibiotics and dispersion (Huan and Tay, 2020). The antibacterial activity of the silver sulfadiazine micelles decreased (MIC values for *S. aureus*) compared to the control solution (Dellera et al., 2014). For instance, one example developed an encapsulating chitosan oleate micelles antibiotic sulfadiazine. This combination of antibiotic and nanoparticle decreases the harmful effects of the drugs on human cells, effectively inhibits bacterial infections, and has the potential to increase the absorption of the antibiotics by body cells, increasing drug distribution throughout the body. It also has the potential to increase the absorption of the antibiotics by the body cells and drug distribution throughout the body as well as its dissolution in water (Lu et al., 2012). The antimicrobial activity of the micelles containing silver sulfadiazine significantly became higher with significantly lower MIC values compared to free drugs. Various bacterial strains are inhibited by the use of chitosan oleate micelles containing silver sulfadiazine (Wang et al., 2019). Stronger antibacterial effects with lower minimum inhibitory concentrations to inhibit various resistant bacteria are exhibited by this complex, making it the best potential approach for the treatment of bacterial infections (Cruz et al., 2020).

11.5.6 Cyclodextrins

Enzymatic breakdown of starch results in cyclodextrins (CDs). CDs are a cyclic oligosaccharide with a hydrophilic outer surface, and the middle cavity is filled with glucopyranose (Bellu et al., 2021). CDs are complexing agents that have been used to enhance a drug's stability, bioavailability, and solubility in water, which are very difficult to work with (Challa et al., 2005). CDs have been utilized in transdermal drug delivery systems to improve stability and solubility of poorly soluble medications. A drug's bioavailability can be enhanced by the inclusion of CD–drug complexes, improving targeted and more effective delivery through the skin (Boomi et al., 2020). Sacran hydrogel films containing CDs reduce inflammation and promote the regulation of cytokines, thus accelerate wound healing therefore it is another effective treatment for wounds that are challenging to treat with standard antibiotics (Cruz et al., 2020; George et al., 2019). The use of CDs incorporated in sacran hydrogels was observed in a study on diabetic foot ulcers. This formulation contains weakly soluble antibiotic medication, and it was observed that the incorporation of CDs increases the solubility and stability of the drugs, as well as improves the sustained and controlled release of the drugs (Maťátková et al., 2022). Cyclodextrin-based hydrogel films leading to quicker wound healing processes, enhanced regulation of cytokines, and reduced inflammation make it a suitable choice for wound healing (Li et al., 2017).

11.5.7 Nitric Oxide-Releasing Nanoparticles

Nitric oxide regulates the growth of keratinocytes and skin homeostasis. It also promotes angiogenesis and extracellular matrix protein deposition during wound healing. Because inducible nitric oxide synthase levels are dysregulated in psoriasis and systemic lupus erythematosus (SLE), nitric oxide has a role in both conditions (Stallmeyer et al., 1999). In excisional wounds, induced nitric oxide-knockout animals exhibit delayed re-epithelialization and fewer keratinocytes (Schwentker et al., 2002). Nitric oxide synthase substrate L-arginine improved collagen deposition, suggesting that nitric oxide is a crucial component in wound closure. The broad-spectrum antibacterial action of nitric oxide is well supported by data. When the microbial pattern recognition receptors of phagocytic cells are stimulated, the inducible nitric oxide synthase transcription is increased, and high-output nitric oxide is produced. Nitric oxide forms reactive nitrogen intermediates by deactivating zinc metalloproteins and thus reacts with ROS, impairing DNA replication and cell respiration (Schapiro et al., 2003). Lastly, the microbial cells are the site of action for the antimicrobial effector molecules (Weller, 2009). Biomaterials with coatings that release nitric oxide reduce the incidence of infections linked to biomaterials. In vitro, *S. aureus* and other pathogen biofilms were eliminated using carbon-based coatings supporting monofilament polypropylene mesh that releases nitric oxide (Engelsman et al., 2009). The proportion of *S. aureus*-infected subcutaneous implants in the mouse model was reduced by 82% by coating medical-grade silicone elastomer implants with sol–gel-derived film (Nablo et al., 2005). The kind and quantity of diamine-containing organosilane precursors, as well as the circumstances of the sol–gel processing, determine the characteristics of the organic–inorganic polymer matrices created by aminosilane-based sol–gel chemistry (DeRosa et al., 2007). In vitro, SiNPs that release nitric oxide suppress the growth of various bacteria (Han et al., 2012). However, according to different research, bacteria exhibit nitric oxide-mediated antibiotic resistance by building up reactive oxygen species and refining toxic substances chemically (Gusarov et al., 2009).

11.5.8 Green Synthesized Nanoparticles

TiO2NPs, AgNPs, AuNPs, and CuONPs produced by green synthesis help heal wounds more quickly while reducing or eliminating infections and negative effects from medication (Makarov et al., 2014). When combined with phytoextracts, green synthesized nanoparticles like AgNPs are economical, ecologically friendly, and have antibacterial and antifungal qualities that aid in wound healing (Kuppusamy et al., 2016). In mice, for instance, the creation of green synthesized AgNPs utilizing *Catharanthus roseus* and *Azadirachta indica* extracts improved excision wound healing and showed efficacy against multidrug resistant bacteria (Lakkim et al., 2020). Sprague Dawley rats with multiple strain infections in their wounds were successfully treated by the manufacture of stable AgNPs utilizing the fungus *Orchidantha chinensis* (Wen et al., 2016). The generated AgNPs decreased the number of scars, re-epithelialized the epidermis, sped up the pace at which wounds healed and contracted, and decreased TNF-α, IL-6, and IL-1β expression (Wen et al., 2016). A chemical made from geniposide known as genipin obtained from the slow-growing fungus *Penicillium nigricans*, as well as PEG and AgNPs when combined with chitosan, enhanced wound healing with more antibacterial activity (Y. Liu et al., 2012). *Coleus forskohlii* root extract in AgNPs was efficient in full thickness excision wound healing in Wistar albino male rats [168].

In Wistar albino rats, the manufacture of AuNPs by extracting the *Coleus forskohlii* root resulted in a considerable acceleration of the re-epithelialization of excision wounds, as well as an increased proliferation rate, development of connective tissue, and migration of epidermal cells. The creation of TiO2NPs decreased the wound area and increased wound contraction in Wistar albino mice in the presence of *Moringa oleifera* leaf extract (Sivaranjani and Philominathan, 2016). In Wistar albino rats, the production of CuONPs using *Ficus religiosa* leaf extract promoted the proliferation of fibroblasts, collagen fibers, and macrophages while also inhibiting human pathogenic microorganisms (Sankar et al., 2015).

11.6 CHALLENGES

There are still a number of serious challenges, apart from the diverse benefits of nanomaterials in the treatments of wound healing. These challenges include toxicity, scalability of manufacturing, regulatory issues, and long-term stability of nanomaterials (Shi et al., 2021). Further research is required to understand how nanomaterials interact with biological systems, and clinical and preclinical investigations should be carried out to enhance their composition and design to guarantee the efficacy and safety of nanomaterials. The key challenges that need to be investigated are cost-effectiveness, safety, regulatory restrictions, and scalability (Kubackova et al., 2020).

11.6.1 Toxicity Issues

Nanomaterials have the potential for transforming the wound healing transformation process through the reduction of infection rates and the promotion of tissue regeneration. However, their safety is at significant risk. The safety of nanomaterials is guaranteed by understanding their interaction with biological systems (Boomi et al., 2020). Subsequent investigations should be conducted to determine the persistent impacts of nanomaterials on the body and mitigation of such harmful effects by applying developing approaches. It is essential to establish regulations and criteria for the safe use of nanoparticles in wound healing (Stoica et al., 2020). Regulatory agencies, scientists, and physicians should work together to evaluate the disadvantages and advantages of nanomaterials in wound healing. By resolving these problems and following safety measures, we can attain the full utilization of nanomaterials for rapid and effective wound healing (Dutta et al., 2012). Safe discovery and application of nanoparticles in wound healing of wound could revolutionize the profession and improve patients' outcomes.

An eye should be kept on the quantity and effects of long-term exposure of nanomaterials on people and the environment. This ongoing assessment will help in recognizing possible hazards or unfavorable conditions over time, enabling modifications to the recommendations and standards (Alvarez et al., 2014). This safety dedication will open the door for further improvements in wound healing and improve patients' outcomes (He et al., 2021). For example, by using nanomaterials, researchers might be able to trace wound healing. For thorough monitoring of the progress of wound healing, smart bandages are being developed to alert healthcare experts in case of any indication of infection and a slow healing process. By lowering long-term problems, patients' outcomes are improved by proactive measures (De Souza et al., 2019).

11.6.2 Regulatory Considerations

There are some challenges and difficulties presented by regulatory considerations for the assurance of the efficacy and safety of the nanoparticles, despite the great benefits of using nanoparticles in wound healing. Wide testing regulations are required for their usage and production (Bellu et al., 2021). The ecological effects and concerns about the toxicity of nanoparticles must be addressed by the regulatory authorities (De Souza et al., 2019). Establishing balance between patient safety and innovation is crucial. Therefore, it is essential to create methods for observing and analyzing the consequences of treatments based on nanoparticles in order to identify possible complications and harmful effects.

A comprehensive approach to regulation requires collaboration among industry stakeholders, scholars, and regulatory bodies (Sandoval et al., 2022). The safety of the environment and patients is protected by this collaboration; it can also promote advancements in treatments based on nanoparticles (Maťátková et al., 2022). Healthy communication among healthcare experts, regulatory agencies, and researchers will quickly address concerns and challenges regarding nanoparticle-based medication. It is critical to keep track of patients receiving nanoparticle-based wound healing treatments in order to evaluate their long-term health outcomes and monitor any potential negative effects or unanticipated consequences (Salvo and Sandoval, 2022). Environment sustainability and patient safety are prioritized as a result of this collaboration, opening the door for the widespread implementation of nanoparticle-based medications in clinical practice. Regular feedback and monitoring will enable healthcare experts to reduce potential risks and increase their efficacy (Nqakala et al., 2021). An extensive structure that promotes accessibility and collaboration should be built for the benefit of patients. Implementing a frequent monitoring and feedback system provides healthcare practitioners with valuable insights regarding the efficacy of various therapies (Boomi et al., 2020).

11.6.3 Production Scalability

One of the biggest challenges in the healing process of wounds is the large-scale production of nanoparticles. Notwithstanding their potential to improve drug delivery and tissue regeneration, large-scale production is a significant challenge (Li et al., 2017). This issue can be addressed by researchers, manufacturers, and regulatory authorities through the optimization of techniques for production, ensuring long-term supply, and standardizing processes for nanoparticle production (Boomi et al., 2020). The difficulties in the commercialization of nanoparticle-based medication for wound healing can also be reduced by these production processes. The efficacy and safety of the nanoparticle-based wound healing treatment can be enhanced by the establishment of guidelines for nanoparticles characterization and evaluation of their use in wound healing (Khatri et al., 2014). Different formulations can be assessed and compared by researchers with the help of these guidelines, thus leading to more enhanced and improved treatments. Possible risks and side effects of therapies based on nanoparticles can be identified and reduced through the collaboration of academia, industry, and regulatory bodies (Maťátková et al., 2022). Multidisciplinary approach and active contribution from various stakeholders are required for the utilization of nanoparticles in wound healing. Customization of nanoparticles, adjustments to their characteristics for specific wound types, and optimizing their healing capabilities of nanoparticles can be attained through technological developments in nanotechnology.

11.6.4 Long-Term Stability

Nanoparticles have the potential to provide long-lasting and powerful medication, which could revolutionize drug delivery methods (De Souza et al., 2019). However, their stability might be affected over time degradation and aggregation. Therefore, researchers are investigating methods that would ultimately enhance nanoparticle stability and shelf life; these methods include surface modification and encapsulation of nanoparticles (Dulińska-Litewka et al., 2019). These methods result in enhanced stable and targeted drug delivery that will ultimately reduce the need for frequent medication. Additionally, these methods will also encourage the release of therapeutic chemicals in a controlled manner over a longer period of time (Stoica et al., 2020). Enhanced stability and longer shelf life of nanoparticles dramatically reduce healthcare expenses by lowering the requirement for frequent medication and replenishment (Bellu et al., 2021). As a result, the overall patient outcomes improve due to decreased adverse effects, which is achieved by the stable and controlled delivery of drugs to the target site. The use of nanomaterials in the healthcare system is increasing day by day as they enhance patients' outcomes and lives globally. Furthermore, nanomaterials increase the bioavailability of the drugs, which results in more effective and enhanced distribution and absorption of drugs in the body. The use of nanoparticles results in better treatment, and patients recover more quickly and efficiently (Singh et al., 2018). Because of their adaptability and bioavailability, nanoparticles are now used in various medical fields, such as imaging, diagnostics, and regenerative medicine. Nanoparticles target specific tissues and cells, thereby reducing the side effects of medication on healthy tissues which results in higher doses of medicine at the target site and thus lower doses of medication are required to achieve the best possible treatment. Patient compliance can be improved by the controlled release of the drug, and thus treatment efficacy can be achieved. Disease treatment and patient overall health can be improved by the use of nanomaterials (Boomi et al., 2020). The risk of side effects on healthy tissues is reduced by doctors precisely targeting specific areas of the body, thereby reducing the overall drug dosages (Singh et al., 2018).

11.6.5 Nanoparticle Limitations in the Healing of Wounds

NPs may heal wounds excellently, and there are still a ton of prospects for their use and advancement in the future. It is, however, noted that the wound surface is not secured by intact skin; nanoparticles, therefore, come into direct contact with wound tissue. Therefore, the safety of NPs must be assured before their use. The most commonly reported side effects of NPs are skin irritation and allergy. For instance, studies reported that skin hypersensitivity is caused by the use of carbon nanotubes and nickel NPs because of the ions released and surface coating from NPs (Palmer et al., 2019). Skin irritation, psoriasis, and inflammation have been observed to be aggravated by the exposure of nanoparticles to the skin (Lai et al., 2021). It has been reported that oxidative stress, programmed cell death in fibroblasts and keratinocytes, and autophagy are caused by NP exposure (M. Wang et al., 2018). Size, shape, steadiness, surface charge, and concentration of the NPs (nickel, gold, and silver, etc.) determine their noxiousness behavior. Therefore, to decrease the harmful effect of new NPs on the skin cells, it is important to adjust their physicochemical properties for use for wound treatment (Hashempour et al., 2019). Stability of the NPs can significantly reduce dermatitis. The utilization of stabilizers, such as surfactants, metal shells, or polymers, is recommended (Pan et al., 2021). In order to reduce skin discomfort, low-sensitization materials have to be utilized for NP surface coating. Furthermore, it has been shown that DNA damage can be caused by NPs and lower methylation of genes, which may indicate a higher risk of cell cancer (Sooklert et al., 2019). However, there is currently little evidence to support the idea that NPs might cause genetic abnormalities and cancerous changes in skin cells. Furthermore, longer exposure studies are necessary, as there is now little data to support the notion that NPs will have significant impacts from chronic percutaneous and skin deposition. After entering the body, NPs can enter the injured blood vessels in lesions and subsequently the bloodstream, where they come in close contact with blood vessels, resulting in hemolysis. It has been demonstrated that hemolysis can be induced by a few metal NPs, including AgNPs and ZnONPs. The physicochemical characteristics of materials can be changed to address the aforementioned issue, or biologically active materials like phospholipids and polysaccharides can be applied to the surface of NPs (Bakshi, 2017).

Another issue is that NPs might cause multisystem abnormalities by dispersing throughout the body and into different organs. After entering the bloodstream, the number of NPs reduces dramatically in comparison to their original concentration and may partially degrade in urine and feces. Death and weight loss have been seen in animal experiments. Nevertheless, there isn't any solid proof that using NPs in real life would cause tumors or organ harm. However, introduction during pregnancy could harm the progeny (Hadrup et al., 2018). Evaluations of NP toxicity in wound healing have typically concentrated on local acute adverse reactions. On the other hand, nanofibers, nanofilms, and other novel NPs are still seldom taken into consideration; instead, a number of studies have focused on metal NPs, carbonaceous NPs, and nanotubes (Teixeira et al., 2020). To address these issues, an urgent NP toxicity assessment is necessary.

11.7 CONCLUSION

The chapter extensively discusses wound healing dressings and highlights the advantages of various nanomaterials. Specifically, it covers inorganic nanomaterials, organic and hybrid nanomaterials, and nanofibers. Customized wound bandages incorporating nanomaterials hold significant potential for individualized wound care. By combining polymeric nanomaterials of natural origin with biomolecules, growth factors, and antibiotics, we can overcome the limitations of current wound healing materials and create scaffolds with improved mechanical resistance. Additionally, a hybrid inorganic/natural polymer/antibiotic-based nanocomposite could enhance wound healing by combining antimicrobial and biodegradable properties. The unique properties of nanomaterials, such as their adjustable physicochemical characteristics and specific targeting capabilities, have facilitated the creation of advanced wound dressings, scaffolds, and various therapies. The broad spectrum of potential applications of nanomaterials in the treatment of both acute and chronic wounds is promising for the future of wound care. The integration of nanotechnology and personalized medicine could revolutionize healthcare by reducing costs and improving treatment durations. Overall, nanomaterials pave the way for a new era of more efficient, patient-centric, and innovative methods of wound healing.

REFERENCES

Adair, T. H., & Montani, J.-P. (2011). *Angiogenesis*. Morgan & Claypool Publishers.

Ahmad, M. Z., Alasiri, A. S., Ahmad, J., Alqahtani, A. A., Abdullah, M. M., Abdel-Wahab, B. A., Pathak, K., Saikia, R., Das, A., Sarma, H., & Alzahrani, S. A. (2022). Green synthesis of titanium dioxide nanoparticles using ocimum sanctum leaf extract: In vitro characterization and its healing efficacy in diabetic wounds. *Molecules*, 27(22). https://doi.org/10.3390/molecules27227712

Alavi, M. (2020). Applications of chitosan and nanochitosan in formulation of novel antibacterial and wound healing agents. In M. Rai (Ed.), Nanotechnology in skin, soft tissue, and bone infections (pp. 169–181). Springer. https://doi.org/10.1007/978-3-030-35147-2_10

Alavi, M. & Varma, R. S. J. C. (2021). Antibacterial and wound healing activities of silver nanoparticles embedded in cellulose compared to other polysaccharides and protein polymers. *Cellulose*, 28(13), 8295–8311

Alberti, T., Coelho, D. S., Voytena, A., Pitz, H., de Pra, M., Mazzarino, L., Kuhnen, S., Ribeiro-do-Valle, R. M., Maraschin, M., & Veleirinho, B. (2017). Nanotechnology: A promising tool towards wound healing. *Current Pharmaceutical Design*, 23(24). https://doi.org/10.2174/1381612823666170503152550

AlBuhairn, B., Hind, D., & Hutchinson, A. (2008). Antibiotic prophylaxis for wound infections in total joint arthroplasty : A systematic review. *Journal of Bone and Joint Surgery - Series B*, 90(7). https://doi.org/10.1302/0301-620X.90B7.20498

Alvarez, G. S., Hélary, C., Mebert, A. M., Wang, X., Coradin, T., & Desimone, M. F. (2014). Antibiotic-loaded silica nanoparticle-collagen composite hydrogels with prolonged antimicrobial activity for wound infection prevention. *Journal of Materials Chemistry B*, 2(29). https://doi.org/10.1039/c4tb00327f

Anandhi, P. & Rajeshkumar, S. (2023). Copper nanoparticles in wound healing: A review. *Journal of Survey in Fisheries Sciences*, 10(1S), 49–63.

Andonegi, M., Peñalba, M., de la Caba, K., & Guerrero, P. (2020). ZnO nanoparticle-incorporated native collagen films with electro-conductive properties. *Materials Science and Engineering C*, 108. https://doi.org/10.1016/j.msec.2019.110394

Ansari, S. A., Satar, R., Jafri, M. A., Rasool, M., Ahmad, W., & Zaidi, S. K. (2016). Role of nanodiamonds in drug delivery and stem cell therapy. *Iranian Journal of Biotechnology*, 14(3). https://doi.org/10.15171/ijb.1320

Arakha, M., Pal, S., Samantarrai, D., Panigrahi, T. K., Mallick, B. C., Pramanik, K., Mallick, B., & Jha, S. (2015). Antimicrobial activity of iron oxide nanoparticle upon modulation of nanoparticle-bacteria interface. *Scientific Reports*, 5. https://doi.org/10.1038/srep14813

Arantes, V. T., Faraco, A. A. G., Ferreira, F. B., Oliveira, C. A., Martins-Santos, E., Cassini-Vieira, P., Barcelos, L. S., Ferreira, L. A. M., & Goulart, G. A. C. (2020). Retinoic acid-loaded solid lipid nanoparticles surrounded by chitosan film support diabetic wound healing in in vivo study. *Colloids and Surfaces B: Biointerfaces*, 188. https://doi.org/10.1016/j.colsurfb.2019.110749

Archana, D., Dutta, J., & Dutta, P. K. (2013). Evaluation of chitosan nano dressing for wound healing: Characterization, in vitro and in vivo studies. *International Journal of Biological Macromolecules*, 57. https://doi.org/10.1016/j.ijbiomac.2013.03.002

Arit, M. D., Rahman, M. A., Milu, M. M. H., Siddik, A. B., & Hogue, M. E. (2021). Green nanomaterials: Synthesis, properties and spectroscopic applications. In Nanomaterials for spectroscopic applications (pp. 213–272). Jenny Stanford Publishing.

Bakshi, M. S. (2017). Nanotoxicity in systemic circulation and wound healing. *Chemical Research in Toxicology*, 30(6). https://doi.org/10.1021/acs.chemrestox.7b00068

Barroso, A., Mestre, H., Ascenso, A., Simões, S., & Reis, C. (2020). Nanomaterials in wound healing: From material sciences to wound healing applications. *Nano Select*, 1(5). https://doi.org/10.1002/nano.202000055

Batt, J., Milward, M., Chapple, I., Grant, M., Roberts, H., & Addison, O. (2018). TiO2 nanoparticles can selectively bind CXCL8 impacting on neutrophil chemotaxis. *European Cells and Materials*, 35. https://doi.org/10.22203/eCM.v035a02

Behm, B., et al. (2012). Cytokines, chemokines, and growth factors in wound healing. *Journal of the European Academy of Dermatology and Venereology*, 26(7), 812–820.

Bellu, E., Medici, S., Coradduzza, D., Cruciani, S., Amler, E., & Maioli, M. (2021). Nanomaterials in skin regeneration and rejuvenation. *International Journal of Molecular Sciences*, 22(13). https://doi.org/10.3390/ijms22137095

Biswas, A. K., Islam, M. R., Choudhury, Z. S., Mostafa, A., & Kadir, M. F. (2014). Nanotechnology based approaches in cancer therapeutics. *Advances in Natural Sciences: Nanoscience and Nanotechnology*, 5(4). https://doi.org/10.1088/2043-6262/5/4/043001

Boomi, P., Ganesan, R., Poorani, G. P., Jegatheeswaran, S., Balakumar, C., Prabu, H. G., Anand, K., Prabhu, N. M., Jeyakanthan, J., & Saravanan, M. (2020). Phyto-engineered gold nanoparticles (AuNPs) with potential antibacterial, antioxidant, and wound healing activities under in vitro and in vivo conditions. *International Journal of Nanomedicine*, 15. https://doi.org/10.2147/IJN.S257499

Chakrabarti, S., et al. (2019). Aspects of nanomaterials in wound healing. *Current Drug Delivery*, 16(1), 26–41.

Challa, R., Ahuja, A., Ali, J., & Khar, R. K. (2005). Cyclodextrins in drug delivery: An updated review. *AAPS PharmSciTech*, 6(2). https://doi.org/10.1208/pt060243

Chaudhari, N. B., Zalte, A. G., & Gulecha, V. S. (2023). Formulation, optimization, characterization and in vitro-ex vivo evaluation of atorvastatin loaded solid lipid nanoparticles using quality by design approach. *Research Journal of Pharmacy and Technology*, 16(3). https://doi.org/10.52711/0974-360X.2023.00236

Chereddy, K. K., Her, C. H., Comune, M., Moia, C., Lopes, A., Porporato, P. E., Vanacker, J., Lam, M. C., Steinstraesser, L., Sonveaux, P., Zhu, H., Ferreira, L. S., Vandermeulen, G., & Préat, V. (2014). PLGA nanoparticles loaded with host defense peptide LL37 promote wound healing. *Journal of Controlled Release*, 194. https://doi.org/10.1016/j.jconrel.2014.08.016

Cherng, J.-H., et al. (2022). Hemostasis and anti-inflammatory abilities of AuNPs-coated chitosan dressing for burn wounds. *Journal of Personalized Medicine*, 12(7), 1089.

Choi, J. U., Lee, S. W., Pangeni, R., Byun, Y., Yoon, I. S., & Park, J. W. (2017). Preparation and in vivo evaluation of cationic elastic liposomes comprising highly skin-permeable growth factors combined with hyaluronic acid for enhanced diabetic wound-healing therapy. *Acta Biomaterialia*, 57. https://doi.org/10.1016/j.actbio.2017.04.034

Chopra, H., et al. (2022). Strategies and therapies for wound healing: A review. *Current Drug Targets*, 23(1), 87–98.

Chu, Y., Yu, D., Wang, P., Xu, J., Li, D., & Ding, M. (2010). Nanotechnology promotes the full-thickness diabetic wound healing effect of recombinant human epidermal growth factor in diabetic rats. *Wound Repair and Regeneration*, 18(5). https://doi.org/10.1111/j.1524-475X.2010.00612.x

Chung, E. J., Leon, L., & Rinaldi, C. (2019). Nanoparticles for biomedical applications: Fundamental concepts, biological interactions and clinical applications. *Nanoparticles for Biomedical Applications: Fundamental Concepts, Biological Interactions and Clinical Applications.* https://doi.org/10.1016/C2017-0-04750-X

Cruz, D. M., Mostafavi, E., Vernet-Crua, A., Barabadi, H., Shah, V., Cholula-Díaz, J. L., Guisbiers, G., & Webster, T. J. (2020). Green nanotechnology-based zinc oxide (ZnO) nanomaterials for biomedical applications: A review. *JPhys Materials*, 3(3). https://doi.org/10.1088/2515-7639/ab8186

Cui, T., Yu, J., Wang, C. F., Chen, S., Li, Q., Guo, K., Qing, R., Wang, G., & Ren, J. (2022). Micro-gel ensembles for accelerated healing of chronic wound via pH regulation. *Advanced Science*, 9(22). https://doi.org/10.1002/advs.202201254

Dam, P., et al. (2023). Wound healing strategies based on nanoparticles incorporated in hydrogel wound patches. *RSC Advances*, 13(31), 21345–21364.

Dave, V., Kushwaha, K., Yadav, R. B., & Agrawal, U. (2017). Hybrid nanoparticles for the topical delivery of norfloxacin for the effective treatment of bacterial infection produced after burn. *Journal of Microencapsulation*, 34(4). https://doi.org/10.1080/02652048.2017.1337249

David, O. M., Lategan, K. L., de Cortalezzi, M. F., & Pool, E. J. (2022). The stability and anti-angiogenic properties of titanium dioxide nanoparticles (TiO2NPs) using caco-2 cells. *Biomolecules*, 12(10). https://doi.org/10.3390/biom12101334

De Souza, R. C., Haberbeck, L. U., Riella, H. G., Ribeiro, D. H. B., & Carciofi, B. A. M. (2019). Antibacterial activity of zinc oxide nanoparticles synthesized by solochemical process. *Brazilian Journal of Chemical Engineering*, 36(2). https://doi.org/10.1590/0104-6632.20190362s20180027

Dellera, E., Bonferoni, M. C., Sandri, G., Rossi, S., Ferrari, F., Del Fante, C., Perotti, C., Grisoli, P., & Caramella, C. (2014). Development of chitosan oleate ionic micelles loaded with silver sulfadiazine to be associated with platelet lysate for application in wound healing. *European Journal of Pharmaceutics and Biopharmaceutics*, 88(3). https://doi.org/10.1016/j.ejpb.2014.07.015

DeRosa, F., Kibbe, M. R., Najjar, S. F., Citro, M. L., Keefer, L. K., & Hrabie, J. A. (2007). Nitric oxide-releasing fabrics and other acrylonitrile-based diazeniumdiolates. *Journal of the American Chemical Society*, 129(13). https://doi.org/10.1021/ja0686864

Dong, H., et al. (2015). Polyol synthesis of nanoparticles: Status and options regarding metals, oxides, chalcogenides, and non-metal elements. *Green Chemistry*, 17(8), 4107–4132.

Dulińska-Litewka, J., Łazarczyk, A., Hałubiec, P., Szafrański, O., Karnas, K., & Karewicz, A. (2019). Superparamagnetic iron oxide nanoparticles-current and prospective medical applications. *Materials*, 12(4). https://doi.org/10.3390/ma12040617

Dutta, R. K., Nenavathu, B. P., Gangishetty, M. K., & Reddy, A. V. R. (2012). Studies on antibacterial activity of ZnO nanoparticles by ROS induced lipid peroxidation. *Colloids and Surfaces B: Biointerfaces*, 94. https://doi.org/10.1016/j.colsurfb.2012.01.046

Eckmann, A., Felten, A., Mishchenko, A., Britnell, L., Krupke, R., Novoselov, K. S., & Casiraghi, C. (2012). Probing the nature of defects in graphene by Raman spectroscopy. *Nano Letters*, 12(8). https://doi.org/10.1021/nl300901a

Engelsman, A. F., Krom, B. P., Busscher, H. J., van Dam, G. M., Ploeg, R. J., & van der Mei, H. C. (2009). Antimicrobial effects of an NO-releasing poly(ethylene vinylacetate) coating on soft-tissue implants in vitro and in a murine model. *Acta Biomaterialia*, 5(6). https://doi.org/10.1016/j.actbio.2009.01.041

Falanga, V., et al. (2022). Chronic wounds. *Nature Reviews Disease Primers*, 8(1), 50.

Fang, J., Lyon, D. Y., Wiesner, M. R., Dong, J., & Alvarez, P. J. J. (2007). Effect of a fullerene water suspension on bacterial phospholipids and membrane phase behavior. *Environmental Science and Technology*, 41(7). https://doi.org/10.1021/es062181w

Fumakia, M., & Ho, E. A. (2016). Nanoparticles encapsulated with LL37 and serpin A1 promotes wound healing and synergistically enhances antibacterial activity. *Molecular Pharmaceutics*, 13(7). https://doi.org/10.1021/acs.molpharmaceut.6b00099

George, D., Maheswari, P. U., Sheriffa Begum, K. M. M., & Arthanareeswaran, G. (2019). Biomass-derived dialdehyde cellulose cross-linked chitosan-basednanocompositehydrogel with phytosynthesized zinc oxide nanoparticles for enhanced curcumin delivery and bioactivity. *Journal of Agricultural and Food Chemistry*, 67(39). https://doi.org/10.1021/acs.jafc.9b01933

Ghafari, Y., et al. (2024). Gold nanoparticles in biomedicine: Advancements in cancer therapy, drug delivery, diagnostics, and tissue regeneration. Scientific Hypotheses, 1(1), 21–35.

Girija, A. R., et al. (2022). Nanomaterials-based drug delivery approaches for wound healing. Current Pharmaceutical Design, 28(19), 711–726.

Godoy-Gallardo, M., Eckhard, U., Delgado, L. M., de Roo Puente, Y. J., Hoyos-Nogués, M., Gil, F. J., & Perez, R. A. (2021). Antibacterial approaches in tissue engineering using metal ions and nanoparticles: From mechanisms to applications. *Bioactive Materials*, 6(12). https://doi.org/10.1016/j.bioactmat.2021.04.033

Gong, C. Y., Wu, Q. J., Wang, Y. J., Zhang, D. D., Luo, F., Zhao, X., Wei, Y. Q., & Qian, Z. Y. (2013). A biodegradable hydrogel system containing curcumin encapsulated in micelles for cutaneous wound healing. *Biomaterials*, 34(27). https://doi.org/10.1016/j.biomaterials.2013.05.005

Gorzelanny, C., Kmeth, R., Obermeier, A., Bauer, A. T., Halter, N., Kümpel, K., Schneider, M. F., Wixforth, A., Gollwitzer, H., Burgkart, R., Stritzker, B., & Schneider, S. W. (2016). Silver nanoparticle-enriched diamond-like carbon implant modification as a mammalian cell compatible surface with antimicrobial properties. *Scientific Reports*, 6. https://doi.org/10.1038/srep22849

Gosain, A., & DiPietro, L. A. (2004). Aging and wound healing. *World Journal of Surgery*, 28(3). https://doi.org/10.1007/s00268-003-7397-6

Gudkov, S. V., Burmistrov, D. E., Serov, D. A., Rebezov, M. B., Semenova, A. A., & Lisitsyn, A. B. (2021). A mini review of antibacterial properties of ZnO nanoparticles. *Frontiers in Physics*, 9. https://doi.org/10.3389/fphy.2021.641481

Gupta, S., Sagar, S., Maheshwari, G., Kisaka, T., & Tripathi, S. (2021). Chronic wounds: Magnitude, socioeconomic burden and consequences. *Wounds Asia*, 4, 8–14.

Gusarov, I., Shatalin, K., Starodubtseva, M., & Nudler, E. (2009). Endogenous nitric oxide protects bacteria against a wide spectrum of antibiotics. *Science*, 325(5946). https://doi.org/10.1126/science.1175439

Hadrup, N., Sharma, A. K., & Loeschner, K. (2018). Toxicity of silver ions, metallic silver, and silver nanoparticle materials after in vivo dermal and mucosal surface exposure: A review. *Regulatory Toxicology and Pharmacology*, 98. https://doi.org/10.1016/j.yrtph.2018.08.007

Han, G., Nguyen, L. N., MacHerla, C., Chi, Y., Friedman, J. M., Nosanchuk, J. D., & Martinez, L. R. (2012). Nitric oxide-releasing nanoparticles accelerate wound healing by promoting fibroblast migration and collagen deposition. *American Journal of Pathology*, 180(4). https://doi.org/10.1016/j.ajpath.2011.12.013

Han, S. K. (2023). Basics of wound healing. In *Innovations and advances in wound healing*, (pp. 1–42). Springer.

Haririan, Y., et al. (2023). Carboxymethyl chitosan-gelatin-mesoporous silica nanoparticles containing Myrtus communis L. extract as a novel transparent film wound dressing. *International Journal of Biological Macromolecules*, 253, 127081.

Hashem, A. H., Shehabeldine, A. M., Ali, O. M., & Salem, S. S. (2022). Synthesis of chitosan-based gold nanoparticles: Antimicrobial and wound-healing activities. *Polymers*, 14(11). https://doi.org/10.3390/polym14112293

Hashempour, S., Ghanbarzadeh, S., Maibach, H. I., Ghorbani, M., & Hamishehkar, H. (2019). Skin toxicity of topically applied nanoparticles. *Therapeutic Delivery*, 10(6). https://doi.org/10.4155/tde-2018-0060

Hayat, M., Nawaz, A., Chinnam, S., Muzammal, M., Latif, M. S., Yasin, M., Ashique, S., Zengin, G., & Farid, A. (2023, July 1). Formulation development and optimization of herbo synthetic gel: In vitro biological evaluation and in vivo wound healing studies. *Process Biochemistry*, 130, 116–126.

He, C., Liu, X., Zhou, Z., Liu, N., Ning, X., Miao, Y., Long, Y., Wu, T., & Leng, X. (2021). Harnessing biocompatible nanofibers and silver nanoparticles for wound healing: Sandwich wound dressing versus commercial silver sulfadiazine dressing. *Materials Science and Engineering C*, 128. https://doi.org/10.1016/j.msec.2021.112342

Huan, S., & Tay, L. (2020). Four curious cases of closed-incision negative pressure therapy (ciNPT). *Cureus*. https://doi.org/10.7759/cureus.8193

Huang, S., & Fu, X. (2010). Naturally derived materials-based cell and drug delivery systems in skin regeneration. *Journal of Controlled Release*, 142(2). https://doi.org/10.1016/j.jconrel.2009.10.018

Husain, F. M., Qais, F. A., Ahmad, I., Hakeem, M. J., Baig, M. H., Khan, J. M., & Al-Shabib, N. A. (2022). Biosynthesized zinc oxide nanoparticles disrupt established biofilms of pathogenic bacteria. *Applied Sciences (Switzerland)*, 12(2). https://doi.org/10.3390/app12020710

Iqubal, M. K., Saleem, S., Iqubal, A., Chaudhuri, A., Pottoo, F. H., Ali, J., & Baboota, S. (2020). Natural, synthetic and their combinatorial nanocarriers based drug delivery system in the treatment paradigm for wound healing via dermaltargeting. *Current Pharmaceutical Design*, 26(36). https://doi.org/10.2174/1381612826666200612164511

Jacob, J., Haponiuk, J. T., Thomas, S., & Gopi, S. (2018). Biopolymer based nanomaterials in drug delivery systems: A review. *Materials Today Chemistry*, 9. https://doi.org/10.1016/j.mtchem.2018.05.002

Janjua, T. I., et al. (2023). Silica nanoparticles: A review of their safety and current strategies to overcome biological barriers. *Advanced Drug Delivery Reviews*, 203, 115115.

Jelinek, M., Voss, A., Kocourek, T., Mozafari, M., Vymětalová, V., Zezulová, M., Písařík, P., Kotzianová, A., Popov, C., & Mikšovský, J. (2013). Comparison of the surface properties of DLC and ultrananocrystalline diamond films with respect to their bio-applications. *Physica Status Solidi (A) Applications and Materials Science*, 210(10). https://doi.org/10.1002/pssa.201228713

Jiang, Q., Zhou, W., Wang, J., Tang, R., Zhang, D., & Wang, X. (2016). Hypromellose succinate-crosslinked chitosan hydrogel films for potential wound dressing. *International Journal of Biological Macromolecules*, 91. https://doi.org/10.1016/j.ijbiomac.2016.05.077

Jones, N., Ray, B., Ranjit, K. T., & Manna, A. C. (2008). Antibacterial activity of ZnO nanoparticle suspensions on a broad spectrum of microorganisms. *FEMS Microbiology Letters*, 279(1). https://doi.org/10.1111/j.1574-6968.2007.01012.x

Juknius, T., Ružauskas, M., Tamulevičius, T., Šiugždiniene, R., Jukniene, I., Vasiliauskas, A., Jurkevičiute, A., & Tamulevičius, S. (2016). Antimicrobial properties of diamond-like carbon/silver nanocomposite thin films deposited on textiles: Towards smart bandages. *Materials*, 9(5). https://doi.org/10.3390/ma9050371

Kalashnikova, I., et al. (2015). Nanomaterials for wound healing: Scope and advancement. *Nanomedicine*, 10(16), 2593–2612.

Kannan, S., Solomon, A., Krishnamoorthy, G., & Marudhamuthu, M. (2021). Liposome encapsulated surfactant abetted copper nanoparticles alleviates biofilm mediated virulence in pathogenic Pseudomonas aeruginosa and MRSA. *Scientific Reports*, 11(1). https://doi.org/10.1038/s41598-020-79976-7

Kell, A. J., Stewart, G., Ryan, S., Peytavi, R., Boissinot, M., Huletsky, A., Bergeron, M. G., & Simard, B. (2008). Vancomycin-modified nanoparticles for efficient targeting and preconcentration of gram-positive and gram-negative bacteria. *ACS Nano*, 2(9). https://doi.org/10.1021/nn700183g

Khalid, A., Ullah, H., Ul-Islam, M., Khan, R., Khan, S., Ahmad, F., Khan, T., & Wahid, F. (2017). Bacterial cellulose-TiO2 nanocomposites promote healing and tissue regeneration in burn mice model. *RSC Advances*, 7(75). https://doi.org/10.1039/c7ra06699f

Khashan, K. S., Sulaiman, G. M., & Mahdi, R. (2017). Preparation of iron oxide nanoparticles-decorated carbon nanotube using laser ablation in liquid and their antimicrobial activity. *Artificial Cells, Nanomedicine and Biotechnology*, 45(8). https://doi.org/10.1080/21691401.2017.1282498

Khatri, V., Halász, K., Trandafilović, L. V., Dimitrijević-Branković, S., Mohanty, P., Djoković, V., & Csóka, L. (2014). ZnO-modified cellulose fiber sheets for antibody immobilization. *Carbohydrate Polymers*, 109. https://doi.org/10.1016/j.carbpol.2014.03.061

Kittana, N., Assali, M., Abu-Rass, H., Lutz, S., Hindawi, R., Ghannam, L., Zakarneh, M., & Mousa, A. (2018). Enhancement of wound healing by single-wall/multi-wall carbon nanotubes complexed with chitosan. *International Journal of Nanomedicine*, 13. https://doi.org/10.2147/IJN.S183342

Kubackova, J., Zbytovska, J., & Holas, O. (2020). Nanomaterials for direct and indirect immunomodulation: A review of applications. *European Journal of Pharmaceutical Sciences,* 142. https://doi.org/10.1016/j.ejps.2019.105139

Kuppusamy, P., Yusoff, M. M., Maniam, G. P., & Govindan, N. (2016). Biosynthesis of metallic nanoparticles using plant derivatives and their new avenues in pharmacological applications – An updated report. *Saudi Pharmaceutical Journal,* 24(4). https://doi.org/10.1016/j.jsps.2014.11.013

Kushwaha, A., Goswami, L., & Kim, B. S. (2022). Nanomaterial-based therapy for wound healing. *Nanomaterials,* 12(4), 618.

Lai, X., Wang, M., Zhu, Y., Feng, X., Liang, H., Wu, J., Nie, L., Li, L., & Shao, L. (2021). ZnO NPs delay the recovery of psoriasis-like skin lesions through promoting nuclear translocation of p-NFκB p65 and cysteine deficiency in keratinocytes. *Journal of Hazardous Materials*, 410. https://doi.org/10.1016/j.jhazmat.2020.124566

Lakkim, V., Reddy, M. C., Pallavali, R. R., Reddy, K. R., Reddy, C. V., Inamuddin, Bilgrami, A. L., & Lomada, D. (2020). Green synthesis of silver nanoparticles and evaluation of their antibacterial activity against multidrug-resistant bacteria and wound healing efficacy using a murine model. *Antibiotics*, 9(12). https://doi.org/10.3390/antibiotics9120902

Lasocka, I., Jastrzebska, E., Szulc-Dabrowska, L., Skibniewski, M., Pasternak, I., Kalbacova, M. H., & Skibniewska, E. M. (2019). The effects of graphene and mesenchymal stem cells in cutaneous wound healing and their putative action mechanism. *International Journal of Nanomedicine*, 14. https://doi.org/10.2147/IJN.S190928

LewisOscar, F., MubarakAli, D., Nithya, C., Priyanka, R., Gopinath, V., Alharbi, N. S., & Thajuddin, N. (2015). One pot synthesis and anti-biofilm potential of copper nanoparticles (CuNPs) against clinical strains of Pseudomonas aeruginosa. *Biofouling*, 31(4). https://doi.org/10.1080/08927014.2015.1048686

Li, D., Nie, W., Chen, L., Miao, Y., Zhang, X., Chen, F., Yu, B., Ao, R., Yu, B., & He, C. (2017). Fabrication of curcumin-loaded mesoporous silica incorporated polyvinyl pyrrolidone nanofibers for rapid hemostasis and antibacterial treatment. *RSC Advances*, 7(13). https://doi.org/10.1039/c6ra27319j

Li, R. et al. (2021). Metal–organic-framework-based materials for antimicrobial applications. *ACS Nano*, 15(3), 3808–3848.

Liao, C. et al. (2019). Bactericidal and cytotoxic properties of silver nanoparticles. *International Journal of Molecular Sciences*, 20(2), 449.

Liao, J. L., Zhong, S., Wang, S. H., Liu, J. Y., Chen, J., He, G., He, B., Xu, J. Q., Liang, Z. H., Mei, T., Wu, S., Cao, K., & Zhou, J. Da. (2017). Preparation and properties of a novel carbon nanotubes/poly(Vinyl alcohol)/epidermal growth factor composite biological dressing. *Experimental and Therapeutic Medicine*, 14(3). https://doi.org/10.3892/etm.2017.4752

Liu, M., Wang, R., Liu, J., Zhang, W., Liu, Z., Lou, X., Nie, H., Wang, H., Mo, X., Abd-Elhamid, A. I., Zheng, R., & Wu, J. (2022). Incorporation of magnesium oxide nanoparticles into electrospun membranes improves pro-angiogenic activity and promotes diabetic wound healing. *Biomaterials Advances*, 133. https://doi.org/10.1016/j.msec.2021.112609

Liu, Y., Chen, W., & Kim, H. Il. (2012). Antibacterial activity of pH-sensitive genipin cross-linked chitosan/poly(ethylene glycol)/silver nanocomposites. *Polymers for Advanced Technologies*, 23(1). https://doi.org/10.1002/pat.1818

Lopes, F. S., Oliveira, J. R., Milani, J., Oliveira, L. D., Machado, J. P. B., Trava-Airoldi, V. J., Lobo, A. O., & Marciano, F. R. (2017). Biomineralized diamond-like carbon films with incorporated titanium dioxide nanoparticles improved bioactivity properties and reduced biofilm formation. *Materials Science and Engineering C*, 81. https://doi.org/10.1016/j.msec.2017.07.043

Lorenzo-Anota, H. Y., Zarate-Triviño, D. G., Uribe-Echeverría, J. A., Ávila-ávila, A., Rangel-López, J. R., Martínez-Torres, A. C., & Rodríguez-Padilla, C. (2021). Chitosan-coated gold nanoparticles induce low cytotoxicity and low ros production in primary leucocytes, independent of their proliferative status. *Pharmaceutics*, 13(7). https://doi.org/10.3390/pharmaceutics13070942

Lu, B., Li, T., Zhao, H., Li, X., Gao, C., Zhang, S., & Xie, E. (2012). Graphene-based composite materials beneficial to wound healing. *Nanoscale*, 4(9). https://doi.org/10.1039/c2nr11958g

Luanpitpong, S., Wang, L., & Rojanasakul, Y. (2014). The effects of carbon nanotubes on lung and dermal cellular behaviors. *Nanomedicine*, 9(6). https://doi.org/10.2217/nnm.14.42

Luo, X., Zhang, H., Cao, Z., Cai, N., Xue, Y., & Yu, F. (2016). A simple route to develop transparent doxorubicin-loaded nanodiamonds/cellulose nanocomposite membranes as potential wound dressings. *Carbohydrate Polymers*, 143. https://doi.org/10.1016/j.carbpol.2016.01.076

Lyon, D. Y., Fortner, J. D., Sayes, C. M., Colvin, V. L., & Hughes, J. B. (2005). Bacterial cell association and antimicrobial activity of a C60 water suspension. *Environmental Toxicology and Chemistry*, 24(11). https://doi.org/10.1897/04-649R.1

Ma, X., Cheng, Y., Jian, H., Feng, Y., Chang, Y., Zheng, R., & Zhang, H. (2019). Hollow, rough, and nitric oxide-releasing cerium oxide nanoparticles for promoting multiple stages of wound healing. *Advanced Healthcare Materials*, 8(16). https://doi.org/10.1002/adhm.201900256

Makarov, V. V., Love, A. J., Sinitsyna, O. V., Makarova, S. S., Yaminsky, I. V., Taliansky, M. E., & Kalinina, N. O. (2014). "Green" nanotechnologies: Synthesis of metal nanoparticles using plants. *Acta Naturae*, 6(20). https://doi.org/10.32607/20758251-2014-6-1-35-44

Makvandi, P., Wang, C. Y., Zare, E. N., Borzacchiello, A., Niu, L. N., & Tay, F. R. (2020). Metal-based nanomaterials in biomedical applications: Antimicrobial activity and cytotoxicity aspects. *Advanced Functional Materials*, 30(22). https://doi.org/10.1002/adfm.201910021

Mallick, S., et al. (2022). Engineered nanotechnology: An effective therapeutic platform for the chronic cutaneous wound. *Nanomaterials*, 12(5), 778.

Man, H. B., & Ho, D. (2012). Front cover: Diamond as a nanomedical agent for versatile applications in drug delivery, imaging, and sensing (Phys. Status Solidi A 9/2012). *Physica Status Solidi (a)*, 209(9). https://doi.org/10.1002/pssa.201290022

Maťátková, O., Michailidu, J., Miškovská, A., Kolouchová, I., Masák, J., & Čejková, A. (2022). Antimicrobial properties and applications of metal nanoparticles biosynthesized by green methods. *Biotechnology Advances*, 58. https://doi.org/10.1016/j.biotechadv.2022.107905

McCarty, S. M., & Percival, S. L. (2013). Proteases and delayed wound healing. *Advances in Wound Care*, 2(8). https://doi.org/10.1089/wound.2012.0370

Min, K. H., et al. (2018). Biomineralized hybrid nanoparticles for imaging and therapy of cancers. *Quantitative Imaging in Medicine and Surgery*, 8(7), 694.

Min, Y., et al. (2023). Metal-based nanoparticles and their relevant consequences on cytotoxicity cascade and induced oxidative stress. *Antioxidants*, 12(3), 703.

Mostafavi, E., Medina-Cruz, D., Vernet-Crua, A., Chen, J., Cholula-Díaz, J. L., Guisbiers, G., & Webster, T. J. (2021). Green nanomedicine: the path to the next generation of nanomaterials for diagnosing brain tumors and therapeutics? *Expert Opinion on Drug Delivery*, 18(6). https://doi.org/10.1080/17425247.2021.1865306

Nablo, B. J., Prichard, H. L., Butler, R. D., Klitzman, B., & Schoenfisch, M. H. (2005). Inhibition of implant-associated infections via nitric oxide release. *Biomaterials*, 26(34). https://doi.org/10.1016/j.biomaterials.2005.05.017

Namdar, R., & Nafisi, S. (2018). Nanodiamond applications in skin preparations. *Drug Discovery Today*, 23(5). https://doi.org/10.1016/j.drudis.2018.04.006

Nandhini, J., et al. (2024). Nanomaterials for wound healing: Current status and futuristic frontier. *Biomedical Technology*, 6, 26–45.

Nasibova, A. (2023). Generation of nanoparticles in biological systems and their application prospects. *Advances in Biology & Earth Sciences*, 8(2), 140–146.

Naskar, A., & Kim, K.-S. J. P. (2020). Recent advances in nanomaterial-based wound-healing therapeutics. *Pharmaceutics*, 12(6), 499.

Nisar, P., et al. (2019). Antimicrobial activities of biologically synthesized metal nanoparticles: An insight into the mechanism of action. *Journal of Biological Inorganic Chemistry*, 24, 929–941.
Niu, A., Han, Y., Wu, J., Yu, N., & Xu, Q. (2010). Synthesis of one-dimensional carbon nanomaterials wrapped by silver nanoparticles and their antibacterial behavior. *Journal of Physical Chemistry C*, 114(29). https://doi.org/10.1021/jp104720w
Nosrati, H., Aramideh Khouy, R., Nosrati, A., Khodaei, M., Banitalebi-Dehkordi, M., Ashrafi-Dehkordi, K., Sanami, S., & Alizadeh, Z. (2021). Nanocomposite scaffolds for accelerating chronic wound healing by enhancing angiogenesis. *Journal of Nanobiotechnology*, 19. https://doi.org/10.1186/s12951-020-00755-7
Nosrati, H., Heydari, M., Tootiaei, Z., Ganjbar, S., & Khodaei, M. (2023). Delivery of antibacterial agents for wound healing applications using polysaccharide-based scaffolds. *Journal of Drug Delivery Science and Technology*, 84. https://doi.org/10.1016/j.jddst.2023.104516
Nqakala, Z. B., Sibuyi, N. R. S., Fadaka, A. O., Meyer, M., Onani, M. O., & Madiehe, A. M. (2021). Advances in nanotechnology towards development of silver nanoparticle-based wound-healing agents. *International Journal of Molecular Sciences*, 22(20). https://doi.org/10.3390/ijms222011272
Opneja, A., Kapoor, S., & Stavrou, E. X. (2019). Contribution of platelets, the coagulation and fibrinolytic systems to cutaneous wound healing. *Thrombosis Research*, 179. https://doi.org/10.1016/j.thromres.2019.05.001
Ostadhossein, F., Mahmoudi, N., Morales-Cid, G., Tamjid, E., Navas-Martos, F. J., Soriano-Cuadrado, B., Paniza, J. M. L., & Simchi, A. (2015). Development of chitosan/bacterial cellulose composite films containing nanodiamonds as a potential flexible platform for wound dressing. *Materials*, 8(9). https://doi.org/10.3390/ma8095309
Pacelli, S., Acosta, F., Chakravarti, A. R., Samanta, S. G., Whitlow, J., Modaresi, S., Ahmed, R. P. H., Rajasingh, J., & Paul, A. (2017). Nanodiamond-based injectable hydrogel for sustained growth factor release: Preparation, characterization and in vitro analysis. *Acta Biomaterialia*, 58. https://doi.org/10.1016/j.actbio.2017.05.026
Palmer, B. C., Phelan-Dickenson, S. J., & Delouise, L. A. (2019). Multi-walled carbon nanotube oxidation dependent keratinocyte cytotoxicity and skin inflammation. *Particle and Fibre Toxicology*, 16(1). https://doi.org/10.1186/s12989-018-0285-x
Pan, Y., Paschoalino, W. J., Szuchmacher Blum, A., & Mauzeroll, J. (2021). Recent advances in bio-templated metallic nanomaterial synthesis and electrocatalytic applications. *ChemSusChem,* 14(3). https://doi.org/10.1002/cssc.202002532
Patel, K. K., Surekha, D. B., Tripathi, M., Anjum, M. M., Muthu, M. S., Tilak, R., Agrawal, A. K., & Singh, S. (2019). Antibiofilm potential of silver sulfadiazine-loaded nanoparticle formulations: A study on the effect of DNase-I on microbial biofilm and wound healing activity. *Molecular Pharmaceutics*, 16(9). https://doi.org/10.1021/acs.molpharmaceut.9b00527
Pati, R., Mehta, R. K., Mohanty, S., Padhi, A., Sengupta, M., Vaseeharan, B., Goswami, C., & Sonawane, A. (2014). Topical application of zinc oxide nanoparticles reduces bacterial skin infection in mice and exhibits antibacterial activity by inducing oxidative stress response and cell membrane disintegration in macrophages. *Nanomedicine: Nanotechnology, Biology, and Medicine*, 10(6). https://doi.org/10.1016/j.nano.2014.02.012
Peng, C. C., Yang, M. H., Chiu, W. T., Chiu, C. H., Yang, C. S., Chen, Y. W., Chen, K. C., & Peng, R. Y. (2008). Composite nano-titanium oxide-chitosan artificial skin exhibits strong wound-healing effect - An approach with anti-inflammatory and bactericidal kinetics. *Macromolecular Bioscience*, 8(4). https://doi.org/10.1002/mabi.200700188
Perumal, S., Ramadass, S. K., & Madhan, B. (2014). Sol-gel processed mupirocin silica microspheres loaded collagen scaffold: A synergistic bio-composite for wound healing. *European Journal of Pharmaceutical Sciences*, 52(1). https://doi.org/10.1016/j.ejps.2013.10.006
Pormohammad, A., (2021). Nanomaterials in wound healing and infection control. *Antibiotics*, 10(5),473.
Poscher, V., & Salinas, Y. (2020). Trends in degradable mesoporous organosilica-based nanomaterials for controlling drug delivery: A mini review. *Materials*, 13(17). https://doi.org/10.3390/MA13173668
Qi, X., et al. (2023). Metal-organic framework-based nanomedicines for the treatment of intracellular bacterial infections. *Pharmaceutics*, 15(5), 1521.
Raffi, M., Mehrwan, S., Bhatti, T. M., Akhter, J. I., Hameed, A., Yawar, W., & Ul Hasan, M. M. (2010). Investigations into the antibacterial behavior of copper nanoparticles against Escherichia coli. *Annals of Microbiology*, 60(1). https://doi.org/10.1007/s13213-010-0015-6
Raghupathi, K. R., Koodali, R. T., & Manna, A. C. (2011). Size-dependent bacterial growth inhibition and mechanism of antibacterial activity of zinc oxide nanoparticles. *Langmuir*, 27(7). https://doi.org/10.1021/la104825u

Rajendran, N. K., et al. (2018). A review on nanoparticle-based treatment for wound healing. *Journal of Drug Delivery Science and Technology*, 44, 421–430.

Ren, Y., et al. (2020). Photoresponsive materials for antibacterial applications. *Cell Reports Physical Science*, 1(11), 100245.

Rizvi, S. I., & Zaid, M. A. (2001). Intracellular reduced glutathione content in normal and type 2 diabetic erythrocytes: Effect of insulin and (-) epicatechin. *Journal of Physiology and Pharmacology*, 52(3).

Rizwan, M., et al. (2021). Types and classification of nanomaterials. In M. B. Tahir, M. Sagir, & A. M. Asiri (Eds.), *Nanomaterials: Synthesis, characterization, hazards and safety* (pp. 31–54). Elsevier.

Rodrigues, M., Kosaric, N., Bonham, C. A., & Gurtner, G. C. (2019). Wound healing: A cellular perspective. *Physiological Reviews*, 99, 665–706.

Salvo, J., & Sandoval, C. (2022). Role of copper nanoparticles in wound healing for chronic wounds: Literature review. *Burns and Trauma,* 10. https://doi.org/10.1093/burnst/tkab047

Sanchez-Moreno, P., Ortega-Vinuesa, J. L., Peula-Garcia, J. M., Marchal, J. A., & Boulaiz, H. (2016). Smart drug-delivery systems for cancer nanotherapy. *Current Drug Targets*, 19(4). https://doi.org/10.2174/1389450117666160527142544

Sandoval, C., Ríos, G., Sepúlveda, N., Salvo, J., Souza-Mello, V., & Farías, J. (2022). Effectiveness of copper nanoparticles in wound healing process using in vivo and in vitro studies: A systematic review. *Pharmaceutics,* 14(9). https://doi.org/10.3390/pharmaceutics14091838

Sankar, R., Baskaran, A., Shivashangari, K. S., & Ravikumar, V. (2015). Inhibition of pathogenic bacterial growth on excision wound by green synthesized copper oxide nanoparticles leads to accelerated wound healing activity in Wistar Albino rats. *Journal of Materials Science: Materials in Medicine*, 26(7). https://doi.org/10.1007/s10856-015-5543-y

Schapiro, J. M., Libby, S. J., & Fang, F. C. (2003). Inhibition of bacterial DNA replication by zinc mobilization during nitrosative stress. *Proceedings of the National Academy of Sciences of the United States of America*, 100(14). https://doi.org/10.1073/pnas.1033133100

Schwentker, A., Vodovotz, Y., Weller, R., & Billiar, T. R. (2002). Nitric oxide and wound repair: Role of cytokines? *Nitric Oxide - Biology and Chemistry*, 7(1). https://doi.org/10.1016/S1089-8603(02)00002-2

Seifi, T., & Kamali, A. R. (2021). Anti-pathogenic activity of graphene nanomaterials: A review. *Colloids and Surfaces B: Biointerfaces*, 199. https://doi.org/10.1016/j.colsurfb.2020.111509

Sharifiaghdam, M., et al. (2022). Chitosan coated metallic nanoparticles with stability, antioxidant, and antibacterial properties: Potential for wound healing application. *Journal of Applied Polymer Science*, 139(10), 51766.

Shi, Y., Ma, X., Fang, G., Tian, X., & Ge, C. (2021). Matrix metalloproteinase inhibitors (MMPIs) as attractive therapeutic targets: Recent progress and current challenges. *NanoImpact,* 21. https://doi.org/10.1016/j.impact.2021.100293

Shnoudeh, A. J., et al. (2019). Synthesis, characterization, and applications of metal nanoparticles. In *Biomaterials and bionanotechnology* (pp. 527–612). Elsevier.

Singh, H., Du, J., Singh, P., & Yi, T. H. (2018). Ecofriendly synthesis of silver and gold nanoparticles by Euphrasia officinalis leaf extract and its biomedical applications. *Artificial Cells, Nanomedicine and Biotechnology*, 46(6). https://doi.org/10.1080/21691401.2017.1362417

Singh, S., et al. (2017). The physiology of wound healing. *Surgery (Oxford)*, 35(9), 473–477.

Sivaranjani, V., & Philominathan, P. (2016). Synthesize of Titanium dioxide nanoparticles using Moringa oleifera leaves and evaluation of wound healing activity. *Wound Medicine*, 12. https://doi.org/10.1016/j.wndm.2015.11.002

Sooklert, K., Nilyai, S., Rojanathanes, R., Jindatip, D., Sae-Liang, N., Kitkumthorn, N., Mutirangura, A., & Sereemaspun, A. (2019). N-acetylcysteine reverses the decrease of DNA methylation status caused by engineered gold, silicon, and chitosan nanoparticles. *International Journal of Nanomedicine*, 14. https://doi.org/10.2147/IJN.S204372

Sorg, H., Tilkorn, D. J., Hager, S., Hauser, J., & Mirastschijski, U. (2017). Skin wound healing: an update on the current knowledge and concepts. *European Surgical Research*, 58(1–2). https://doi.org/10.1159/000454919

Spielman, A. F., et al. (2023). Beyond the scar: A basic science review of wound remodeling. Advances in Wound Care, 12(2), 57–67.

Stallmeyer, B., Kämpfer, H., Kolb, N., Pfeilschifter, J., & Frank, S. (1999). The function of nitric oxide in wound repair: Inhibition of inducible nitric oxide-synthase severely impairs wound reepithelialization. *Journal of Investigative Dermatology*, 113(6). https://doi.org/10.1046/j.1523-1747.1999.00784.x

Stoica, A. E., Chircov, C., & Grumezescu, A. M. (2020). Nanomaterials for wound dressings: An up-to-date overview. *Molecules*, 25(11). https://doi.org/10.3390/molecules25112699

Strudwick, X. L., & Cowin, A. J. (2018). The role of the inflammatory response in burn injury. *Hot Topics in Burn Injuries*, 3(1–23).

Su, Y., Zheng, X., Chen, A., Chen, Y., He, G., & Chen, H. (2015). Hydroxyl functionalization of single-walled carbon nanotubes causes inhibition to the bacterial denitrification process. *Chemical Engineering Journal*, 279. https://doi.org/10.1016/j.cej.2015.05.005

Sukhanova, A., Bozrova, S., Sokolov, P., Berestovoy, M., Karaulov, A., & Nabiev, I. (2018). Dependence of nanoparticle toxicity on their physical and chemical properties. *Nanoscale Research Letters*, 13. https://doi.org/10.1186/s11671-018-2457-x

Sun, Y., Bhattacharjee, A., Reynolds, M., & Li, Y. V. (2021). Synthesis and characterizations of gentamicin-loaded poly-lactic-co-glycolic (PLGA) nanoparticles. *Journal of Nanoparticle Research*, 23(8). https://doi.org/10.1007/s11051-021-05293-3

Sundaram, P., & Abrahamse, H. (2020). Phototherapy combined with carbon nanomaterials (1d and 2d) and their applications in cancer therapy. *Materials*, 13(21). https://doi.org/10.3390/ma13214830

Suramwar, N. V., Thakare, S. R., & Khaty, N. T. (2016). One pot synthesis of copper nanoparticles at room temperature and its catalytic activity. *Arabian Journal of Chemistry*, 9. https://doi.org/10.1016/j.arabjc.2012.04.034

Teixeira, M. A., Paiva, M. C., Amorim, M. T. P., & Felgueiras, H. P. (2020). Electrospun nanocomposites containing cellulose and its derivatives modified with specialized biomolecules for an enhanced wound healing. *Nanomaterials,* 10(3). https://doi.org/10.3390/nano10030557

Tocco, I., et al. (2012). Nanotechnology-based therapies for skin wound regeneration. *Journal ofNanomaterials,* 2012, 4.

Uzair, B., et al. (2020). Green and cost-effective synthesis of metallic nanoparticles by algae: Safe methods for translational medicine. *Bioengineering (Basel),* 7(4), 129.

Veiseh, O., Kievit, F. M., Fang, C., Mu, N., Jana, S., Leung, M. C., Mok, H., Ellenbogen, R. G., Park, J. O., & Zhang, M. (2010). Chlorotoxin bound magnetic nanovector tailored for cancer cell targeting, imaging, and siRNA delivery. *Biomaterials*, 31(31). https://doi.org/10.1016/j.biomaterials.2010.07.016

Verma, N., Pramanik, K., Singh, A. K., & Biswas, A. (2022). Design of magnesium oxide nanoparticle incorporated carboxy methyl cellulose/poly vinyl alcohol composite film with novel composition for skin tissue engineering. *Materials Technology*, 37(8). https://doi.org/10.1080/10667857.2021.1873634

Wahajuddin & Arora, S. (2012). Superparamagnetic iron oxide nanoparticles: Magnetic nanoplatforms as drug carriers. *International Journal of Nanomedicine,* 7. https://doi.org/10.2147/IJN.S30320

Wang, L., et al. (2023). Biomimetic enzyme cascade structural color hydrogel microparticles for diabetic wound healing management. *Advanced Science,* 10(14), 2206900.

Wang, M., Lai, X., Shao, L., & Li, L. (2018). Evaluation of immunoresponses and cytotoxicity from skin exposure to metallic nanoparticles. *International Journal of Nanomedicine*, 13. https://doi.org/10.2147/IJN.S170745

Wang, X., et al. (2021). Inorganic nanomaterials with rapid clearance for biomedical applications. *Chemical Society Reviews,* 50(15), 8669–8742.

Wang, W., Lu, K. J., Yu, C. H., Huang, Q. L., & Du, Y. Z. (2019). Nano-drug delivery systems in wound treatment and skin regeneration. *Journal of Nanobiotechnology,* 17(1). https://doi.org/10.1186/s12951-019-0514-y

Wang, X., Coradin, T., & Hélary, C. (2018). Modulating inflammation in a cutaneous chronic wound model by IL-10 released from collagen-silica nanocomposites: Via gene delivery. *Biomaterials Science*, 6(2). https://doi.org/10.1039/c7bm01024a

Weller, R. B. (2009). Nitric oxide-containing nanoparticles as an antimicrobial agent and enhancer of wound healing. *Journal of Investigative Dermatology*, 129(10). https://doi.org/10.1038/jid.2009.149

Wen, L., Zeng, P., Zhang, L., Huang, W., Wang, H., & Chen, G. (2016). Symbiosis theory-directed green synthesis of silver nanoparticles and their application in infected wound healing. *International Journal of Nanomedicine*, 11. https://doi.org/10.2147/IJN.S106662

Yang, Y., Dong, Z., Li, M., Liu, L., Luo, H., Wang, P., Zhang, D., Yang, X., Zhou, K., & Lei, S. (2020). Graphene oxide/copper nanoderivatives-modified chitosan/hyaluronic acid dressings for facilitating wound healing in infected full-thickness skin defects. *International Journal of Nanomedicine*, 15. https://doi.org/10.2147/IJN.S278631

Yimeng, S., Huilun, X., Ziming, L., Kejun, L., Chaima, M., Xiangyu, Z., Yinchun, H., Yan, W., & Di, H. (2023). Copper-based nanoparticles as antibacterial agents. *European Journal of Inorganic Chemistry*, 26(4). https://doi.org/10.1002/ejic.202200614

Yousefi, M., Dadashpour, M., Hejazi, M., Hasanzadeh, M., Behnam, B., de la Guardia, M., Shadjou, N., & Mokhtarzadeh, A. (2017). Anti-bacterial activity of graphene oxide as a new weapon nanomaterial to combat multidrug-resistance bacteria. *Materials Science and Engineering C*, 74. https://doi.org/10.1016/j.msec.2016.12.125

Yu, W., et al. (2023). Biofunctionalized decellularized tissue-engineered heart valve with mesoporous silica nanoparticles for controlled release of VEGF and RunX2-siRNA against calcification. *Bioengineering*, 10(7), 859.

Yu, Y., et al. (2023). Double-or triple-tiered protection: Prospects for the sustainable application of copper-based antimicrobial compounds for another fourteen decades. *International Journal of Molecular Sciences*, 24(13), 10893.

Yun, Y. H., Goetz, D. J., Yellen, P., & Chen, W. (2004). Hyaluronan microspheres for sustained gene delivery and site-specific targeting. *Biomaterials*, 25(1). https://doi.org/10.1016/S0142-9612(03)00467-8

Zeng, R., Lin, C., Lin, Z., Chen, H., Lu, W., Lin, C., & Li, H. (2018). Approaches to cutaneous wound healing: Basics and future directions. *Cell and Tissue Research*, 374. https://doi.org/10.1007/s00441-018-2830-1

Zhang, Y., Wang, B., Meng, X., Sun, G., & Gao, C. (2011). Influences of acid-treated multiwalled carbon nanotubes on fibroblasts: Proliferation, adhesion, migration, and wound healing. *Annals of Biomedical Engineering*, 39(1). https://doi.org/10.1007/s10439-010-0151-y

Zheng, Y., Li, S., Han, D., Kong, L., Wang, J., Zhao, M., Cheng, W., Ju, H., Yang, Z., & Ding, S. (2021). Eco-friendly preparation of epoxy-rich graphene oxide for wound healing. *ACS Biomaterials Science and Engineering*, 7(2). https://doi.org/10.1021/acsbiomaterials.0c01598

Zhou, K., et al. (2022). Hybrid Ag nanoparticles/polyoxometalate-polydopamine nano-flowers loaded chitosan/gelatin hydrogel scaffolds with synergistic photothermal/chemodynamic/Ag+ anti-bacterial action for accelerated wound healing. *International Journal of Biological Macromolecules*, 221, 135–148.

Zhou, Z., Joslin, S., Dellinger, A., Ehrich, M., Brooks, B., Ren, Q., Rodeck, U., Lenk, R., & Kepley, C. L. (2010). A novel class of compounds with cutaneous wound healing properties. *Journal of Biomedical Nanotechnology*, 6(5). https://doi.org/10.1166/jbn.2010.1157

Index